The Gospel of Germs

The
Gospel
of
Germs

*Men, Women, and the Microbe
in American Life*

Nancy Tomes

HARVARD UNIVERSITY PRESS
Cambridge, Massachusetts
London, England
1998

Library of Congress Cataloging-in-Publication Data

Tomes, Nancy, 1952–
    The gospel of germs : men, women, and the microbe in American life /
Nancy Tomes.
        p. cm.
    Includes bibliographical references and index.
    ISBN 0-674-35707-8 (cloth : alk. paper)
    1. Germ theory of disease—Public opinion—History—19th century.
2. Germ theory of disease—Public opinion—History—20th century.
3. Hygiene—United States—History—19th century.   4. Hygiene—
United States—History—20th century. I. Title.
RM214.T66 1998
616'.01'0973—dc21   97-28734

For Annie, Chris, and the families who love them

Contents

IV *The Gospel in Retreat*

Illustrations

Preface: Memories of Disease Past

In the late 1990s, the subject of germs has become a major societal preoccupation. The perils of infection are heralded everywhere: in the memoirs of AIDS sufferers, in best-selling books with titles such as *The Coming Plague,* in movies with names like *The Last Stand* and *Outbreak,* in magazine articles titled "Revenge of the Killer Microbes"—even in *A Field Guide to Germs.* Over the last ten years, there has been a remarkable resurgence of interest in the changing ecology of human-microbial relationships and its implications for life in the twenty-first century.

I must admit that until very recently, I did not share this fascination with germs. Like most middle-class baby boomers who came of age after the antibiotic revolution of the mid-twentieth century, I have never been very concerned about germs. My urge to write about the history of germ beliefs did not grow out of any close encounters with exotic diseases or life-threatening bouts of illness. As anyone who has visited my home can attest, I am hardly a stickler for antiseptic standards of cleanliness. Until I began working on this project, I regarded people who put paper covers on toilet seats as hopelessly neurotic and viewed housecleaning as a trivial pursuit that kept women from leading productive lives.

My formerly condescending attitude toward the perils of infection reflects the unprecedented protection from deadly contagious diseases that my generation has enjoyed—at least until recently—due to the availability of twentieth-century wonder drugs such as penicillin. By virtue of my age and middle-class upbringing, I escaped what was once a common occurrence: namely, seeing relatives and friends stricken with potentially mortal infections.

At the same time, my historical consciousness has been shaped by a family history steeped with memories of a disease environment quite unlike the one I knew as a child. Being a late-born child of two southerners, I grew up listening to stories that made it clear that the past was once a far more dangerous place. Over the dinner table and during visits to relatives, my parents and other kinfolk passed on memories of disease that conjured up fears and discomforts I had never known. Their stories made me dimly aware of the bad old days, when the chances of sudden death from disease, and the hazards of unsanitary conditions, were ever present.

My awareness of this difference was heightened when I was taken as a child to visit my paternal grandparents, who lived on a farm in south-central Kentucky where my father was born in 1911. Like many rural southern areas, Edmonson County remained outside the web of modern conveniences—such as electricity, running water, and indoor toilets—well into the twentieth century. Even in the 1960s, my father's childhood home resembled the pioneer homesteads that I loved to read about in Laura Ingalls Wilder's novels. My grandparents had an outdoor well, whose dark, sulphurous depths I regarded with considerable awe. Their only toilet was an outdoor privy built over a steep hillside (a style known colloquially as the "long drop"), which offered a terrifying view to a small child. Never had using the toilet seemed so perilous or so thrilling.

Along with these exotic features of my grandparents' house, the tales of life long ago in the country, which were told and retold as the family congregated on the big front porch, revealed other elements of danger. My relatives reminisced about my aunt Stelsa, who cut her lip one day and died from "blood poisoning" the next, and my second cousins, a set of triplets improbably named Zeta, Zuma, and Zula, who were broken up by the death of one from "the typhoid." These tales became all the more poignant when I saw the gravestones of these mythic characters near the tiny country church down the road.

In a more subtle yet painful way, visits to my mother's hometown about an hour's drive from Nashville reinforced my childhood awareness of disease hazards of the past. My mother, who was born in 1912, came from a more genteel family than my father's. My maternal grandfather worked as a clerk for a copper mining company and served as the town postmaster—he kept his records in a fine boiler-plate script that I was once taken to the courthouse to admire. Yet despite their greater degree of comfort, my mother's immediate family suffered more from disease than did my father's farmer-clan. When my mother was five, her father died from kidney failure of unknown cause, and when she was thirteen, her favorite brother died of peritonitis after his appendix ruptured. Growing up, I sensed vividly how those deaths had permanently darkened my mother's life. Contemplating my grandfather's picture, which sat in the living room, or the snapshots of my uncle Hamilton as a boy, which my mother kept in the cedar chest with the family Bible, made me fear suddenly losing my father or brother to disease as she had.

My parents' lives were marked by perils that did not exist in my own safe, clean childhood of the 1950s and 1960s. Of course, death came to my world, too; in my youth, I saw relatives and acquaintances die, but from heart disease, cancer, and old age, rather than from blood poisoning, typhoid, or peritonitis. Although polio lingered as a source of parental concern in the 1950s, infections were largely a matter of such routine, minor ailments as childhood bouts with measles and mumps and adult encounters with colds and the flu.

I knew in some vague way that good health and good manners required that I wash my hands after using the toilet (especially my grandparents' privy) and cover my nose and mouth when I sneezed or coughed. But those little rituals had no connection in my mind with the scarlet fever or consumption that carried off little girls in my favorite books about the olden days. I had much more concrete fears of death from a nuclear attack or a tornado, two unlikely disasters that my public school prepared for with frequent drills. Beyond one trip to the local high school to take the polio vaccine in the early 1960s, avoiding infection required no precautions as dramatic as those I learned for how to minimize radiation poisoning or stock a bomb shelter.

Although few of my readers may share my formative encounters with outhouses, my experiences are otherwise representative of most middle-class baby boomers. By the 1960s and 1970s, a significant generation gap in disease experience had developed between Americans of different age groups. Those of us born after World War II, especially into the middle classes, were able to take for granted one of the most remarkable developments in modern history—what demographers refer to as the "great mortality transition." The diseases that had killed the majority of people in my grandparents' era—that is, tuberculosis, pneumonia, and typhoid—were defanged. Ours was a charmed existence, protected from diseases that had decimated families and communities for millennia.

Growing up with the security of the antibiotic miracle drugs, we baby boomers were released from the anxieties about casual contact that had shaped personal hygiene during the preceding century. Without any experience of the deadly power of infectious diseases, we regarded the older generation's obsessions with germs and disinfectants as mildly amusing. To the "peace and love" generation, the propensity of our mothers and grandmothers to douse everything in sight, from doorknobs to toilet seats, with Lysol was but more evidence of their old-fashioned, uptight approach to life.

But in the last fifteen years, that youthful confidence has been badly shaken, not only by the appearance of AIDS, but also by a whole new generation of "superbugs," from drug-resistant tuberculosis to the Ebola virus. Newspapers, magazines, television, and movies are filled with stories about menacing germs. Suddenly, we feel vulnerable again to the invisible world of the microorganism. Although I am not certain that the "coming plague" will be as catastrophic as some commentators fear, I am convinced that the achievements of the early twentieth-century public health movement are not nearly so secure as we once thought.

With these contemporary issues in mind, I have discovered a greater appreciation for the disease hazards that my parents and grandparents once faced so stoically. Their experiences hardly seem so distant or quaint anymore. The stories of my aunt Stelsa or my uncle Hamilton, who were cut down in "the bloom of youth," once served to reinforce my sense of distance from the past but now evoke a feeling of kinship. This awareness has spurred my efforts to understand the landscape of disease that they once knew so well and that we may find increasingly familiar.

This book asks a simple set of questions: When and how did ordinary Americans come to believe in the existence of germs? How did they first imagine and understand their coexistence with a world of living microbes unseen by the naked eye? How did coming to believe in the existence of these invisible enemies change the way men and women lived their everyday lives? In answering these questions, I touch on some familiar aspects of medical history and social history, including the development of laboratory science, the expansion of the public health movement, and the implications of the new hygiene for different groups in American society. But I also have a new story to tell about the origins of germ consciousness itself and the ways in which it moved from the realm of scientific investigation and discourse into the American household.

From this perspective, I am most interested in the germ theory of disease as a source of popular beliefs and behaviors about how to avoid illness. In other words, what fascinates me is how bacteriology was applied to the dinner table, the parlor, and the bathroom. *The Gospel of Germs* looks at the educational crusades that brought women and men from all walks of life to believe in the existence of germs and to alter fundamental aspects of their daily lives to avoid them. From 1880 to 1920, reformers of many stripes promoted this code of behavior with religious fervor and made believing in germs part of the credo of modern living—hence the title of the book. These beliefs remain the core of our understandings about how to avoid "catching" diseases from other people and continue to shape our reactions to new infectious diseases such as acquired immune deficiency syndrome, or AIDS.

In exploring the origins of our modern-day beliefs about germs, the reader will be introduced to many curious details of toilet design, dusting, food preparation, and the like, which may at first seem arcane or trivial. But I hope to show how many everyday behaviors we now see as inconsequential, such as disinfecting a toilet, drinking from a water fountain, or screening a window, began out of a fervent desire to evade a grisly death. In the process, I want to call attention to the sanitary underpinnings of everyday life that we take for granted and to remind us what a privilege it initially was—and still is—for Americans to use a flush toilet, to drink an uncontaminated glass of water, and to enjoy a salmonella-free meal.

I hope also that by exploring the historical origins of our beliefs about the germ, we will be better able to face the new challenges that infectious diseases pose to our society. As in my family, the disease memories of one generation are handed on and reinterpreted by the next. Today our experience with AIDS, the Ebola virus, and all the other microbial dangers that now perturb us are shaping the beliefs and behaviors of our children. With greater awareness of the historical dimension of our own health beliefs, perhaps we can better calculate the influence of contemporary crusades to inculcate a new "gospel of germs."

Introduction: The Gospel of Germs

In 1989, a woman wrote the columnist Ann Landers asking for advice. Her fiancé had insisted on inviting a friend who had AIDS to their wedding, and even though she knew none of her guests could catch the virus simply by being in the same room with him, still, she worried, "What if someone should accidentally use R's fork, or drink out of the same glass? What if he should sneeze across the table or, heaven forbid, give me a kiss of congratulations?" The bride-to-be was nervous that if news leaked out that a person with AIDS was attending her wedding, other guests would be afraid to attend.

Like many public health authorities confronted with popular anxieties about HIV, Ann Landers showed no patience with the woman's confusion; she responded testily, "You need to educate yourself."[1] Landers was quite right, of course; HIV is not spread by what public health authorities call "casual" infection, that is, nonintimate contact between the sick and the well. But despite repeated assurances from experts, many Americans continue to worry that any exposure to people with AIDS, or even to the objects they touch, could infect them with the disease. For those who carry the virus, this unshakable faith in the dangers of casual infection has caused untold misery and cruelty.

With no intention of excusing that cruelty, this book nonetheless attempts to understand the fears that have possessed so many Americans in the face of the AIDS epidemic. Although homophobia and racism have played a huge role in their genesis, there is more to those fears than simply ignorance, irrationality, or prejudice. They stem

1

from fundamental beliefs about contagion and its association with certain behaviors and objects—attitudes that even the most enlightened and humane individuals share. People with AIDS and those who love and care for them also experience the dread that seemingly insignificant forms of contact will spread the disease. This book examines how and why that collective sense of apprehension came to be.[2]

When Ann Landers's bride-to-be expressed anxiety about forks and cups and kisses, she unwittingly gave testimony to what I term the "gospel of germs," that is, the belief that microbes cause disease and can be avoided by certain protective behaviors. Today, we recognize that a wide variety of organisms, referred to colloquially as "germs," are capable of causing disease; they include bacteria, viruses, rickettsiae, parasites, and fungi. Beliefs concerning the existence of germs are among the most widely shared scientific precepts governing everyday life in modern Western societies. Although some may still disagree about the link between a specific microbe and a particular ailment, the general principle that pathogenic microorganisms can cause sickness is rarely disputed. Like the law of gravity or the solar-centered planetary system, the so-called germ theory of disease has the aura of a timeless and universal truth.

As a result, we are taught from a very young age to believe in disease agents that we cannot discern with our own senses and to shun certain contacts with other people—including their sneezes, coughs, and feces—as a way to avoid encountering germs. Parents, teachers, health care professionals, and advertisers all continually reinforce the association between practices such as hand washing or refrigerating food and the preservation of health. The rituals of germ avoidance are so many and so axiomatic that we scarcely can remember when or where we first learned them.[3]

Yet far from being timeless or universal, our beliefs and fears about germs are a relatively recent acquisition. Only a century ago, our grandparents and great-grandparents had no idea that the agents of infectious diseases were microorganisms. The reality that we now take for granted—that we share our bodies and homes, our air and food, with a multiplicity of microorganisms, some of which are quite dangerous—they had to be carefully taught. How Americans came to believe in the existence of germs, and how that understanding changed their lives, is the subject of this book.

We can best appreciate the magnitude of this transformation by contemplating the way that ordinary Americans conceived of disease prior to the late 1800s. Long before the germ theory had gained wide acceptance, Americans were aware that people suffering from certain diseases, such as smallpox or bubonic plague, gave off some sort of intangible substance capable of making others sick. Common wisdom held that the sick person's breath, skin, evacuations, and clothing could all harbor the "seeds" of disease and spread them to those who were well. But the nature of this infective substance remained mysterious. The fact that many diseases spread without any known contact with the ill led many physicians to suspect a more generalized, atmospheric source of infection. This suspicion was often referred to as the "miasma" theory.[4]

Although they avoided those who were ill from smallpox or cholera, most nineteenth-century Americans showed little concern about those forms of casual contact with other people, or contamination of water and food, that are today shunned in the name of health. They shared beds, at home with relatives or in hotels with strangers, without inquiring deeply about their bed partner's health. They exchanged combs, hairbrushes, and even toothbrushes, and fed babies from their mouths and spoons, with no sense of hazard. They coughed, sneezed, and spit with blithe disregard for the health consequences to those around them. They stored and cooked their meals with scant concern for foodborne illness. They drank unfiltered water from wells and streams, often using a common dipper or drinking cup. Last but not least, they urinated and defecated in chamber pots and outdoor privies with little regard for where the contents ended up in relation to the community water supply.[5]

To be sure, some Americans began to shun these promiscuous minglings with other people's bodies long before the germ theory of disease was introduced, but they did so for reasons other than avoiding disease. Ever since the Renaissance period, etiquette books had counseled personal cleanliness and had prohibited behaviors such as spitting and coughing among those seeking social distinction. In the eighteenth century, educated and genteel Americans began to cultivate habits of cleanliness to enhance their well-being and to project a pleasant social persona. The pursuit of gentility and politeness, not the fear of disease, fueled a revolution in cleanliness that began among the colonial elite and gradually spread to the urban middle classes.[6]

Still, for all their growing dedication to soap and water, few Americans prior to the Civil War worried about the daily hazards of infection except during times of epidemic. This lack of concern reflected in part how relatively few illnesses physicians and lay persons classified as "catching" prior to the mid-1800s. Many common diseases that we now know to be communicable were thought to be constitutional in origin, that is, the result of poor heredity complicated by unhealthy living habits. Consumption is a case in point. The disease we now recognize as tuberculosis was endemic during the first two-thirds of the nineteenth century: it caused perhaps as many as one in four deaths and took a particularly heavy toll among young adults. Until the 1880s, consumption was widely attributed to an inherited weakness of the lungs, which could be aggravated by overwork, a damp climate, a neglected cold, or overindulgence in alcohol. Those who suffered from the disease had no conception that the droplets they expelled when spitting or coughing carried the bacteria responsible for the disease.[7]

In retrospect, Americans' lack of awareness about the potential deadliness of their secretions is chilling. In his biography, the physician Edward Trudeau, who later became one of the first American converts to the germ theory of tuberculosis, recalled nursing his consumptive older brother through his final illness in the 1860s. For weeks he tended his brother in a hot, close room, sleeping beside him at night to be close at hand, unaware of the risk to himself. The brother's physician instructed Trudeau to keep the windows tightly shut, a recommendation that only increased his likelihood of infection. Within a few years of his brother's death, Trudeau too developed tuberculosis. Not until 1882, when he read of Robert Koch's discovery of the tubercle bacillus, did Trudeau realize that he had most likely contracted the disease by being such a faithful nurse to his dying brother.[8]

Food handling provides another telling example of how different hygiene standards were before people became schooled in the germ theory. Food producers and consumers alike routinely handled food in ways guaranteed to ensure frequent microbial contamination. Milk was left exposed to air and flies, perishables were stored at uneven temperatures, and cooked food was left on the table between meals. With no knowledge of the healthy carrier concept, seemingly robust cooks and waiters passed on the microorganisms of typhoid and other ailments to those they served.

Given how hard it is to prevent food poisoning even today, we can hardly begin to imagine how frequently Americans suffered from foodborne disease in the nineteenth century. To give but one hair-raising example: in 1891, over sixty people attending a wedding in Lyndon, Kentucky, succumbed to a violent gastrointestinal illness. Six people died, including the bridegroom. Suspicions that a spurned admirer of the bride had poisoned the guests with arsenic led to a full-scale medical investigation. The culprit turned out to be the chicken salad, which had been made from meat cooked two days before the wedding and stored in broth at room temperature.[9]

This innocence, or lack of awareness, regarding practices that we know now can spread deadly disease characterized many aspects of daily life prior to the mid-1800s. In those days, Americans neither thought of casual contact or food contamination as an omnipresent source of infection nor assumed that safety from contagion required constant, unrelenting discipline of their bodies and households. In the absence of that awareness, death rates from infectious diseases such as typhoid and tuberculosis rose precipitously as more and more people moved to cities, lived in closer proximity to one another, and depended upon common supplies of water and food.

The growing threat of mysterious fevers and wasting diseases left no class or ethnic group unscathed. Even the most powerful and respected figures in Anglo-American society knew the humbling experience of deathbed scenes. For example, typhoid fever robbed a war-weary President Lincoln and his wife of their adored son Willie; left Queen Victoria grief-stricken over the death of her husband and the near-death of her eldest son; and smote the mother of a future president, Theodore Roosevelt, the same day that his wife died in childbirth. Although then as now the poor suffered more, wealth and social status offered no protection against what contemporaries often referred to as the "invisible enemies" of disease.[10]

In this climate of loss and anxiety, the germ theory of disease began to attract popular interest even before physicians agreed that it was valid. The idea that living organisms had a role in causing disease had a long and venerable history dating back to classical times, but as of the mid-1800s, what was sometimes referred to as the "animacular hypothesis" was distinctly unpopular among medical men, as I will discuss in Chapter 1. During the 1860s and 1870s, however, experimentalists such as the French chemist Louis Pasteur and the German physician Robert Koch compiled increasingly convincing proof that distinctive species of microbes were linked with the most deadly diseases of the era. Starting in the late 1870s, the new scientific discipline of bacteriology scored a succession of dramatic discoveries by rapidly identifying the bacteria responsible for cholera, tuberculosis, gonorrhea, typhoid, and scarlet fever. Although many physicians continued to have reservations about the germ theory of disease, the general principle that microorganisms played a central role in causing communicable diseases had by 1900 achieved widespread acceptance in both Europe and America.

Initially bacteriologists had high hopes that their new understanding of disease would yield effective cures and preventive vaccines. But although a few useful measures were discovered—the diphtheria antitoxin and a vaccine for rabies in the 1890s, the arsenical-derivative Salvarsan for syphilis and a vaccination for typhoid in the early 1900s—hopes for reliable chemotherapies were not realized until the 1930s and the 1940s, when sulfa compounds, penicillin, and other antimicrobial drugs were discovered.

For the first fifty years after its acceptance, the germ theory provided its greatest utility as a guide to the *prevention* of disease through modification of individual and collective behavior. Bacteriologists not only identified the specific agents of infection, but also tracked how they spread from the sick to the well. With an increasingly detailed and accurate road map of the circulation of germs, they could better direct public health efforts to interrupt the way the organisms were spread. Gradually, older theories of atmospheric infection gave way to a more modern understanding of how diseases are transmitted by casual contact, food and water contamination, insect vectors, and healthy human carriers.[11]

These revelations prompted a radical expansion of collective public health practices, including municipal sewerage systems, water purification, garbage collection, and food inspection. Our modern conceptions of governmental responsibility for public health date from this period, 1890 to 1930, which is often referred to as the "golden era" of the American public health movement. These same years also represented a period of intense interest in what I think of as the "private side of public health," that is, the reformation of individual and household hygiene. Between the 1880s and the 1920s, Americans of all ages were subjected to aggressive public health campaigns that taught them the new lessons of the laboratory: that microscopic living particles were the agents of contagion, that sick bodies shed germs into the environment, and that disease spread by seemingly innocuous behaviors such as coughing, sneezing and spitting, sharing common drinking cups, or failing to wash hands before eating.

In retrospect, the task of convincing ordinary Americans that they coexisted with an invisible world of microorganisms—like "Gulliver among the Lilliputians," as one commentator termed it—appears daunting. Germs could not be seen, smelled, or touched. Confirming their existence required gazing down the lens of a microscope, a privilege that many Americans a century ago never enjoyed. The first apostles of the germ faced a considerable challenge in convincing their contemporaries that such an intangible being as a germ existed, much less that it caused potentially deadly illnesses.[12]

At the same time, this leap of faith was perhaps less extreme than it might at first appear. By virtue of their religious heritage, ordinary Americans had been conditioned to believe in an "invisible world" dominated by unseen forces that held the power of life and death; as public health reformers often noted, there were striking similarities between traditional fears of malign spirits and the new views of the germ. In addition, rational or naturalistic explanations for epidemic disease had long assumed the reality of intangible disease agents borne in the unseen miasma or the sick person's breath. What science had done, apostles of the germ argued, was to demonstrate the "true" identity of these invisible, malign agents and to show that they were part of a natural order and thus controllable by human action. Thanks to bacteriology, T. Mitchell Prudden, one of the first Americans trained in the new experimental methods, asserted in 1890, "We no longer grope after some mysterious, intangible thing, before which we must bow down or burn something, as if it were some demon which we would exorcise."[13]

The identification of dread disease with a concrete enemy piqued popular interest in the germ theory from its earliest days. As one commentator observed in *Popular Science Monthly* in 1885, "The germ theory appeals to the average mind: it is something tangible; it may be hunted down, captured, colored, and looked at through a microscope, and then in all its varieties, it can be held directly responsible for so much damage." At the same time, the tones of awe and apprehension so frequently apparent in early accounts of the microbial world suggest the lingering influence of religious and magical views of disease. It is no wonder, then, that when the science fiction writer H. G. Wells needed to rescue humankind from an invasion of Martians in his celebrated 1898 short story "The War of the Worlds," he turned to bacteria as the most powerful deus ex machina at his fictional command.[14]

For all its novelty, however, the new germ theory of disease did not immediately displace established ways of thinking about and warding off contagion. Just as people today respond to the AIDS epidemic in terms of current understandings of infection, Americans a century ago tried to incorporate the microbe into old and familiar explanations for disease. In particular, initial understandings of the germ theory were deeply indebted to an older scientific discipline known as "sanitary science," which stressed the ubiquity of airborne infection and the disease-causing properties of human wastes and organic decay.

The first three chapters of *The Gospel of Germs* explore the marriage of sanitary science and germ theory, which shaped the first generation of preventive lessons aimed against the microbe and taught in the 1880s and 1890s. The initial understanding of germ diseases reflected sanitarian beliefs in the existence of what were termed "house diseases," that is, illnesses caused by defective plumbing, ventilation, and housekeeping. As a result, Americans first understood the chief menace of microbes to arise from their toilets and washbasins, which they feared as portals for dangerous bacteria-laden "sewer gas" to enter their homes. This fear led to an obsessive concern with domestic plumbing and ventilation.[15]

With the maturation of bacteriology in the 1890s and early 1900s, there came a second, more expansive version of the gospel of germs, described in Chapter 4. The recognition that consumptives' coughs and sneezes contained the tubercle bacillus intensified attention to the infective discharges of the mouth and nose. Correspondingly, turn-of-the-century preventive education focused increasingly on the exchange of germs through unguarded coughing and sneezing, shared drinking cups, and other common practices that facilitated the transfer of saliva. But the new anxieties about infection via casual contact by no means eclipsed the older focus on "house diseases." For example, bacteriologists found that they could culture the tubercle bacillus from common house dust; thus ridding the home of that dust became a fundamental tenet of tuberculosis prevention. (Only later did further investigations reveal that the germs cultured from dust had little infective power.) Theories of "fomite infection," which held that objects could harbor the dried microbes of disease for months and even years, heightened concerns about household furnishings and clothing. Evidence that houseflies carried the tubercle bacillus and other germ diseases on their feet led to crusades to get window screens in every home. As a result, the house disease concept remained an integral part of the turn-of-the-century gospel of germs.

Bacteriologists' fine-tuning of the preventive lessons of everyday life coincided with the mass dissemination of the gospel of germs during the Progressive period. During the 1880s and 1890s, avoiding germs had been primarily the obsession of prosperous urban families. In the early 1900s, however, reformers sought to bring hygienic enlightenment to all Americans, in order to emancipate the whole society from the fear of infectious diseases. To that end, the gospel of germs was taken up by an impressive array of Progressive-era institutions, including municipal and state health departments, life insurance companies, women's clubs, settlement houses, Boy Scouts and Girl Scouts, YMCAs and YWCAs, labor unions, and agricultural extension programs.

In Chapters 5 and 6, I look at two of the most influential conduits of popular germ-consciousness in the early 1900s, the antituberculosis crusade and the domestic science movement. To get across the message that T.B. was a communicable disease, the antituberculosis movement pioneered methods of health education that were copied widely by other public health workers and that with surprisingly little modification remain in use today. Borrowing overtly and enthusiastically from the new advertising culture of the early 1900s, antituberculosis workers turned out posters, slogans, and other forms of propaganda to "sell" their message of protection against the ubiquitous tubercle bacillus. The domestic science movement, which originated at about the same time as did the antituberculosis crusade, focused more specifically on educating housewives and mothers about germ life in order to make the American home more healthy and productive. During the early twentieth century, an army of new female professionals, including home economists, visiting nurses, and social workers, dedicated themselves to bringing the insights of "household bacteriology" to every homemaker in the nation.

Although it was portrayed as a set of obligations that both sexes had to honor, the kind of cleanliness required by the gospel of germs clearly had more profound implications for women, both as private citizens and as professionals. Men certainly had their roles to play under the new germ credo, but on a day-to-day basis the bulk of the worry and work fell on women, most of whom were housewives. The gospel of germs turned even humble chores such as dish washing and sweeping into "a fine action, a sort of religion, a step in the conquering of evil, for dirt is sin," to quote the pioneer home economist Ellen Richards. The association between house diseases and housekeeping simultaneously ennobled women's work in the home and made it more physically and emotionally burdensome.[16]

The third section of the book looks more closely at how different groups of Americans began to understand and act upon the gospel of germs. In Chapter 7, I survey the extraordinary range of behaviors that affluent Americans changed between the 1890s and the 1920s in order to evade the germ. Men shaved their beards and women shortened their skirts to eliminate potentially germ-catching appendages. They stripped their homes of allegedly microbe-laden furnishings and embraced as necessities for a germ-free life such household institutions as the white china toilet, the vacuum cleaner, and the refrigerator. They purchased, stored, and cooked their food in new ways designed to retard bacterial contamination. They learned to avoid other people's sneezes and coughs, and to shy away from familiar social customs such as handshaking and baby kissing. This domestic revolution carried over into "homes away from home" such as hotels, railway cars, restaurants, and even funeral parlors. As more and more germ-conscious Americans dined out and traveled, the institutions that served them strove to cater to their sanitary scruples. Hotels began to supply individual cakes of soap and to use extra long sheets so that sleepers might fold them back over potentially germ-ridden blankets. Churches adopted individual communion cups, and cities installed sanitary water fountains to replace the contagion-spreading common cup.

Marketplace forces and mass advertising played a central role in fostering this more "antisepticonscious America." We rarely think of corporate America, particularly advertising agencies, as having much to do with the dissemination of scientific ideas. Yet as both Chapters

3 and 7 make clear, reformers and educators were not the only ones to interpret the germ theory's relevance for the habits of everyday life. From the 1880s onward, entrepreneurs and manufacturers of all sorts realized that the fear of the microbe could be effectively exploited to sell a wealth of goods and services. Under the aegis of the first, sanitarian-dominated gospel of germs, entrepreneurs promoted safeguards against the dangers of sewer gas and polluted water, such as special toilet attachments and household water filters. In response to the second, more bacteriologically informed gospel, the range of aids to combat the microbe expanded dramatically between 1895 and 1915 to include everything from antiseptic floor coverings and wall paint to sanitary dish drainers and fly traps.

Germ-conscious advertising campaigns became an educational force, yet they did not represent a simple extension of the work of the antituberculosis crusaders or the domestic scientists. For all their invocations of laboratory authority to sell their products, manufacturers and advertisers displayed no deep allegiance to the scruples of science. By keeping alive aspects of the gospel of germs that public health experts wished increasingly to jettison, such as the fear of sewer gas, advertising served other, more profit-oriented motives.

Although hygiene reformers and manufacturers both cast the principles of germ protection as universal goods in American society, the expense of sanitary protections placed these aids out of the reach of many Americans well into the 1930s. Working-class families could ill afford even the most basic prerequisites for practicing the gospel of germs, such as flush toilets, clean running water, and a safe milk supply. The ability to conform to "antiseptic" standards of cleanliness differentiated rich from poor, educated from unschooled, American-born from foreign-born. In Chapter 8, I look at some of the consequences of the uneven spread of sanitary boons as it affected two different groups of women on the periphery of middle-class America: immigrant housewives and farm women.

For many middle-class Americans in the early 1900s, the association of poor, immigrant, and non-white citizens with disease germs only deepened their feelings of class prejudice, nativism, and racism. By harping on the menace of contagion, the apostles of the germ inevitably increased the stigmatization of the sick and the poor. The specter of infection served nativists and racists well in their efforts to legitimate immigration restriction and racial segregation.

But simultaneously, the gospel of germs gave rise to other, counter-vailing pressures toward inclusion and reform. Many converts to the germ theory believed deeply in a "chain of disease," a "socialism of the microbe" that linked all members of American society together. If not for simple humanity, then for this reason alone, they argued, the health problems of the poor and the newcomer had to be addressed. The great disease crusades of the early twentieth century created a common set of assumptions about contagion that crossed class and race lines and became intellectual capital in movements for broad-ranging social change. Chapter 9 explores how labor unionists and African American community activists tried to turn the fear of disease into weapons to do battle for economic and social justice.

The main body of the book ends in the 1920s, when the intensity of efforts to spread the gospel of germs began to fade. With further bacteriological scrutiny, the frequency of infection by dust and fomites came into question; instead, experts placed increasing emphasis on the importance for disease control of contact infection and healthy carriers of disease. Practitioners of the so-called new public health advocated more attention to identifying and isolating the sick and distanced themselves from the expansive social concerns and evangelical fervor of their predecessors' educational crusades.

At the same time, many of the basic tenets associated with the gospel of germs remain the foundation of public health practice to this day. The idea that germs may survive months and even years in dust and on fomites may have been discarded, but most other principles of the preventive code have been preserved. As one reads through the latest edition of the American Public Health Association's standard handbook on the control of infectious diseases, it is evident that droplet infection via sneezing and coughs, casual contact, fecal contamination of water and food, insect vectors, and improper food handling are still recognized as the most common ways that infectious diseases spread.[17]

But after World War I, personal and household hygiene practices gradually came to be less essential to the control of deadly disease. The gospel of germs declined in importance, largely due to the steady decline in mortality rates from infectious diseases and a strengthening of collective protections against germs, such as water filtration and food regulation. By the late 1920s, heart disease, kidney disease, and cancer had replaced respiratory and gastrointestinal infections as the leading causes of death. Understandably then, both scientific interest and public health initiatives turned increasingly to the prevention of chronic, noninfectious ailments that now consti- tuted the chief threat to American health. Still, as Chapter 10 makes evident, germ consciousness remained strong until the 1950s, largely due to the polio menace and the influence of advertising. Not until the widespread availability of antibacterial drugs after World War II did the gospel of germs truly fade into insignificance as a road map for avoiding deadly disease—at least until the AIDS epidemic. I sug- gest in the Epilogue that the gospel of germs has taken on new relevance since 1980, as Americans have been confronted with a new generation of "superbugs," chief among them the human immuno- deficiency virus.

The "triumph" of the germ theory of disease has long been a central theme in both medical and social history. Reams of paper have been devoted to the scientific antecedents and experimental discoveries that gave rise to the modern view of infection. There are many fine studies of the changing thought of individual scientists and of con- ceptions of specific diseases, such as cholera and tuberculosis. Histo- rians of public health, of women, of advertising, and of architecture have all noted the fact that Americans became extraordinarily germ conscious at the turn of the century. Yet surprisingly little has been written about the collective dimensions of this great watershed in thinking about disease, and the ways in which the "lessons of the laboratory," as I term them, became part of the fabric of everyday life.[18]

In tracking that collective consciousness, I have been inspired by historians of science such as John Burnham, Roger Cooter, Bruno Latour, and Martin Pernick, who have drawn new attention to the historical processes by which popular understandings of disease are constructed. Like them, I prefer to think of popularization not as a hierarchical, top-down process where the focus is on what the public gets "right" or "wrong," but as a dynamic where ideas and images are traded among different audiences, including laboratory scientists, practicing physicians, hygiene reformers, and interested lay people. Instead of treating popular views as merely pale, distorted images of the "real" knowledge generated by "real" scientists, such a model allows for ideas to travel in more than one direction, to accommo-

date, for example, the influence of sanitarian thought on early formulations of the germ theory. This approach also helps to describe
what interests me most, that is, how scientific precepts become a part
of the working hypotheses of everyday life, or what is sometimes
termed "ethnoscience."[19]

Although I have sought to avoid a hierarchical view of popularization, I have tried not to gloss over the simplifications and distortions
that inevitably occurred as bacteriological knowledge moved from
the laboratory to the parlor, so to speak. Judged as a rational process
of education, there is no doubt that the great disease crusades of the
early twentieth century fell short. To the extent that many public
health reformers ceased trying to convey any comprehensive understanding of disease and focused only on teaching health habits, the
gospel of germs became what John Burnham has termed the "func-
tional equivalent of superstition." Shorn of its scientific underpinnings, the new germ credo all too easily turned into another form of
magic to be followed blindly and mechanically. When that magic
failed and illness could not be avoided, its practitioners were left with
a heavy load of anxiety and pain, which easily became fodder for the
sorts of irrational hatreds and prejudice directed toward consumptives a century ago, or toward AIDS patients today.[20]

At the same time I acknowledge the dark consequences of the
gospel of germs, I also tell another side of the story that has been
given less attention by historians. The great disease crusades of the
early twentieth century did not succeed in giving every American
an accurate understanding of the germ theory of disease, yet they
nonetheless did a remarkable job of inducing people from varied
backgrounds to change fundamental personal behaviors. Even if
practiced only as a form of health "superstition," the rules of germ
avoidance still served a useful function; a family need not understand
the principles of the germ theory of disease in order to abandon
household practices that spread fecal contamination and thereby
enjoy less risk of cholera or typhoid.

My interest in individual and household *practices*—that is, the habits that ordinary Americans associated with disease prevention—takes
this book in directions rather different from those followed by previous historians of public health. For many years, the golden age of
public health has been equated with the collective, state legislated
measures enacted at the turn of the century. The popular education

campaigns of the same time have been comparatively neglected, in large part due to the assumption that what they taught did not matter very much. That dismissive view dates back to the 1910s and 1920s, when proponents of the new public health began to extol the triumph of "real" science over the "bogus" precepts of their predecessors—a view enshrined in popular works such as Sinclair Lewis's novel *Arrowsmith* and Paul de Kruif's *The Microbe Hunters*.[21]

Until recently, historical demographers tended to reinforce the assumption that changing personal and household practices had little to do with the decline in mortality rates from infectious disease that began in the nineteenth century. Here the work of the English physician Thomas McKeown was especially influential. In the 1960s and 1970s, McKeown argued that the great mortality transition was an unintended outcome of better nutrition and higher living standards, rather than a result of the interventions of either organized medicine or the public health movement.[22]

When I first began this book, I will admit that I too was inclined to see the personal health practices that I was tracking as having little demographic significance. But I soon came to accord my subject more respect, for two reasons. First, I was continually reminded by current events how these same practices were still promoted as fundamental forms of disease control. In the precautions urged on New Yorkers during the T.B. mini-epidemic of the early 1990s, and the rules of safe meat handling issued by the U.S. Department of Agriculture, I could see the direct descendants of the lessons in tuberculosis control and household bacteriology that I was studying.

I was also profoundly influenced by the growing interest in personal health practices shown by a new generation of historical demographers and medical historians. Scholars such as Gretchen Condran, Anne Hardy, Samuel Preston, and Simon Szreter have begun lately to suggest that voluntary changes in health behavior contributed more to the mortality decline than the so-called McKeown thesis allowed. This new work points out that death rates began to diminish decades before citywide public health works had ensured safe sewage systems, water purity, and food supplies. Had individuals and families begun to practice some of the sanitarians' recommendations, such as avoiding fecal contamination, boiling drinking water, and practicing isolation procedures while nursing sick relatives in the home, it is conceivable that they could have improved their families' chances of survival.[23]

Prior to 1900, the demographic effect of such household-level changes probably remained modest, especially in reducing rates of infant and child mortality. But the work of Douglas Ewbank and Samuel Preston suggests that between 1900 and 1930 popular health crusades aimed at teaching principles of milk purification and childhood disease management to mothers contributed to a striking drop in death rates among the young. The demographic data, they conclude, suggest that "changing personal health practices may have been an important contributor to the decline in infant and child mortality" in both the United States and Britain.[24]

My study does not pretend to make any definitive contribution to this complex demographic debate. Knowing that those debates are going on, however, has strongly affirmed my conviction that the revolution in personal hygiene behavior is far more worthy of serious analysis than the work of Lewis and de Kruif allowed. This conviction has been further strengthened by my appreciation of the gendered dimensions of this revolution. As the new demographic work suggests, much of the germ theory's relevance for personal health practices fell in the realm of housecleaning, childcare, and food preparation, domains traditionally designated women's work and consequently ignored or trivialized. One of my goals in writing this book has been to challenge the implicitly gendered division of knowledge that regards as significant what Pasteur did in the laboratory but dismisses as inconsequential what a public health nurse or housewife did with his insights.

Taking personal hygiene behaviors seriously as forms of disease prevention inevitably puts my interpretation at odds with another long scholarly tradition, that of analyzing germ fears primarily as cultural artifacts. In their influential accounts of cleanliness behaviors, the German sociologist Norbert Elias and the English anthropologist Mary Douglas formulated a position that many scholars have followed ever since: that the apprehensions about disease expressed in the pursuit of cleanliness are a mere rationalization for, in Douglas's words, "gestures of separation and classification" that serve other, more powerful needs to create and maintain social order. As Douglas put it in her 1966 classic *Purity and Danger,* "In chasing dirt, in papering, decorating, tidying we are not governed by anxiety to escape disease, but are positively re-ordering our environment, making it conform to an idea."[25]

This perspective has led to a much needed appreciation of the cultural dimensions of cleanliness and has certainly shaped my own understanding of the gospel of germs. I agree with the premise that no disease is ever observed in a totally unbiased way; there is always a scrim of culture affecting our perceptions of and attempts to treat illness. Yet taken too far, the Douglas approach too easily discounts fear of disease as a motivation for specific cleanliness behaviors. The perspective of *Purity and Danger* reflects the confidence of the 1960s, the same era in which the U.S. surgeon general decreed that infectious diseases no longer constituted a serious threat to the public's health. Without denying that the cultural construction of dirt reflects more than just the fear of disease, my interpretation emphasizes how everyday encounters with illness and death reinforced the lessons of germ avoidance.[26]

To use a modern parallel, imagine an anthropologist writing a hundred years from now about the AIDS epidemic and interpreting such preventive measures as using condoms and disinfecting needles with bleach purely as "gestures of separation and classification" aimed at homosexual men and intravenous drug users. Such a position would strike us as ludicrous. Even the most ardent proponents of the cultural construction of disease are unlikely to deny that safe sex and clean needles save lives. Dismissing late-nineteenth-century reformers' efforts to convince people to prevent fecal contamination of their water supply or to eliminate pathogenic microorganisms from their food as merely attempts to act out middle-class status anxieties or to stigmatize certain groups of Americans is equally simpleminded.

My historical analysis seeks to strike a balance, then, that honors both the cultural construction of cleanliness and the biological dimension of disease. Ailments such as typhoid and tuberculosis do have a biological reality, a set of distinctive pathological features that shape the cultural meanings attached to them. Fears of their dangerousness are founded in painful experiences of illness and death that must not be overlooked. Bringing this "biological body" back into the historical narrative is essential to understanding the transformations wrought by changing disease theories a century ago.[27]

My narrative also questions the widespread tendency to blame the acceptance of the germ theory for the limitations of the modern biomedical model of disease. The growing authority of bacteriology as a scientific discipline is usually portrayed as a conservative development that inevitably narrowed the scope of American medicine and encouraged new forms of discrimination. The germ theory is often linked with images of inspection, exclusion, and incarceration: the Ellis Island inspectors using buttonhooks to check immigrants' eyelids for trachoma, the health officials sending poor consumptives off to the sanitarium to die, or the sad figure of Typhoid Mary banished to North Brother Island for over thirty years.[28]

These developments were certainly one consequence of the germ theory's acceptance, but they do not represent its sole legacy to American politics and culture. Although historians have traditionally highlighted their invocation in campaigns for immigration restriction and racial segregation, the "truths" of the germ theory were also invoked in less conservative movements for economic justice and social equality. By looking closely at these latter efforts, I want to emphasize that the discovery of the germ had no fixed moral or social message; there was nothing inherently narrow or discriminatory in the germ theory of disease itself. Its meanings for everyday life were susceptible to multiple interpretations and were deployed in competing arguments about the problems of American society. If certain views gained more influence than others, we must look to the political and cultural context of the debate, and not to the theory itself, for an explanation of that fact.[29]

Although this study focuses only on the United States, it is important to acknowledge that similar public health crusades occurred during the same time period in other Western countries such as England and France. Through the influence of colonial rule, the gospel of germs was also exported to many non-Western nations such as China, India, and the Philippines. Any systematic comparison of these movements is beyond the scope of this already overly ambitious study. But on the basis of my limited forays into other national scenes, I would like to emphasize two aspects of the American health crusades that I believe to be distinctive.[30]

First, I would point to the heavy influence of advertising methods and consumer-oriented approaches. Recent studies of advertising and popular culture have underlined their precocious and dominant effects on American society. My work suggests that this influence carried over into the public health movement as well. American hygiene reformers displayed a precocious interest in and talent for exploiting the new forms of mass communication and persuasion available in the early twentieth century. Conceptions of "selling" health were central to their programs of popular education. Second, I believe that crusades against disease have played a special role in American political culture. Although other nations have certainly invoked health as a common civic goal above the fray of party politics, that vision has loomed especially large in the United States. In a democratic society riven by gender, racial, ethnic, and class differences, notions of public health citizenship have offered a seemingly neutral ground for building consensus, for purposes of both exclusion and inclusion.[31]

Although I seek to challenge some of the conventional academic wisdom about the effects of the germ theory on American medicine and popular culture, let me be the first to point out that I commit some major sins of overgeneralization of my own. It is impossible to write about the evolution of ideas about germs without imposing more order on that process than actually existed. When I interpret the meanings assigned to phrases such as "the germ theory of disease" or the "gospel of germs," I endow them with more coherence and consistency than they ever really had. Rather than place those phrases in quotation marks throughout the whole text, I warn the reader now that I use them as a form of historical shorthand to track a very untidy set of ideas.

I have tried also to avoid using terms such as "popularization," "mass education," or "the public" as if any such unitary processes or collective entities existed. In the time period of my study, the 1870s to the 1920s, Americans were, as they are now, a highly heterogenous people. Their diversity requires that we look carefully at how different groups responded to the same set of ideas. I have thus tried to balance sweeping generalizations with in-depth case studies that bring a wide range of voices into the narrative.

Likewise, I want to acknowledge the importance of *individual* variance. Members of the same gender, ethnic, class, or racial group may hold very different opinions on matters of health. One member of a family might be very anxious about the threat of radon gas, whereas another might dismiss it as an overblown fear. The same was undoubtedly true in the period of my study. Some individuals shrugged off the dire warnings about germ dangers, whereas others worried about them incessantly. These varying degrees of germ consciousness were shaped by factors such as parental views about cleanliness, personal experiences with illness, and basic traits of personality that I as a historian cannot easily assess. At best what I have done here is to track a cluster of beliefs and behaviors concerning infection that were held by many, if not all, Americans in the decades between 1870 and 1930.

One final note on terminology: technically, the terms "infectious," "contagious," "communicable," and "epidemic" have different meanings when applied to disease. "Infectious" denotes a disease that may spread from person to person without actual contact between them; in contrast, a contagious disease is directly transmitted from person to person. "Communicable" is a more general term that covers both infectious and contagious diseases. Epidemic diseases spread rapidly from a few cases to a large number of people, then gradually disappear; endemic diseases exist more or less constantly in the population.

These terms had similar meanings in the late nineteenth century. In actual use, however, the distinctions are and were difficult to maintain. In the time period of my study, medical authorities frequently admitted the hopelessness of precision in the usage of these terms, especially the futility of maintaining the distinction between infectious and contagious diseases. In the *Transactions of the American Medical Association* for 1866, one physician expressed the general sense of frustration: "The three appellatives, Epidemic, Contagion, and Infection, not infrequently confuse the investigator, and the boundary line between them seems even more imaginary than the equator."[32] With all due respect to the technical differences between infectious and contagious, which I know to exist, I have also chosen to "ignore the equator" and use those terms interchangeably in this book.

· I ·

The Gospel Emergent, 1870–1890

1 · Apostles of the Germ

On a cold, damp day in February 1884, the children of Martha Bulloch Roosevelt were summoned to attend at her deathbed. A few days earlier, Martha, the widow of the prominent New York City philanthropist Theodore Roosevelt, Senior, had developed what had appeared to be a slight cold. As her condition rapidly worsened, her doctor diagnosed her illness as typhoid fever, an often fatal gastrointestinal disorder. Telegrams were sent to her daughter Corinne in Baltimore and to her son Theodore in Albany, urging them to come quickly to their mother's elegant home on West Fifty-Seventh Street. For Theodore, the summons had a second somber purpose: in the same house lay his wife, Alice, who had just given birth to their first child and was gravely ill from a kidney ailment known as Bright's disease. At three o'clock in the morning on February 14, Martha Roosevelt died, surrounded by her children. Alice Roosevelt soon followed, dying in her husband's arms less than twelve hours later. Years afterward, Corinne could recall the words that her brother Elliott spoke to her as she arrived home that night: "There is a curse on this house."[1]

The affluent mourners who crowded the Fifth Avenue church for the double funeral of Martha and Alice Roosevelt must have pondered the "strange and terrible fate," as Theodore put it, that so prematurely took their lives. Martha was only forty-eight and Alice but twenty-two years old. Family and friends no doubt regarded Alice's demise as the more tragic, given the newborn daughter she left behind, yet in one sense it was the more understandable. The rigors of childbirth claimed many women's lives, and one who had an undiagnosed kidney disease, as Alice did, would be especially vulnerable.[2]

Martha Bulloch Roosevelt's death was a different matter. She succumbed to what by the early 1880s had been clearly typed as a "filth disease," that is, an illness spread by fecal contamination. As such, typhoid fever was considered eminently preventable by the practice of proper domestic sanitation. We now know that Martha Roosevelt could easily have contracted typhoid outside her home from eating contaminated seafood in a restaurant or from coming in contact with a healthy person who carried the bacillus, such as the infamous cook "Typhoid Mary" Mallon. But as of 1884, these modes of infection had yet to be discovered, and public health authorities thought that typhoid spread via defective household plumbing, which tainted air and water with the disease "poison."[3]

At the time of Martha Roosevelt's death, the precise nature of the typhoid poison was in dispute. The majority of physicians thought the disease agent was a chemical substance produced by decaying fecal matter. In contrast, a small but growing number of public health experts believed that the chief danger came not from the fecal matter but from the living microorganisms it contained. In 1880, two German physicians had isolated a species of bacteria, subsequently referred to as "Eberth's bacillus," that they believed caused typhoid fever. The same year that Martha Bulloch Roosevelt died, the New York City Department of Health issued the *Handbook of Sanitary Information for Householders,* which warned that "the greatest danger . . . in the breathing of sewer-air is that of inhaling with it the living particles (bacilli, etc.) contained or developed in the excreta of diseased persons."[4]

But although experts argued over the causal agent, no one questioned that typhoid fever's appearance in any home should prompt close scrutiny of its sanitary arrangements. Therein lay the mystery of the Roosevelts' tragedy. The mansion on West Fifty-Seventh Street, which Theodore Senior had built in the 1870s, had been designed by the leading architects of the day and equipped with the finest fixtures and appointments. Moreover, Martha Roosevelt, familiarly known as "Mittie," was excessively concerned about cleanliness, to a point that family and friends considered almost obsessive. She moved from rural Georgia to New York City after her marriage in 1853 and was never reconciled to the extraordinary filth of that rapidly growing metropolis. To combat it, Mittie developed a rigorous set of cleanliness rituals, which included bathing daily with two changes of water, putting a sheet on the floor when she knelt to pray at night, and wearing white clothing even in winter so that no speck of dirt could escape her scrutiny. With the assistance of a small army of servants, she kept her home equally pristine. A family friend recalled reading a many-paged set of housekeeping instructions prepared for her daughter-in-law Alice, outlining an exacting regimen of polishing, scrubbing, sweeping, and dusting. For example, each morning the cook was required to greet the ash man with a bucket of boiling water to scald his ash can before it entered the house.[5]

Yet for all these heroic efforts to keep herself and her home spotlessly clean, Martha Roosevelt succumbed to what public health experts knew to be a "filth disease." Such a fastidious woman would surely have died of shame, had the typhoid fever not killed her, to discover that she had contracted a disease spread by fecal contamination. Martha Roosevelt's death from typhoid epitomized the uncertainties that beset even the most conscientiously clean households of the Gilded Age: no one, not even the most careful, seemed to be safe from the invisible agents of disease.[6]

The heated scientific debates of the 1870s and 1880s over the identity of those invisible enemies coincided with a period of intense anxiety about rising disease rates in both the United States and Europe. The mortality statistics collected by municipal authorities confirmed what personal experiences such as the Roosevelts' had suggested—that rates of illness and death had risen alarmingly in the nineteenth century, making large cities very unhealthy places to live. Not only were inhabitants thrown into periodic crisis by recurrent epidemics of diseases like cholera and smallpox, but they were also harried by endemic diseases such as typhoid and pneumonia that killed steadily year after year. The young were particularly at risk: in Martha Roosevelt's New York City, for example, one-fifth of all infants died before the age of one, often of the dreaded "summer complaint," or infant diarrhea, and those lucky enough to survive until adulthood still faced nearly a one in four chance of dying between the ages of twenty and thirty.[7]

As a result, late-nineteenth-century Americans of all classes had an intimate knowledge of ailments that are rarely seen today, even by specialists in infectious disease. They were familiar, for example, with the blue skin and rice-water discharges of cholera and the high fever and rash that signaled typhoid. They could recognize the characteristic skin eruptions of smallpox, and the sore throat, strawberry tongue, and sunburn-like rash of scarlet fever. They could differentiate the coughs associated with whooping cough, pneumonia, and consumption. They knew too well the chronic diarrhea and wasting that indicated the "summer complaint" and the labored respiration and blocked airways produced by diphtheria.

This all-too-everyday experience with disease and death left many urban Americans with a profound sense of dread. Even the "best" households seemed under siege from mysterious fevers and wasting diseases that came and went with little predictability. On the surface, the everyday life of affluent Victorians appeared comfortable and well ordered; compared to their grandparents and parents, for example, Martha Bulloch Roosevelt's generation had achieved an unprecedented degree of gentility. Yet they still fell prey to diseases of uncleanness. This sense of vulnerability created the backdrop for the growing debate over the germ theory of disease, a theory that its advocates felt explained the mystery of why the determined pursuit of cleanliness had failed to protect Martha Roosevelt and her contemporaries.

A World of Unseen Dangers

In an 1880 paper delivered to the San Francisco Medical Society, imposingly titled "On the Supposed Identity of the Poisons of Diphtheria, Scarlatina, Typhoid Fever, and Puerperal Fever," a physician named William H. Mays began emphatically, "I will state at the outset that I am an ardent germ-theorist, viewing any doctrine that conflicts with that theory much as I would an attempt to controvert Newton's law of gravitation." In catechism style, he spelled out the tenets of his faith in these terms:

> I hold that every contagious disease is caused by the introduction into the system of a living organism or microzyme, capable of reproducing its kind and minute beyond all reach of sense. I hold that as all life on our planet is the result of antecedent life, so is all specific disease the result of antecedent specific disease. I hold that as no

germ can originate *de novo,* neither can a scarlet fever come into existence spontaneously. I hold that as an oak comes from an oak, a grape from a grape, so does a typhoid fever come from a typhoid germ, a diphtheria from a diphtheria germ; and that a scarlatina could no more proceed from a typhoid germ than could a sea-gull from a pigeon's egg.[8]

Read over a century later, when the germ theory of disease is viewed as a scientific truth on the order of Isaac Newton's law of gravitation or Charles Darwin's theory of evolution (to which Dr. Mays also subscribed), this declaration of faith sounds decidedly strange. Whatever controversies may exist today about the nature of infectious diseases, scientists no longer debate whether the streptococcus responsible for scarlet fever can generate spontaneously, much less turn into the bacillus that causes typhoid fever. But May's litany of beliefs captures precisely what it meant to believe in this new theory of disease when the majority of doctors in Europe and the United States still believed that a seagull could hatch from a pigeon's egg, so to speak.[9]

As of 1880, the majority of Anglo-American physicians found these radical ideas about disease causation hard to accept. The medical world was firm in its allegiance to another explanation, the so-called zymotic theory of disease, which rested on a different set of convictions: that the disease agents were chemical ferments produced by decaying filth, and that they could generate spontaneously given the right atmospheric circumstances. Moreover, they were more than satisfied with the progress of preventive medicine, or "sanitary science" as it was often called, in suggesting ways that the zymotic diseases might be brought under better control.[10]

Given the widespread satisfaction at the time with the zymotic theory and sanitary science, it is little wonder that advocates of the germ theory expressed their new faith with such bravado. In the 1870s, believing in the germ theory was often likened to a religious conversion. Its adherents referred to themselves as "converts" to the new "doctrine" and presented the tenets of their creed in catechism form, as did William Mays. Like born-again Christians, ardent germ theorists saw the world with new eyes, as a place where air, water, and soil teemed with invisible life and their own skin and secretions swarmed with microbes. As the microscopist Lionel Beale put it, "the higher life is everywhere interpenetrated by the lower life," locked in a microscopic survival of the fittest.[11]

The English word "germ," which derives from the Latin verb "to sprout," had long been used to refer to the intangible "seeds" of contagion. Advocates of the new theory adopted the term to signify any microscopic organism capable of causing human or animal disease. Researchers would eventually be able to distinguish among the larger and more complex organisms, such as bacteria, fungi, and parasites, and the much smaller viruses and rickettsiae. But in the 1870s, the individual agents lumped under the category of "germ" or "microbe" were not so well known, and early expositions of the germ theory employed a bewildering variety of terms to describe them, including "vibrio," "algae," "cryptograms," "microzymes," and "schizophytes."[12]

Conversion to this way of looking at infectious disease required faith in a new mode of scientific investigation. Traditionally, physicians had based their theories about disease on observations of illness in the individual and the community, or what today we would term clinical and epidemiological evidence. In contrast, belief in the germ theory rested on evidence derived from the laboratory. Experimental methods for linking germs and disease began to develop in the 1860s and 1870s, most brilliantly in the work of the French chemist Louis Pasteur and the German physician Robert Koch. For adherents of the germ theory, the evidence that experimentalists gathered from microscopic examinations, test-tube cultures, and animal experiments provided a new kind of divination into the fundamental nature of disease.

But for most of their contemporaries, the new experimentalism seemed little reason to abandon insights derived from decades of clinical and epidemiological observation, insights that supported the validity of the zymotic theory and sanitary science. As a result, from 1865 to 1895 Western medicine underwent a virtual civil war over the truth of the germ theory. In both Europe and the United States, the profession divided into hostile camps that jousted across countless pages of medical journals and textbooks. In the end, advocates of the germ theory triumphed: by the 1890s, medical students were being educated to revere the germ theory as scientific orthodoxy and to regard Pasteur and Koch as heroes.[13]

But in the 1870s and early 1880s, when the new view of disease was first introduced to both medical and popular audiences, it had yet to ascend to that privileged status. Instead, the germ theory was often linked to an ancient and discredited tradition in medicine referred to as the "animacular hypothesis." As its advocates knew well, the proposition that the agents of infection were living beings had a long history. In a widely read 1874 article on the germ theory, Karl Liebermeister noted that "positive indications of such an idea are to be found among the writers of antiquity." In succeeding centuries, observers periodically hypothesized that a mysterious *contagium vivum* might account for the spread of epidemic diseases such as the bubonic plague.[14]

With Antoni van Leeuwenhoek's invention of a simple microscope in the late 1600s, Liebermeister continued, "some sort of an actual basis for such theories was furnished by the microscopical demonstration of very minute living organisms, invisible to the naked eye." In a series of widely reported observations, the Dutch merchant detailed a world of microscopic characters, many of which lived in or on the human body, who seemed likely candidates for the elusive contagium vivum. Unfortunately, eighteenth century believers got carried away with their microscopic imaginations. Despite the crudeness of their instruments, they devised elaborate identities and family trees for the different microorganisms and produced detailed drawings of creatures with "crooked bills and pointed claws," which some proposed shooting out of the sky with cannons. It was entirely understandable, Liebermeister observed, "that such fantastic ideas should bring down ridicule upon the whole theory," and the weight of medical opinion turned in favor of an atmospheric theory of infection.[15]

The cholera epidemics of the 1830s revived interest in the contagium vivum theory, especially after the English physicians John Snow and William Budd demonstrated that the disease was spread by water polluted with the bowel evacuations of the sick. In 1840, Budd declared his belief that the cholera poison was a living organism. Still, few of his contemporaries were converted to that view, and Liebermeister noted that of this older generation, the German physician Friedrich Gustav Jacob Henle, writing in 1853, was "perhaps the last who elaborated the theory of a *contagium vivum*." During the middle decades of the nineteenth century, the zymotic theory of disease held virtually undisputed sway in Western medicine.[16]

Still, many naturalists and physicians continued to study microorganisms. Their ability to do so was greatly enhanced in the late 1820s by the introduction of the achromatic compound microscope, invented by an English wine merchant named Joseph Jackson Lister; the new instrument eliminated the problems of distortion at high magnification that had long hampered microscopic observations. As better and cheaper instruments became available, microscopy became a popular pastime among physicians and lay people in both England and the United States. Their growing familiarity with the microscopic world helped set the stage for the rebirth of the old animacular hypothesis as the new germ theory of disease.[17]

This metamorphosis began in the late 1850s and early 1860s with the work of Louis Pasteur. At first glance, it may seem curious that a chemist rather than a physician played the pivotal role in starting a revolution in medical thinking, yet chemistry and medicine had long been intimately related. The zymotic theory of disease was associated with the German chemist Justus von Liebig, whose work had helped popularize the analogy between disease and fermentation. In taking up his research on fermentation, Pasteur knew that his studies had potential significance for theories of disease.[18]

Trained in chemistry at the Ecole Normale Supérieure in Paris, Pasteur first established his scientific reputation in the field of crystallography. In the mid-1850s, while teaching chemistry at a university in Lille, a center of the beet-sugar distilling industry, he became increasingly interested in the process of fermentation. Pasteur's microscopic researches convinced him of an observation that his countryman Cagniard de la Tour had advanced as early as 1835: that the agents of both fermentation and putrefaction were different species of living microorganisms. Working with brewers, vintners, and vinegar makers, all of whose livelihoods turned out to hinge on the successful management of these curious microscopic creatures, he became an expert on the applied science of fermentation. Using his microscope to examine cultures grown in flasks of clear broth, he learned to distinguish between microbial species that produced good beer, fine wine, and flavorful vinegar and those that produced nothing but slimy, revolting messes. He discovered that some species were aerobic, or needed air to live, whereas others were anaerobic and throve in its absence.

Pasteur immediately perceived that his research on fermentation had a significance beyond its usefulness to French industry. The leading medical authorities of the day believed that infectious diseases were caused by chemical ferments, and he had shown the agents of fermentation to be living microorganisms. The implications seemed clear: infectious diseases might be caused by these same microbes. As early as 1859, Pasteur wrote in a paper on microorganisms and fermentation that "everything indicates that contagious diseases owe their existence to similar causes." A few years later, in an 1861 treatise, he suggested that microscopic examination of airborne dust and dirt might provide valuable insight into the spread of epidemics.[19]

Pasteur's work embroiled him in a long-standing scientific controversy over the possibility of spontaneous generation, a debate closely linked to theories about epidemic disease. For centuries, philosophers had debated whether living creatures could originate from nonliving matter. The observation of microscopic life became a central element in the debate; commentators reasoned that if these most primitive forms of being could be generated in sterile flasks of broth, life could arise spontaneously. By analogy, the same reasoning suggested that epidemics could originate de novo—without any connection with a prior outbreak of disease.[20]

In a famous experimental duel with the naturalist Félix-Archimède Pouchet, the most prominent French advocate of spontaneous generation, Pasteur challenged the truth of this ancient doctrine. By an ingenious series of investigations, Pasteur proved that if "ordinary air" (air laden with common dust and dirt) was excluded from contact with a flask of nutritive broth, the broth remained pure and clear. But within a short period of exposure either to unpurified air or to a drop of water filled with microorganisms, the same sterile solution was soon teeming with life. Pasteur suggested that Pouchet and other exponents of spontaneous generation achieved contrary results because their experimental methods were not exacting enough to keep out the ever-present germ matter in the air.[21]

In retrospect, the connections between Pasteur's early work and the germ theory of disease that emerged around 1870 seem obvious. But in the late 1850s and early 1860s, Pasteur was primarily interested in the general problems of fermentation and spontaneous generation, not the specific relationship between microbes and disease. Only in the mid-1860s did he begin to investigate an actual disease, an ailment of silkworms called *pébrine*; not until the 1870s, after the germ theory of disease had already been articulated, did he start his celebrated research on anthrax and rabies.[22]

Although it is often credited to Pasteur, the modern germ theory of disease actually emerged through a far more collaborative sharing of ideas and research. In the 1860s and early 1870s, a small group of natural scientists and physicians, following their own interests or inspired by early reports of Pasteur's work, began to investigate the relationship between microbes and disease. As reports of their work appeared in medical and scientific journals, they gradually came to be seen, and to see themselves, as proponents of a cohesive doctrine regarding the agency of microbes in causing human and animal diseases.

The most numerous group of researchers, which included the French physician Casimir Davaine, the English physician John Burdon Sanderson, and the German physician Robert Koch, used experimental methods to study the process of infection. From blood or other matter extracted from a person or animal suffering from a disease, they tried to isolate the infective agent and then inject it into a healthy animal, a procedure that they hoped would produce the same ailment. By the mid-1870s, experiments of this sort suggested that tuberculosis, diphtheria, septicemia, cattle plague, and anthrax were "inoculable," meaning they could be passed from one creature to another.[23]

The germ theory also gained legitimacy from previous research that had convincingly demonstrated that living parasites caused muscardine, a disease of silkworms, as well as localized skin diseases such as favus and scabies. In the 1850s, investigators showed that an intestinal worm, *Trichinella spiralis,* was able to enter the human digestive tract via partly cooked pork, where it reproduced and sent colonies to burrow into other parts of the body. This "new revelation," as one expositor of the germ theory explained, "showed that the whole system, as well as a particular organ or tissue, might suffer from the effects of parasitic contamination." The model of parasitic behavior provided a useful way of understanding the relationship between microbe and host.[24]

Yet curiously, in the earliest accounts of the germ theory, investigations of actual diseases were often overshadowed by experiments that verified Pasteur's observations about the infective properties of air. Here the work of the English physicist John Tyndall was particularly important in shaping Anglo-American opinion. While conducting research on gases and radiant light, Tyndall became aware of the enormous amount of "floating matter" in the air. Upon reading of Pasteur's work, he became convinced that this matter contained disease germs. To prove it, he did a number of his own experiments and engaged in a long debate with the leading English proponent of the spontaneous generation theory, Henry Charlton Bastian. Tyndall, an accomplished popular lecturer and author, became one of the most important English-speaking advocates of the germ theory during the 1870s.[25]

Another extremely important type of evidence adduced in favor of the fledgling germ theory came not from the laboratory but from the operating room. Like Tyndall, the surgeon Joseph Lister, son of the Lister who invented the achromatic compound microscope, read of Pasteur's speculations about the infective matter in the air and wondered if it might be the source of the postoperative infections that made surgery such a risky enterprise. To neutralize the air's infective properties, Lister began to use carbolic acid as an antiseptic spray and wound dressing; the result was a dramatic reduction in his rates of postoperative infections. Although skeptics claimed that the so-called antiseptic method worked simply because it counteracted the infective properties of the air itself, not the living germs it contained, Lister presented his surgical experience as proof of Pasteur's theory.[26]

The phrase "germ theory of disease," which came into common use in the English-language medical literature around 1870, was scientific shorthand for propositions associated with the work not only of Pasteur, but also of Koch, Tyndall, Lister, and other investigators. Put simply, the germ theory consisted of two related propositions: first, that animal and human diseases were caused by distinctive species of microorganisms, which were widely present in the air and water; and second, that these germs could not generate spontaneously, but rather always came from a previous case of exactly the same disease.[27]

It should be noted that not everyone who swore allegiance to the germ theory of disease endorsed the second proposition. Many early converts accepted the causal link between microbes and disease while continuing to believe that under the right environmental conditions, disease germs might originate de novo and then spread from person to person. As the British physician Thomas J. MacLagan insisted in *Lancet*, "That every germ must originate from a pre-existing one may be true; but such a belief forms no essential part of the germ theory of disease." In addition, many early adherents of the germ theory assumed that disease particles required specific conditions to develop, or germinate; thus the disease germ and the mature pathogenic microorganism were not necessarily the same. This assumption was subsequently reinforced by the discovery that some species of bacteria form spores, that is, hardy reproductive cells that under the right environmental circumstances will grow into the mature organisms. Thus there remained considerable diversity of opinion on these points even among professed believers in the new theory.[28]

Early Criticisms of the Germ Theory

As first articulated around 1870, the germ theory was truly a theory, a radical extrapolation from a limited set of experimental observations. Indeed, if one tries to read the early debates over the germ theory impartially, without favoring the side that eventually proved correct, the antigerm theorists appear armed with some formidable arguments. Despite its advocates' appeal to a higher order of experimental evidence, the early laboratory "proofs" offered in favor of the germ theory were few and unconvincing. Even its most fervent advocates freely admitted that essential aspects of the hypothesis remained unproven. Believing in the germ theory, as it was initially formulated from the experimental evidence available in the 1870s, required a considerable leap of faith that most physicians simply could not make.[29]

Objections came not just from poorly educated or marginal physicians, but also from some of the most intelligent, systematic thinkers of the period. Many physicians committed to making medicine more scientific were deeply suspicious of overly simplistic theories of any sort, which they felt harkened back to the sterile hypothesizing of eighteenth-century medicine. Reducing the whole complex origin of an epidemic to the agency of a microbe struck them as a step backward, not forward, in medical thinking. Others objected to the premises of experimentalism itself. To their way of thinking, the behavior of test tube cultures or experimental animals bore no useful analogy to human disease; close observation of many cases of illness provided a much more authoritative body of evidence about the nature of illness. Still others objected not to the validity of laboratory evidence, but rather to its interpretation. Skeptics such as Félix Pouchet and Henry Charlton Bastian fought fire with fire by devising their own experiments to show that microbes could be generated in fluids even after boiling, a process known to kill microorganisms. Particularly in the 1870s, when experimental methods were still relatively crude, the antigerm theory camp could offer experimental results that seemed no less authoritative than those provided by the theory's supporters.[30]

Thoughtful observers also raised a host of objections to the germ theory that could not be easily answered given the available research methods. The very ubiquity claimed for the germ made it difficult for physicians to accept its causal role in disease. Microscopists routinely found many microbial forms on the body and in the secretions of healthy people, so it seemed obvious that the presence of germs alone did not cause illness. As Massachusetts physician Edward P. Hurd remarked in an 1874 review of the evidence for the germ theory, "All the higher organisms seem to be indifferent to them," at least so long as they remained in good overall health.[31]

Moreover, skeptics argued, the growth of unusual bacteria in the secretions of the sick could be the consequence rather than the cause of their illness. The zymotic theory held that when people ingested the chemical ferments of disease, their bodies began to manufacture the by-products of decay that such bacteria needed to grow. As Hurd put it, "There is no proof" that the "lower cryptograms," as he called them, "are not accompaniments, or effects, and not causes of the diseased conditions with which they are found associated." The same problem was raised concerning the animal experiments offered in favor of the germ theory. Early investigators could not easily separate the microorganisms from the blood or tissue of the diseased animal; thus critics argued that some other chemical substance in the inoculated material might have caused the symptoms. As Hurd noted, "It is quite impossible to introduce bacteria into the blood of a healthy animal without at the same time introducing with them septic or putrescent matters which might initiate disastrous changes in the blood, and become the elements of contagion."[32]

Those assumptions about the aerial spread of disease germs common to the early work of Pasteur, Lister, and Tyndall also met with skepticism. If disease germs floated in atmospheric clouds, it was logical to assume that the air surrounding sick people would be heavily laden with distinctive germs. But when investigators sampled sickroom air or exposed microscopic slides to the breath and saliva of patients sick from highly contagious diseases, they recovered microbes that looked no different from those found in their parlors or offices. A Chicago physician who compared the air of sick rooms and ordinary habitations concluded in 1871, "We were unable to detect the slightest particle of any kind in one, which was not equally present in the other."[33]

Moreover, opponents of the germ theory could point to numerous well-publicized cases in which early microscopical enthusiasts supposedly isolated the living agent of a deadly disease from secretions of the ill, only to have the germ in question turn out to be some innocent organism. A case in point was the American physician James H. Salisbury, who claimed in the 1860s to have found the fungal causes of measles, typhoid, malaria, and other fevers. Salisbury developed some ingenious proofs to associate the microscopic *palmella* plant with malaria; for example, he had volunteers sleep in rooms with boxes of palmella-infused soil on their window sills and observed that they soon fell ill of the fever. Other investigators found it easy to disprove the palmella thesis by showing that it existed in regions that had no malaria. Anticipating Max von Pettenkoffer's famous cholera cocktail of the 1890s, the Philadelphia physician Horatio C. Wood even drank a glass of water infused with the microscopic organism to show that it caused no ill effect.[34]

Salisbury's claims about palmella and malaria represented but one example of many unconvincing attempts to link specific microorganisms with specific diseases. Liebermeister noted regretfully in 1874 that contemporary enthusiasts had done as much harm to the germ theory with their premature claims as their eighteenth-century predecessors had done with their fanciful drawings and cannon shootings. "The utter lack of critical discernment and method which have characterized some of the works in this field, and, on the other hand, the recklessness with which facts of uncertain significance have been proclaimed certain proofs, have also in our time driven away many an earnest investigator," he lamented.[35]

Skeptics like Edward Hurd insisted, quite understandably, that "till, then, more convincing experiments shall have been performed, the poison theory of the older pathologists will hold against the living ferment theory of the newer." He concluded, "In rejecting the Germ

Theory as untenable, we have either to confess our ignorance of the causes of all febrile and inflammatory contagious diseases . . . or, guided by analogy, to accept the alternative that the principle of contagion is a subtle chemical ferment, an organic poison, generated in the body of the diseased individual."[36]

The experimental work on anthrax, also called splenic fever, proved crucial to resolving these objections to the germ theory of disease. Anthrax was the first disease that experimenters could convincingly link to a specific microorganism. Primarily a disease of cattle, sheep, and horses that occasionally spread to humans, it caused painful boils, fever, and congested lungs. In 1876, while still a country doctor, Robert Koch identified the *Bacillus anthracis,* a large and relatively easy to grow rod-shaped bacillus. Using microorganisms cultured in a special medium, in this case the aqueous humor of cattle eyes, Koch showed by repeated experiments that the anthrax bacillus did not exist in the blood of healthy animals but when injected into them consistently produced the disease's distinctive symptoms.[37]

In addition, Koch discovered that the anthrax bacillus had two forms. The mature bacillus, a slender filament, did not survive long after the death of its host, but it produced spores—small, black, seed-like capsules—that were capable of surviving extreme cold or heat. Koch's findings helped to explain why anthrax was confined to certain localities and appeared and disappeared so mysteriously: the anthrax spores remained in the soil and only ripened into maturity under a precise set of environmental conditions. This discovery helped to explain why competent experimentalists could get fermentation in boiled solutions; the heat killed the bacteria but not the hardier spores.[38]

As it turned out, this cycle of bacillus and soilborne spore proved unusual among pathogenic microorganisms; besides anthrax, it was found only in the family of organisms responsible for tetanus, gas gangrene, and botulism. The microbes that caused the vast majority of common communicable diseases, including typhoid and tuberculosis, formed no such resilient spores. Yet in the late 1870s and early 1880s, the anthrax model was widely invoked to explain the origin and spread of germ diseases in general. By chance, the first disease-causing bacteria clearly identified by experimentalists confirmed the perception that the microbial "seeds" of disease were widely dispersed in the air and soil and required only the right conditions to germinate. The soilborne anthrax spore powerfully reinforced the association of dirt and disease germs. These assumptions, that pathogenic microorganisms were extremely hardy and widely broadcast in the environment, strongly colored the first generation of preventive strategies advocated in the name of the germ theory.

Going Public

Although increasingly sophisticated experimental methods gradually filled in the gaps of what researchers could prove about the germ theory, its advocates did not wait for incontrovertible laboratory evidence before trying to convert others to their views. From the outset of the debate, critics and champions of the germ theory alike actively sought out public forums for rehearsing their arguments. In so doing, they followed a long tradition of "public science" in which even the most elite scientists courted legitimacy by giving public demonstrations of their ideas and experiments. The terminology and style used in these public discourses were far less formal and abstruse than they would become even a few years later. Early commentaries on the germ theory of disease were often delivered in simple language and embellished with colorful imagery that an educated lay person as well as a physician could understand. In the pages of medical journals, in public lectures reprinted for wide distribution, and in magazines such as the *Popular Science Monthly*, early converts to the germ theory explained the new lessons of the laboratory using everyday experiences of baking and brewing, spoiled food, and dust motes dancing in a sunbeam.[39]

These imaginative modes of describing the microscopic world were particularly useful for introducing the germ theory of disease to a wider audience. Well in advance of winning the divisive medical battle over the issue, advocates of the germ theory sought out audiences of mostly middle-class, city-dwelling men and women who took an avid interest in matters of health. In the popular health literature, favorable notices of the germ theory began to appear in the 1870s, when many physicians were still either hostile to or unacquainted with it. Thus the scientific arguments and "proofs" initially offered on the germ theory's behalf were quickly incorporated into popular writings on health and disease.

Explaining the significance of the experiment was an important feature of these early accounts of the germ theory. Although the germ theory's advocates used all sorts of reasoning and evidence to make their case, they presented the laboratory as the source of a new and special kind of knowledge. At the same time, there were relatively few experiments presented on the germ theory's behalf in the 1870s that a serious amateur could not replicate. Early accounts of laboratory life in the writings of such popularists as John Tyndall were a curious mix of the familiar and the awesome. The experimental materials and methods described in them had a homely cast; investigators lovingly recounted how they constructed the air chambers for their test tubes from household materials; prepared culture media from turnips, herring, or beef tea; and warmed their microbial broths over a kitchen stove. (Tyndall once reported that he took a set of test tubes to a Turkish bath in order to incubate them.)[40]

Yet such ordinary, everyday materials as a basin of turnip slices or the juice from a mutton chop produced dramatic results. Commentators used vivid language to describe the bacteria's transformative effects on liquid or solid media: meat or soup initially described as "sweet," "pure," or "limpid" became "slimy," "putrid," or "turbid." One physician experimenter wrote, "My wonder never ceases when I take up one of the flasks and bulbs which have remained barren in my chamber for three or four years, though supplied with air (filtered through cotton-wool) and suitable heat." It was equally amazing, he added, to withdraw the cotton plug and allow the germ-laden dust or water access to the broth: "In a few hours the stillness of years gives place to life and activity."[41]

Experimental accounts also emphasized how one seemingly inconsequential mistake—not rinsing a pipette with sterile water, or failing to sterilize a flask before filling it with the culture medium—could introduce the fertile hordes into a barren environment. By stressing the importance of having an exacting technique, advocates of the germ theory had an all-purpose explanation for why their critics seemed unable to replicate their results: they simply were not careful enough.[42]

Laboratory proofs of the germ theory depended on a willingness to see what happened in the test tube or the experimental animal as a model for what happened in a disease epidemic. The comparisons made were often quite simple. Tyndall pointed out, for example, that the interval of time between introducing the airborne germs to the culture medium and their multiplication into abundant life neatly corresponded to the latency period physicians had long observed between an individual's exposure to contagion and subsequent development of sickness. Likewise, he noted that different broths nourished the germs to different degrees, just as individual constitutions provided more or less resistance to disease. Watching how a hundred test tubes filled with varied infusions of herring and turnip became turbid at different rates, Tyndall observed how "the whole process bore a striking resemblance to the propagation of a plague among a population, the attacks being successive and of different degrees of virulence."[43]

Advocates of the germ theory continually appealed to their audiences to see the parallels between laboratory experiment and everyday observation. As Pasteur's work on fermentation so well exemplified, the insights of the germ theory had much in common with familiar domestic processes such as bread making and beer brewing. In a Glasgow address reprinted in the *Popular Science Monthly*, Tyndall urged his audience to "observe how these discoveries tally with the common practices of life" and offered examples from his own household, such as his housekeeper's use of brief applications of heat to keep pheasants and milk "sweet." To illustrate the prevalence of germs in the air, he asked his listeners to think about the molds that grew on wet boots or a piece of fruit left exposed to the air and about the dust that appeared in a beam of sunshine after the housemaid cleaned a room. Using his neighbor's efforts to make alcohol from sour cherries as an analogy for the disease process, he explained, "We began with the cherry-cask and beer-vat; we end with the body of man."[44]

Expositors of the new germ theory often struggled to find the right words to describe the "milky way of lower organisms," as the botanist Christian Ehrenberg once called it, that the microscope revealed. First there was problem of terminology; as mentioned before, observers used an exotic array of terms to describe these organisms, such as "monad," "cryptogram," and "infusoria." Individual species had their own strange names such as the "micrococcus" and the *Bacillus subtilis*. Then there was the challenge of describing their various shapes and movements. Commentators resorted to all sorts of analogies to convey the vagaries of microbial forms: this species resembled a eel, that one a "string of beads," another a "twirling wheel"; their movements across the microscope's field of vision were described as "leaping," "darting," and "springing."[45]

In the *Popular Science Monthly,* the botanist Ferdinand Cohn, whose scheme for classifying bacteria according to their shape (spherical, oblong, rodlike, and spiral) gradually became the accepted standard, commented on the organisms' antics: "When they swarm in a drop of water, they present an attractive spectacle, similar to that of a swarm of gnats, or an ant-hill." Their patterns of movement were endlessly fascinating. "At one time they advance with the rapidity of an arrow, at another they turn upon themselves like a top; sometimes they remain motionless for a long time, and then dart off like a flash," he observed.[46]

Those who had seen these microscopic marvels continually marveled at how small and fertile they were. In an 1878 lecture delivered to the Philadelphia Social Science Association, the microscopist Joseph Richardson invented some dramatic numbers to convey their minuteness. Disease "spores" were "so small," he wrote, "that 20,000 of them placed end to end, would measure less than one inch in length, and a mass the diameter of one of the periods (.) upon this printed page might contain 50,000,000." Each of these 50 million seeds, he added, was "capable, under favorable circumstances, of reproducing its own kind with almost inconceivable rapidity."[47]

Commentators sought to fit these minute beings into the classifications that naturalists had already developed to describe animals higher up on the evolutionary scale. In their great biological chain of being, microbes occupied the lowest niche as the most "primitive" form of life. They were so primitive, writers often noted, that they reproduced not by the mating of male and female, but by budding, dividing, or producing spores. Their physical structures were remarkably simple—a cell wall enclosing a largely undifferentiated mass of protoplasm—and when not moving, they often could not be distinguished from crystals or other inanimate forms of matter.

In compiling their microscopic bestiaries, early chroniclers divided the microbial world into friends and foes. Much as Victorian naturalists characterized the lion as a noble beast and condemned the wolf as a savage predator, late-nineteenth-century commentators sorted the various species of microorganism into good and bad microbes.

The good species enriched human society, making it possible for people to enjoy bread, wine, and beer, and they played an essential role in breaking down dead matter into elements that could be used by new forms of life. As the American physician George Sternberg wrote dramatically, "But for the power of these little giants to pull to pieces dead animal matter, we should have dead bodies piled up on all sides of us in as perfect a state of preservation as canned lobster or pickled tongue."[48]

Only a few species of "bad" microbes preyed upon humans and animals, yet their potential for creating havoc was impressive. Converts to the germ theory often painted a chilling picture of an environment saturated with these invisible enemies. Bacterial clouds floated about in the atmosphere, carried along by shifting air currents and dropping into the water supply, until eventually they found a receptive media, the human equivalent of the turnip infusion or the beef tea. As Ferdinand Papillon wrote in the *Popular Science Monthly*, "Our atmosphere . . . is the receptacle for myriads of germs of microscopic beings, which play an important part in the organized world." These "penetrating agents of decay, baneful toilers for disease," he observed, "lie ever in wait for the chance to pierce the internal machinery of animals and plants, and create slight or grave disturbances within it."[49]

To describe how germs found a suitable host, germ theorists frequently resorted to comparisons between germs and seeds. Since antiquity, physicians had used the "seed and soil" metaphor, drawn from the New Testament parable of the sower, to describe how the interplay between one's individual constitution and an external disease agent determined one's susceptibility to disease. Exponents of the germ theory found that image particularly well suited for their purposes: the germ or "seed" required suitable "soil"—that is, a weakened constitution—for its full development. The seed and soil metaphor also worked well to underline the *specificity* of disease agents. Just as a farmer expected to get wheat when he sowed wheat and corn when he sowed corn, the scarlet fever germ only gave rise to scarlet fever and the smallpox germ to smallpox. Nor did the farmer anticipate a crop of wheat or corn to grow where he had sowed no seed at all, as advocates of spontaneous generation had asserted.[50]

Commentators also likened microbes to insects and worms, using the examples of the tiny insect responsible for scabies, or the trichinae worms carried in uncooked pork, to explain the parasitical nature of germs. In a lecture on "the origin and propagation of disease" delivered at the New York Academy of Medicine in 1873, the physician John Dalton developed these examples at some length to help his audience comprehend the unfamiliar world of bacteria and disease. The great potential of the germ theory, he argued, lay in its ability to harmonize with natural science as a whole, "for it will show how large a part of human pathology is connected with the general physiology of vegetative life."[51]

Other expositions of the germ theory used a more feral imagery to describe the microbial parasite. William Mays told his audience that germs "hunt in packs," and another physician referred to them as "atmospheric vultures." Microbes were often described in martial terms as attacking, invading, and conquering their human hosts. Joseph Richardson combined both the botanic and feral images in his 1878 speech to the Social Science Association when he explained that contagious diseases were caused by "the transplanting of microscopically visible spores, or seeds, which have a separate vitality of their own, each after its kind, and which are to be escaped, just as we would escape hordes of animal[s], or swarms of insect pests, by shutting them out or killing them before they can succeed in fastening upon our bodies."[52]

The Microbial Survival of the Fittest

Early accounts of the germ theory, with their frequent use of the terms "higher" and "lower" organisms and their references to microscopic predators and parasites, purposefully conjured up images of a microbial survival of the fittest. To a generation of medical and lay readers familiar with evolutionary theory—Charles Darwin's *Origin of Species* had been published in 1859, and Herbert Spencer's *Principles of Biology* had introduced the term "survival of the fittest" in 1861—these were potent analogies. The strong overlap in language between Darwinian theory and germ theory was not accidental. Many of the leading figures in the English debate were committed supporters of Charles Darwin. John Scott Burdon Sanderson was a good friend and frequent correspondent of Darwin. John Tyndall first gained national renown for his defense of evolutionary theory; his rival, Henry Charlton Bastian, also was a Darwinist.

As Bastian's case suggests, professed Darwinians could be found on both sides of the germ theory debate. The implications of evolutionary theory for microbial behavior and vice versa were by no means clear-cut, especially in regard to the vexed subject of spontaneous generation. Still, although advocates of the germ theory had no exclusive claim on evolutionary theory, its growing popularity probably did more for their cause than for their opponents' because the image of a "microbial survival of the fittest" proved to be such a powerful model for the relationship between microbe and host.[53]

At the simplest level, many commentators likened the species of microbes to the distinctive species evident among more complex plant and animal life forms. The popularity in early accounts of the germ theory of adages about seagulls not hatching from pigeons' eggs or horses not foaling donkeys invoked a broader conception of natural law that limited miraculous or unexpected transformations. For physicians intent on making medicine more scientific, this evolutionary perspective on disease had enormous appeal. As Henry Gradle, a professor of physiology at Chicago Medical College, noted, "It eliminates the factor 'accident' from the consideration of disease, and assigns disease a place in the Darwinian programme of nature."[54]

Gradle made explicit the "survival of the fittest" themes that ran through many early accounts of the germ theory. "In the light of the germ theory," he wrote in 1883 in *Bacteria and the Germ Theory of Disease*, "Diseases are to be considered as *a struggle between the organism and the parasites invading it*." The contest between microbes and higher life forms was similar to parasitical relations throughout nature in which a smaller species preyed upon the body of a larger creature. He concluded, "We are again ignorant as to the weapons of the contending armies, we do not know yet how the warfare is carried on between the hostile vegetable and animal cells, but that the struggle exists is evident, and it must terminate in the victory of one or the other side."[55]

Although Darwin himself resolutely avoided seeking moral meanings in the workings of evolution, many of his contemporaries observed no such restraint and implied a conscious malevolence to disease germs. Using highly charged adjectives such as "foreign," "base," "murderous," and "cunning," they endowed microbes with a frightening will to destroy their biologically superior competitors. The recognition that "these lowest of created things" worked out their destiny by wreaking disease and death on the human race was both humbling and terrifying.[56]

Yet on the whole, the tone of these early accounts of the germ theory was overwhelming optimistic. Converts portrayed the new discipline of the laboratory as a royal road to safety: by identifying the true agents of disease, their modes of travel, and their sure methods of destruction, the insights of the germ theory would make it easier to outwit the invisible agents of disease. A Philadelphia medical student writing in 1885 captured that sense of optimism, asserting that "after centuries of silent resignation, mankind enlightened by science at last begins to recognize its relentless and hitherto mysterious enemies." He asked rhetorically, "Shall we then continue indefinitely yielding up the inumerable [sic] victims that yearly succumb to the attacks of foes whose only force lies in their minuteness?" and answered dramatically, "No! Man is no more made to become their prey than that of the wild beasts among whom he had to fight his way in the infancy of the race and whom he has conquered or destroyed by his industry, intelligence, and work."[57]

Paths to Conquest

The most glamorous of these vistas of progress was the potential for discovering new vaccines and drugs. Inspired by the known value of the smallpox vaccination, converts to the germ theory dreamed of devising concoctions of tamed germs that would confer similar protection against other deadly diseases. In the 1870s and 1880s, Louis Pasteur devoted himself to developing vaccines against anthrax, chicken cholera, and rabies. In the 1890s, Robert Koch touted his "tuberculin" as a cure for tuberculosis. Many lesser-known researchers and clinicians experimented with "internal antiseptics," or chemical substances that when ingested would kill microbial invaders.[58]

But from the 1870s to the early 1900s, such hopes were repeatedly dashed. With a few exceptions, such as the rabies vaccine and the diphtheria antitoxin, none of the measures developed in the first flush of enthusiasm for the germ theory stood the test of time. Laboratory scientists continued searching for the fabled "magic bullet," which did eventually materialize—first in the discovery in 1909 of Salvarsan, which cured syphilis, and several decades later in the

discovery of sulfa drugs and penicillin. Yet prior to 1900, the therapeutic promise of the germ theory remained elusive.

Far more immediate and useful were the insights about hygiene
and sanitation derived from the germ theory. Here the apostles of
the germ did not have to break such hard, new ground as they did in
searching for magic bullets. Early understandings of the germ, which
emphasized its ubiquitous presence in air and water and its hardiness
outside the body, neatly harmonized with already accepted modes of
protection against zymotic disease. As a result, the first version of the
gospel of germs represented a surprisingly successful marriage between the old sanitary science and the new germ theory.[59]

That such a happy union could come about was not immediately
apparent to the older generation of public health reformers. Such
eminent figures as Benjamin Richardson, Florence Nightingale, and
Elizabeth Blackwell expressed fears that acceptance of the germ
theory would undercut the achievements of sanitary science. They
were profoundly uncomfortable with the moral randomness they
perceived in the germ theory; if contact with a microbe was the sole
cause of disease, then living a virtuous, clean life did not necessarily
protect one from its ravages.[60]

In response, early advocates of the germ theory sought to reassure
the older sanitarians that the new disease faith only verified the
great "truths" of sanitary science. Indeed, for all the controversy that
the germ theory engendered in medical circles, its implications for
preventive action initially seemed to be consistent with existing tenets of private and public hygiene. In 1873, after reviewing the debates over the germ theory, the president of Columbia University,
F. A. P. Barnard, concluded gratefully that when it came to preventive medicine, "The champions of conflicting theories, however
freely they may splinter lances in the arena of controversy," could
be found "in the face of the common enemy, marching harmoni-
ously side by side."[61]

Certainly when it came to the practice of personal and domestic
cleanliness, the new experimental evidence about germs counte-
nanced no slackening off in the zeal required by traditional sanitary
science. The reconceptualization of disease ferments as minute living
creatures able to replicate in the millions from a single speck only
hcightened the importance of exacting precautions against their
spread. Likewise, the anthrax model of bacterial spores capable of

surviving high heat and normal disinfectant processes pointed up the need for increasingly rigorous forms of cleanliness.

Pasteur's own reputation for meticulous cleanliness exemplified how acceptance of the germ theory went hand in hand with vigilance to hygienic detail. Having lost two daughters to typhoid fever, he knew intimately the havoc that so-called filth diseases could wreak on a household. Perhaps for this reason, he carried over into his personal life the rituals of cleanliness that he practiced in the laboratory. His son-in-law and biographer René Vallery-Radot wrote that whether dining at home or out "he never used a plate or a glass without examining them minutely and wiping them carefully; no microscopic speck of dust escaped his short-sighted eyes." He was "more than difficult to please in that respect," causing him to be a terror to his hostesses. It was Pasteur's long acquaintance with the microbial world that caused him to bring "such minute care into daily life," concluded Vallery-Radot. If a speck of dust could generate a veritable horde of bacteria in a flask of beef broth, what could it do in a soup tureen?[62]

The lore of the laboratory reinforced the point that seemingly inconsequential actions, such as failing to sterilize a flask or a pipette, could bring about rampant germ life. Carried over into everyday life, that mentality pointed toward an even more exacting practice of domestic cleanliness. The practical lessons that advocates of the germ theory derived from the laboratory only underlined the urgency of the sanitarians' warnings that utmost care needed to be taken to evade the domestic sources of disease. Here at last was an explanation for tragedies such as the death of Martha Bulloch Roosevelt. Her rituals of cleanliness had simply not been precise enough to counteract the depredations of the wily typhoid germ.

The new lessons of the laboratory thus contributed to a widening effort by public health reformers to "pathologize" the home because they invested ordinary behaviors and objects with the capacity to cause or prevent deadly illnesses. Acceptance of the germ theory fed into a revolution in personal and domestic behavior already under way by the 1870s. Commandeering the established truths of sanitary science, apostles of the germ synthesized old and new beliefs about contagion into a new code of protection for the American home. When it came to domestic rituals of purification, the dramatic insights of the germ theory turned out to be so much new wine in old bottles.

2 · Whited Sepulchers

In 1883, the year before Martha Roosevelt's death, an article that read almost like a prognostication of her fate, entitled "The Unsanitary Homes of the Rich," appeared in the *North American Review.* "Much has been written and said of late years about the wretched homes of the poor of New York, their squalor, their filth, and the moral and physical degradation of their occupants," the author, Charles Wingate, noted. Few of the city's inhabitants realized that the sumptuous mansions being built all over Manhattan were filled with hidden dangers that made them as unhealthy as the worst tenement house. Wingate might have been describing the Roosevelts' home on Fifty-Seventh Street when he referred to dwellings "of imposing dimensions, palatial in their adornments, and seeming to lack nothing to promote comfort, enjoyment, and health." Yet due to their faulty plumbing, he concluded, "a large number of these houses are mere whited sepulchers, and their luxurious inmates are exposed to constant risk of disease and death."[1]

Assuming that his readers would be well acquainted with the New Testament, Wingate's use of the phrase "whited sepulcher" was meant to conjure up the image of the Pharisees, who, like a whitewashed tomb that disguised the decaying bodies inside, appeared righteous, but actually were riddled with sin. It was a shocking allusion, but one frequently used by late-nineteenth-century health reformers for it drove home the point, literally as well as figuratively, that sanitary sins were to be found among the highborn as well as the lowborn, and that the whole of American society needed hygienic redemption.[2]

The "whited sepulcher" image also underlines how deeply concerned Wingate's generation was about the domestic origins of disease. Long before the agency of the germ was suspected, popular belief held that the sick left behind the "seeds" of their disease in the houses where they had died and on the objects that they had touched. In the nineteenth century, sanitarians expanded on the association of houses and disease, blaming damp cellars, poor ventilation, and defective plumbing for the fearsome increase in zymotic diseases. Under their relentless proddings, many middle-class Americans began in the mid-1800s to make major changes in the design of their homes and in how they performed household chores.

The new revelations about microbes and disease were thus introduced against the backdrop of a household revolution already in progress. The germ theory of disease entered the popular discourse about disease prevention at a time when the majority of Americans, physicians and lay people alike, believed quite fervently in the reality of what we would today call "sick buildings." The first gospel of germs, which emerged gradually in the 1880s, simply superimposed the menace of the microbe onto existing mappings of disease dangers in the household. The initial measures encouraging Americans to "germ-proof" their homes therefore owed much of their success to a pregerm gospel of domestic disease prevention widely disseminated in the 1860s and 1870s.[3]

The Origins of Domestic Disease Prevention

The code of scrupulous household cleanliness that emerged in the late nineteenth century, first under the aegis of sanitary science, then in tandem with the new germ theory of disease, built upon traditional methods for warding off infection during times of epidemic. Essentially, hygiene reformers took the exacting measures once demanded only during a visitation of dread disease and recast them as everyday protections against the growing menace of endemic, or "ordinary," fevers.

Nineteenth-century Americans were heir to a long tradition of precautions developed to combat epidemic diseases such as bubonic plague and smallpox. For centuries, outbreaks of disease had prompted rituals of household purification designed to guard against the dangers of both a corrupt atmosphere and direct contagion from the sick. When an epidemic threatened, individuals tried to stay warm, well fed, and rested, the better to resist the disease; they also cleaned their homes and yards of any filth or stagnant water that might breed an infective atmosphere. As soon as the epidemic broke out, they avoided contact with other people, and if possible, fled the area.[4]

Simple observation suggested that contact with someone already sick often preceded an illness such as smallpox. Popular belief held that the breath, spit, skin particles, and bodily evacuations of the sick were all capable of spreading the disease. Invisible bits of contagious matter could supposedly adhere to bed linen, clothing, papers, and even household pets; these objects, or "fomites," seemed to retain the power to infect others for months and even years. Lay people were more likely than physicians to believe that epidemics spread by direct contagion and fomite transmission, rather than by the more elusive agency of atmospheric infection. As a result, they frequently shunned the sick and their belongings during times of unusual illness.[5]

Containing the spread of this contagious matter necessitated exacting home nursing practices, especially because so few hospitals existed prior to 1860 (and even those often refused to take patients with contagious diseases). During an epidemic, every household functioned essentially as a hospital, and the measures taken there to reduce the spread of infection were regarded as vitally important. Families were expected to isolate sick members and to disinfect their sick rooms and personal belongings in order to destroy the seeds of disease. Those who failed to exercise such care faced severe censure from neighbors and municipal authorities.[6]

Although epidemics certainly reinforced the association between disease avoidance and domestic prevention, once the crisis had passed, the heroic efforts at personal and household reformation it had inspired soon dissipated. Starting in the eighteenth century, advocates of a new science of public health began to advocate more long-term changes such as land drainage, sewer construction, and the like, to reduce the threat of disease. But there remained little an individual or private household could do to stave off epidemics. In eighteenth- and early nineteenth-century personal hygiene manuals, there were many more pages devoted to avoiding "constitutional" diseases such as dyspepsia or gout than there were to preventing the spread of infectious diseases. The Philadelphia physician Benjamin

Rush perhaps summed up conventional wisdom best when he wrote his wife during the 1793 yellow fever epidemic, "There is but one preventative that is certain, and that is 'to fly from it.'"[7]

Beginning in the mid-1800s, this calculus of responsibility began to change. By attributing the rising incidence of fevers to the extraordinary filthiness of cities and habitations, sanitary science brought the control of zymotic diseases within the sphere of individual as well as collective action. The experience of sanitary reformers during the Civil War strengthened the conviction that higher standards of cleanliness contributed directly to the saving of lives. At first, reformers focused primarily on the poorest homes and neighborhoods of the city, assuming, as a sanitary survey of New York City during the Civil War concluded, that the "nuisances" found there were sufficient to "pollute the atmosphere of the entire city." But by the 1870s, hygiene reformers were increasingly questioning the presumed sanitary superiority of the prosperous classes. As they traced the complex routes of atmospheric and water pollution, they found that dangerous defects in domestic hygiene were common in even the best homes.[8]

The chief culprit, sanitarians agreed, was household plumbing. In the 1840s, affluent families began to move their toilets indoors and to use water to flush the contents away. The increasing volume of discharges from these new "water closets" had overwhelmed existing cesspools and earthenware sewer lines, impregnating cellars and yards with liquid wastes. Improving municipal sewer lines solved the soil saturation problem, but it presented a new hazard: unless households installed airtight "sewer traps" on every toilet or washbasin, the connection with the public sewer provided multiple points of ingress for potentially deadly gases given off by decaying fecal matter.[9]

The frequency with which damp cellars, foul odors, and untrapped drains could be found in even the best of homes convinced sanitarians that the crusade against zymotic diseases had to enlist the rich as well as the poor. To make that point, public health reformers often invoked the tragedies that beset the British royal family. Queen Victoria's husband, the prince consort, died of typhoid in 1861, and their son Edward, the prince of Wales, nearly succumbed to the same disease in 1872. Investigations into both cases blamed faulty domestic plumbing in royal residences for the outbreaks. In the United States, a comparable sanitary scandal arose in 1881 over the condition of the White House, a controversy that will be looked at more closely in

Chapter 3. Such well-publicized cases of sanitary negligence in the symbolic first homes of the land underscored the same message inherent in Martha Bulloch Roosevelt's death: no one was safe.[10]

The sanitarian crusade was couched in religious as well as scientific terms. Equating God's law with natural law, hygiene reformers called on all citizens to embrace a sanitary gospel of redemptive cleanliness. In an 1875 lecture to the graduating class of the University of Michigan Medical School, Professor R. C. Kedzie explained, "The old superstitions which connected unusual sickness with the wrath of offended Deity have faded in the light of science." He continued, "The 'mysterious providences' about which we have heard so much are resolving themselves into 'defective drainage,' 'sewage contamination,' 'unwholesome food,' 'poisoned walls,''no ventilation,' etc." With this knowledge in mind, Kedzie exhorted his listeners not to "flout our filth in the face of Deity, and say that these afflictions come from His hand," but to be "clean in your person, and homes, the food you eat, the water you drink and the air you breathe."[11]

But however religiously sanitarians promoted the private observance of cleanliness, they never conceived of it as sufficient in and of itself. The reform of individual households was always linked with the need for strong public health boards and municipal sanitary reforms. Sanitarians nevertheless recognized that as of the 1870s and 1880s, the state's public health powers were still very rudimentary, so that even the wealthiest homeowners could not count on having safe municipal water supplies or sewerage systems. In an era of undependable municipal services, the installation of household-level protections, such as sewer traps or water filtration systems, restored a sense of control to the individual homeowner. As Joseph Edwards put it in his 1882 manual, "You cannot look into the [public] sewer and see whether it is clean or not. But, into all the arrangements of your own individual house you can peer at all times, and can plainly see whether they are right."[12]

Channels of Influence

The commitment to popular education so pronounced among late-nineteenth century domestic reformers coincided with changes in technology and marketing that made available increasingly inexpensive forms of print culture. From the mid-1860s onward, the multiplying host of medical and lay authors seeking to instruct the Victorian pater and mater in domestic disease prevention found a variety of channels at their command. The volume and variety of publications devoted to home hygiene in the last quarter of the nineteenth century testifies to the extraordinary interest in the topic.[13]

As the physician and public health authority Henry Bowditch remarked in 1876, "Sanitary discussion seems to interest many persons as much as the pages of the novel attract others." At a time when genteel folk customarily avoided plain talk about bodily parts or functions, he marveled at their eagerness to purchase and read detailed accounts of toilets, sewer gas, and fecalborne disease. Referring to the success of a series on household plumbing in the *Atlantic Monthly*, Bowditch was astonished at the way the sanitary engineer George Waring "discusses sewerage . . . with infinite gusto, and apparently to the satisfaction of all readers of this popular monthly."[14]

The treatment of home hygiene ranged from highly technical tomes that thoroughly expounded the rationale for their recommendations to short, simple summaries that supplied only the most rudimentary explanations for the suggested precautions. At one end of the spectrum were treatises on specialized subjects, such as William Eassie's *Sanitary Arrangements for Dwellings* (1874), an English volume frequently cited in the United States and "intended for the use of officers of health, architects, builders, and householders." More general manuals such as Henry Hartshorne's *Our Homes* (1880) treated a wider range of topics, from building or choosing a home to creating a "home hospital" for the care of contagious illness, and were conveniently sized for armchair or bedside use. Bulkier domestic encyclopedias and family medical guides, such as *Wood's Household Practice of Medicine, Hygiene, and Surgery* (1880), intended for the use of "families, travelers, seamen, miners, and others," contained greatly condensed versions of sanitarian doctrine.[15]

The interest in domestic hygiene carried over into the periodical literature, including ladies' magazines, popular science journals, and even literary reviews. The venerable *Godey's Ladies Book* contained short homilies on home health matters, and the more up-to-date *Ladies' Home Journal*, which began publication in 1883, had regular features on the prevention and management of infectious diseases. From its first issue in 1872, the *Popular Science Monthly* carried frequent articles on sanitary plumbing, disinfection, and the germ theory of disease. Periodicals of general interest such as *Frank Leslie's Monthly* and *Scribner's* soon followed suit. In 1875, the *Atlantic Monthly* scored its great success with George Waring's series on domestic plumbing gas, and by the early 1890s, even weekly religious newspapers such as the *Independent* had columnists covering public health issues.[16]

The gospel of home hygiene also reached poorer audiences, albeit less frequently and in greatly condensed form, through broadsides and circulars. When epidemics of cholera, smallpox, or diphtheria threatened, local and state health departments often printed instructions about disease prevention that were distributed in poor neighborhoods and published in newspapers. These two- or three-page circulars, which presented short, simple versions of the sanitarian gospel, were no doubt the chief way that detailed information about disease prevention reached working-class families prior to 1900. In addition, local "sanitary associations" in large cities sold pamphlets on home hygiene for a few cents. In the South, the Hampton Institute, a Virginia school established for former slaves, prepared "Tracts for the People," which included one on "preventable diseases" written by Mary Armstrong, wife of the school's superintendent.[17]

But given that even a nickel was a prohibitive sum for many working-class families, we may safely assume that the readership for domestic hygiene writings remained limited prior to 1900. Although reformers' rhetoric emphasized the universality of their gospel, their educational methods inevitably circumscribed distribution of their message to the literate, affluent, and leisured. If and when circulars did come into their hands, working-class families often lacked both the time and the money to implement their suggestions. Many did not own their own homes and thus had little power to improve them. As a result, in the late 1800s sanitary knowledge and practice remained largely the province of middle- and upper-class families.

The People's Germ Theory

Given the growing interest in the subject of house diseases, domestic hygiene writers were quick to report on the new scientific speculations linking microbes and disease. Long before physicians had reached any consensus on its validity, discussions of the germ theory had begun to appear in popular advice literature. These accounts suggest that the authors had some familiarity with the scientific debates described in Chapter 1. Yet even into the 1880s, when researchers began to identify and describe specific microorganisms such as Eberth's bacillus for typhoid, or the comma bacillus for cholera, popular advice givers still tended to refer simply to "disease germs" as an undifferentiated group. Most also ignored the battle against spontaneous generation fought by Louis Pasteur and John Tyndall and assumed that germs could originate de novo from decaying matter.

One of the earliest accounts of the germ theory to be found in the American advice literature, the 1875 *Household Manual,* suggests how complex arguments were simplified for a general audience. In language similar to that used by John Tyndall, the anonymous author of the manual distinguished between good and bad microbes and appealed to commonsense observations. The text defined germs as "little animals—animalculae—and the seeds of microscopic plants" that entered the body through the lungs and stomach and found their way into the bloodstream. "Here they develop and multiply to a remarkable extent, and with astonishing rapidity. Some kinds are very deadly; others are much less so, and some again seem to be almost harmless." In the spirit of Tyndall, the author advised the reader to look at the "motes, dancing in the sunbeam" to see the more benign germs, and added, "These are the agents which cause bread to rise, malt and wine to ferment, and the housewife's carefully hoarded canned food to spoil." In contrast, "All kinds of putrid matter, whether of animal or vegetable origin, send out immense quantities of germs which are of a much more poisonous character than those just referred to."[18]

The "seed and soil" metaphor also found special favor among popular interpreters of the germ theory. In an 1874 article the *New York Times* observed, "The thought itself is an impressive and natural one, that there may be a 'cholera seed' or 'scarlet-fever germ' or 'typhoid sporule' floating through the air, just as there are floating seeds of thistles or dandelions, or germs of tulip-trees or limes, or scores of the nameless plants which sow themselves wherever there is the slightest bit of soil or moisture favoring." Echoing Tyndall's image of bacterial clouds, the article concluded, "No doubt in this City there is an invisible cloud of 'scarlet-fever germs,' 'typhoid seeds,' and cholera or 'diphtheria spores,' always drifting over from the densely-crowded poor quarters into those of the wealthy, filling the houses and garments, and lying *perdu* until the favorable moment in the organism of some child or delicate person gives them the chance to spring up into vigorous growth."[19]

The germ theory harmonized well with traditional beliefs that the bodies of the ill, as well as the objects they touched, had the capacity to spread disease. In her 1878 tract, Mary Armstrong assured her readers, "Many great men have, of late, devoted themselves to studying the causes of diseases, and they have almost beyond question, established the fact that all contagious diseases, that is all diseases which can be taken by contact with the sick person, or carried in clothing, or left in bedding, furniture, etc. are caused by germs, that is infinitesimally small living organisms which are thrown off from the body of the sick person." She explained further, "They are found in the breath from the lungs and in the secretions and excretions of the body, in fact the whole atmosphere surrounding a person who is ill with what we call a contagious disease, as for example, scarlet fever, diphtheria, or measles, is full of these germs which possess the power of multiplying themselves with inconceivable rapidity." Armstrong concluded, "Now it seems to me, that as soon as this is understood, the first impulse of every reasonable person will be to get rid of these poisonous organisms, to kill them and cleanse the air from their dangerous presence."[20]

Although popular authors' renderings of the germ theory remained rather crude, the precocity of their interest still bears emphasis given the reservations still common among physicians well into the 1880s. Whereas doctors debated among themselves about the sufficiency of microorganisms to cause disease, lay hygiene writers had little trouble accepting the equation of dirt, infection, and germs. As a simple rule of thumb, the further removed the author was from the medical establishment, the more likely he or she was to accept the germ theory as a credible and important scientific discovery. In a typical burst of excitement, Emma Hewitt reported in her text, *Queen of the Home,* "The study of the theory of germ proliferation has yielded amazing results in the way of furnishing the means of checking epidemics." In contrast, most physicians penning hygiene advice in the 1870s and early 1880s presented the germ theory as a controversial thesis. For example, in his 1880 manual, physician George Wilson characterized the germ theory as an unproven hypothesis and concluded tepidly, "Of far greater importance is to know that, whatever be the origin or mode of propagation of these diseases, they are to a very large extent controllable."[21]

Invisible Enemies in the Home

In the late 1870s and 1880s, the germ theory was easily incorporated into popular advice literature precisely because it seemed to justify widely accepted precautions of ventilation, disinfection, isolation of the ill, and general cleanliness. The first incarnation of the gospel of germs simply bestowed germicidal rationales on already trusted strategies of protection: the same practices adopted to defend against the dangers of damp cellars and impure air also worked to defeat malevolent germs.

This union of the old sanitary science and the new germ theory of disease was facilitated by a shared vision of the human body as a potent source of pollution. As Mary Armstrong explained, "Everything which is thrown out from the human body is unclean, and becomes at once dangerous to human life." Common wisdom held that human bodies polluted, or "vitiated," indoor air with gases respired by the lungs such as "carbonic acid" (carbon dioxide) and sulfureted hydrogen, as well as with the "sewer gas" arising from decaying excrement. Human fecal matter also polluted the water directly. Under normal circumstances, the environment contained sufficient natural disinfectants—that is, sunlight, air, and soil—to purify these bodily by-products. But when too many people were packed into too little space, these natural processes of purification were overwhelmed. Dangerous accumulations of filth thus poisoned the air and water, providing the ideal breeding ground for germs. This imbalance was potentially dangerous wherever it occurred, but particularly so in the home, where people spent so much of their time.[22]

Henry Hartshorne, the first professor of hygiene at the University of Pennsylvania Medical School, eloquently summed up the problem in his domestic manual, *Our Homes,* published in 1880. "Apart from human interference, there is in nature a balance of formation and destruction, of life and death, food and waste, making a perfect natural economy everywhere," he wrote. Then "Man comes in with his artificial constructions, and sweeps away much of this economy of nature," resulting in "foulness of the earth, water, and air; stench, miasma, pestilence." Hartshorne concluded, "A guerilla warfare seems to be waged all around the invader of nature." In order to become healthy again, communities must "restore the original balance of primeval nature, by providing for the reappropriation of the products of life and the results of death and decay around us."[23]

Early converts to the germ theory simply added microbes to this already dangerous state of pollution. Wherever foul air and water were allowed to accumulate, they argued, germs were sure to follow. The result was a two-pronged assault on human health: prolonged exposure to vitiated air and impure water weakened the body's overall resistance, making it easier prey for microbial enemies; then, when disease germs were introduced, the inhabitants had little chance to escape illness and death.[24]

The first generation of preventive strategies promoted in the name of the germ theory simply commandeered the sanitarians' long-standing obsessions with domestic architecture, particularly as it related to ventilation and plumbing. In this regard, disease prevention began literally with the home's foundations; Victorian homeowners were urged to build their dwellings on dry soil, to prevent the dampness that encouraged decay and germ life, and to maximize their exposure to the "natural disinfectants" inimical to bacteria, namely fresh air and sunshine.[25]

In the older manuals, the need for ventilation was presented as a defense against accumulations of the gases thrown off from the lungs. Domestic guides contained precise directions concerning the proper ventilation and proportioning of rooms, especially sleeping chambers, to ensure that enough cubic feet of air per occupant was available to dilute organic matter in the air. Many included instructions on how to rig windows with simple "ventilators" made of wedges and boards to circulate fresh air without creating dangerous drafts. The newer, more germ-oriented manuals contained exactly the same recommendations regarding ventilation, but their instructions were endowed with the additional virtue of diffusing the bacterial "clouds" in the air. T. J. MacLagan put the premise simply in the *Popular Science Monthly:* "Keep the windows shut, and you keep the germs in; open them, and they pass out with the changing air."[26]

The obsession about ventilation went hand in hand with a preoccupation with household plumbing. "Ours is the Age of Plumbing," emphasized Henry Hartshorne, and hygiene writers continually emphasized that if a family did not attend to its toilets, the rest of its protective rituals would be rendered worthless. Even the simplest domestic hygiene manuals included lengthy, detailed discussions of the complexities of traps, water closets, and soil pipes needed to prevent fecal contamination of the domestic water and air supply. Initially advocated as a defense against dangerous sewer gases, these measures were subsequently transformed into protections against the bacteria that these gases supposedly carried into the home.[27]

Late-nineteenth-century hygienists advocated as the bare minimum of precautions against sewer gas and its accompanying germ life: complete separation of the drinking and waste water systems; water closets that flushed thoroughly to prevent the accumulation of wastes; watertight pipes to conduct wastes into the sewer; traps on all drains to prevent the discharge of sewer gas back into the room; and a special soil pipe running up the side of the house and venting above roof level to allow the safe conduct of gases away from the home. Writing in a Staten Island missionary paper in 1878, one author confidently claimed of the soil pipe, "With one of these simple appliances out-of-doors, a cellar tight and dry, and indoor drain pipes without material leakage, domestic life would be secure from the worst of its invaders."[28]

Plumbing of this complexity required outside assistance. Sanitary authorities recommended that homeowners hire only the best plumbers and carefully supervise their work. J. Pridgin Teale designed his popular "pictorial guide to domestic sanitary defects" so that the homeowner might "test every sanitary point, one by one, and as he goes round book in hand, . . . catechise his plumber, his mason, or his joiner." Those renting or buying a home were advised to use the "peppermint test," which involved introducing oil of peppermint into a water closet and sniffing to see if the aromatic odor leaked out elsewhere in the house. The appearance of its telltale scent indicated the presence of plumbing defects that might taint the domestic air with sewer gas or germs.[29]

Domestic guides included other instructions for testing and purifying the household water supply. Warning that appearance, taste, and smell were not enough to determine water's safety, they urged householders to filter or boil all drinking water in order to remove noxious organic matter and kill disease germs. Many manuals included instructions for constructing simple home filters of sand, charcoal, and cloth. "But really suspicious water should, before using it for drinking or cooking, be boiled as well as filtered," advised Hartshorne in 1880.[30]

In general, a high level of household cleanliness was recommended as a safeguard for domestic health. Under the old sanitarian code, any accumulation of decaying matter or dirt, especially in association with damp and darkness, was regarded as a potential breeding ground for the invisible agents of disease. The association of airborne dust, soilborne spores, and pathogenic microbes in early conceptions of the germ theory only intensified the menace of household dust and dirt. As a domestic manual compiled in 1887 for the Association of College Alumnae explained, "The general acceptance of the germ theory of disease makes it imperative for every housekeeper to guard against all accumulations of dust, since such accumulations may harbor dangerous germs." Echoing earlier hygiene reformers, the ACA manual urged women to break with the dominant Victorian aesthetic and choose less dust- and germ-friendly home furnishings: "To propitiate the goddess of health we can well afford to sacrifice on her altar the superfluous draperies, carpets, and ornaments of our living and sleeping rooms."[31]

Even more extreme precautions were required when nursing family members through an infectious ailment. Hygiene manuals routinely included a chapter spelling out the proper conduct of a "home hospital." To aid the patient's recovery as well as protect others from contagion, homeowners were urged to choose a light and airy chamber, strip it of carpeting and drapes, and hang a sheet drenched in a strong disinfectant, preferably carbolic acid, at the doorway. Manuals stressed that the patient's evacuations should be immediately disinfected and removed from the house. The careful observance of such isolation measures was portrayed as a duty not only to one's own family but also to the whole community. As T. J. MacLagan concluded in his article on typhoid fever, "It rests with those who have such ailments in their houses to carry into effect the measures calculated to destroy and get rid of the poison, before it has had time or opportunity to be a source of danger to those around."[32]

Traditional concerns about "fomites," a term applied to any object thought capable of conveying infectious material, were greatly heightened by the germ theory of disease. Hygiene reformers warned that many common objects could harbor the particles of disease for long periods of time. For example, a circular on scarlet fever distributed by the Massachusetts State Board of Health cautioned that the contagion could be transmitted by "air, food, clothing, sheets, blankets, whiskers, hair, furniture, toys, library-books, wallpaper, curtains, cats, [and] dogs." The discovery of disease "spores" only reinforced the popular belief that disease particles could lie dormant for many years. In his book *Health in Our Homes* (1887), Joseph Perry told a standard cautionary tale about a cap that had hung in the room of a boy who died of scarlet fever and had been "put away in a closely covered tin box without disinfecting." When his younger brother wore the cap two years later, he promptly fell sick of the same disease, furnishing conclusive proof of the contagion's "tenacity," in Perry's words. To avoid such tragedies, parents were warned to destroy clothing and toys used by sick children, a practice immortalized in Margery Williams's classic children's story, *The Velveteen Rabbit*.[33]

At a time of high mortality rates for children, disease prevention in the nursery consumed whole chapters of many domestic advice manuals. The importance of location was illustrated by a story in Emma Hewitt's 1888 manual about a woman whose children kept having "recurring diphtheric symptoms" because she kept them in a basement workroom during the day; after the family physician advised her to move the children's daytime nursery upstairs, "the change was almost magical." Hygiene writers instructed parents to place the nursery on an upper floor, to furnish it sparsely, and to keep their children out-of-doors as much as possible. At night, they were to be placed in single beds so that their "exhalations" would not mingle (a recommendation also given for adult sleeping arrangements). As a precaution against illness, particularly the dreaded "summer complaint," or infant diarrhea, feeding utensils, especially for infants, had to be kept scrupulously clean and their drinking water boiled or filtered. Because milk was thought to be particularly susceptible to absorbing airborne impurities, including germs, parents were admonished to seek a clean, fresh milk supply.[34]

Before and after germs entered the preventive equation, disinfectant use figured prominently in domestic advice as both a special protection in the sick room and a daily precaution against disease. Circulars and home hygiene manuals described at length the properties of various disinfectants and supplied recipes for inexpensive solutions to be made up as needed. Disinfectants were advocated for a multiplicity of uses: to purify the air of a sickroom, bathe the patient's skin, disinfect excreta, fumigate clothing and linens, wash down a corpse, and cleanse household plumbing systems. Frequent, systematic disinfection was presented as one of the most important precautions that householders could practice, lest, according to an 1885 New Hampshire State Board of Health circular, "by neglect the health of the family may suffer."[35]

Victorian Women and the Cultural Work of Cleanliness

In the late 1800s, public health experts and lay hygiene reformers wrote countless pages elaborating on the need for domestic disease prevention. How widely or deeply their message cut into the popular consciousness is not easy to determine; the volume of advice on a given subject is no sure guarantee that it was followed. But the gospel of germs developed at a time when affluent Americans set great store by cleanliness and when the image of the "whited sepulcher" had deeper cultural meanings that resonated with the dire warnings about house diseases. Hygienists' laments about the hidden sources of infection in the home were heard by a peculiarly body-conscious and house-proud generation of Americans. Over the course of the nineteenth century, a meticulous attention to personal and domestic cleanliness had become an important marker of high class standing. By assaulting the middle and upper classes' sense of sanitary superiority, hygiene reformers had attacked a cherished aspect of their identity.[36]

The ability to maintain an exacting level of cleanliness, both of the body and of the home, was an essential prerequisite of gentility. A popular 1878 etiquette manual informed its female readers concerning personal hygiene, "On this head, fastidiousness cannot be carried too far. Cleanliness is the outward sign of inward purity." At a time when housing was frequently invoked in class designations, as in references to the "shanty Irish," the "tenement house poor," or the "brownstone classes," social standing was closely tied to a certain style of domestic life, which necessitated not only the latest in modern plumbing but also careful attention to household cleanliness.[37]

In pursuit of gentility, middle- and upper-class families invested heavily in their houses in terms of time, money, and anxiety. The home functioned as a stage for both the intense family life and the elaborate social customs so characteristic of the Victorian era; the necessary props included many difficult-to-clean items, including ornate furniture, plush upholstery, heavy draperies, and wall-to-wall carpeting. The service of formal meals involved maintaining extensive collections of china, glassware, silver, and table linen. Personal hygiene required the installation and maintenance of what were discretely referred to as "modern conveniences," such as water closets, bathtubs, and bedroom washstands. In sum, by the 1870s and 1880s, Victorian Americans had invested heavily in a way of life that stressed the surface cleanliness and order of the home. Thus they were very disturbed when public health experts informed them that their dwellings, particularly the new modern conveniences of toilet and washbasin, were harboring death-dealing agents of disease.[38]

The house disease concept exposed the Achilles' heel of Victorian gentility: by dint of great effort and a horde of servants, individuals and households could acquire the appearance of cleanliness. Yet underneath the most meticulous exterior there always existed the disorderly body, constantly exuding different sorts of "filth." Whatever one's class standing, the physiology of living required the "coarseness" of producing sweat, feces, urine, saliva, and for women, menstrual blood. Although they were quite willing to blame the poor for epidemics, the Victorian gentleman and lady knew in their heart of hearts that the unsanitary enemy was also themselves. Their fears of pollution ultimately stemmed from a basic distrust of all human bodies, including their own.

With these twin themes—the deceptiveness of appearances and the ubiquity of filth—the Victorian social system resonated in powerful ways with the dominant views of infection and disease. Just as the smooth-talking stranger's fine clothes and manners might give no hint of his potential for social perfidy, the cleanest-looking gentleman or lady might also harbor some seed of disease. Likewise the finely kept mansion might give no sign, on the surface, of the dangers that lay within its plumbing or "hidden work." Both the dark and damp parts of the body, and the areas of the house most closely associated with them, figured as natural breeding grounds for disease.[39]

The looming specter of infectious disease turned the sanctuary of the middle-class home into a space fraught with peril. As the New York physician Robert Tomes wrote in the *Bazar Book of Health,* the man who felt secure in his unsanitary "castle" only "shuts himself up in it with his worst enemies." Similarly, an 1874 editorial in the *Sanitarian* presented a frightening antithesis to the image of "Home, Sweet Home" by warning, "From the cellar, store-room, pantry, bedroom, sitting room and parlor; from decaying vegetables, fruits, meats, soiled clothing, old garments, old furniture, refuse of kitchen, mouldy walls, everywhere, a microscopic germ is propagating."[40]

The "whited sepulcher" image had a particularly keen edge for women of the genteel classes. For centuries male artists and moralists had used that trope to describe the female body, to claim that woman's exterior beauty masked hidden filth and disease inside. Although the nineteenth-century "cult of true womanhood" promoted a radically different image of middle-class women as exemplars of moral authority and purity, traditional conceptions of women's bodies as dirty and dangerous lingered. As society's appointed guardians of moral virtue and class standards, middle- and upper-class women had to struggle to rise above their "baser" physical nature, with all its anxiety-laden associations to sexual desire and its disruptive consequences. Through rituals of personal and household cleanliness such as those that Martha Roosevelt embraced, many sought to distance themselves from the image of the "whited sepulcher." To a disproportionate degree, the work of cleanliness, both in symbolic and practical terms, landed squarely in women's cultural domain.[41]

Hygiene manuals placed enormous importance on small details of housekeeping. Stressing the need for exacting sanitary observances, Harriette Plunkett stated in her 1885 text, "Eternal vigilance is the price of everything worth the having or keeping." The careless woman, she warned, only came to comprehend the importance of hygiene "when, too late, she stands beside the still form of some precious one, slain by some one of those preventible diseases that, in the coming sanitary millennium, will be reckoned akin to murders." Joseph Perry echoed this sentiment in his 1887 manual: "Many, very likely, will say that it is too much trouble to take the preventive measures advised." But "in the lives of all who study convenience so closely," he replied rhetorically, "there may come a time when, had those simple hints been observed, serious illnesses would have been averted, and possibly lives been spared."[42]

Such heavy-handed attempts to inspire guilt and anxiety were not likely to affect the two sexes equally. To be sure, hygiene reformers emphasized that men had important hygienic roles to perform. The use of masculine and feminine pronouns in the texts suggests that men were expected to attend to the general construction and repair of the home, including the plumbing and the cellar, whereas women had charge of home nursing, child care, and general housekeeping. Of course, in middle- and upper-class households, both sexes had ways that they could delegate these responsibilities. Men could employ plumbers or architects to make the house safe, and their wives had domestic servants to do the heavy cleaning. But on a daily basis, the oversight of sanitary practice fell most heavily and unrelentingly on the women in the family.[43]

Both men and women advice givers assumed that female readers had a greater responsibility for and interest in preserving health because their lives were more closely tied to home and family. Even in the more technical areas of plumbing, women were urged to take an active role. In an 1885 treatise, the sanitary engineer George Waring explained that the housewife's work did not end "when her husband has paid a good round sum to the engineer and to the plumbers." Domestic plumbing, he wrote, was either "a means of safety, or an engine of destruction, according as she performs her duty, or neglects it." Along similar lines, Harriette Plunkett's 1885 text, aptly named *Women, Plumbers, and Doctors*, urged her female readers to conquer their fears of "dark, damp spaces" and take charge of the cellar, that "weird, forbidding, and uncanny" region that "belonged naturally to the gentlemen of a household."[44]

Given the continuing scientific controversy about how infectious diseases actually originated and spread, domestic hygiene authors in the 1870s and 1880s placed a surprisingly high level of responsibility on individual households to prevent disease. To the extent that they accepted this responsibility, middle-class women likely found the gospel of domestic prevention to be a mixed blessing. On the one hand, hygiene reformers promised wives and mothers that the scourge of illness could be greatly lightened by scrupulous adherence to a detailed hygienic code of behavior. On the other hand, it potentially laid the blame for illness and death squarely on their shoulders.

As their frequently repeated tales of germ-laden caps and diphtheric rooms made evident, failure to follow hygienists' recommendations could have the most terrible of consequences for mothers: a sick or dead child. In an era when one-fifth of all children died before their first birthday, this emphasis on the preventability of disease had the potential to generate a powerful sense of culpability. The heightened intensity with which Victorian parents mourned their children, a trend often attributed to the smaller size of their families, may also have reflected these determined efforts to cultivate greater parental, and chiefly maternal, responsibility to prevent "house diseases."[45]

The equation of meticulous housekeeping with disease prevention also helps to explain the often strident tone privileged women brought to their already tense relations with their working-class servants. Lapses in cleaning not only transgressed the code of gentility and endangered the family's social standing; they also put the household at risk from potentially mortal "filth" diseases. Underlying the late-nineteenth-century "servant problem" so endlessly bemoaned in the popular women's literature lay the special burdens of safeguarding the home against infection that the concept of "house diseases" placed on both mistress and servant.[46]

For some middle-class women, this sense of responsibility for guarding the home ultimately impelled them to forsake its limited domain; the gendered division of labor that cast men as experts and lawmakers and women as vigilant wives and mothers proved too confining. Like Plunkett, some began writing about public health issues in domestic manuals and newspaper columns as a way of educating other women about disease prevention; others became active in neighborhood sanitary associations and reform politics. For a few, the emphasis on women's obligations to prevent disease became a rationale for pursuing professional careers, as I will discuss further in Chapter 6. As Plunkett correctly foresaw in 1885, for women, "a new sphere of usefulness and efficiency opens with the knowledge that in sanitary matters an ounce of prevention is worth a ton of cure."[47]

Through varied types of print media, the basic concepts of domestic disease prevention were widely disseminated among the affluent urban classes by the late 1880s and early 1890s. The sanitarians' longstanding efforts to pathologize the home led to the relatively rapid notice of the germ theory, which in turn reinforced the basic tenets of the sanitary gospel. Although hygienic measures of ventilation, sanitary plumbing, disinfection, and water purification were widely accepted before the invocation of the germ theory, the added threat of the microbe gave them renewed force in the late 1800s. Contrary to the worst fears of older sanitarians, the gospel of germs did not weaken the connection between a clean, moral life and safety from disease. Instead, in the words of Emma Hewitt, the new scientific discoveries "placed in the hands of every one, if not the power of destroying these germs, at least the power to prevent their proliferation" by the practice of "antiseptic cleanliness" in the home. For Victorian women, this new knowledge brought both a new sense of power and a heavy load of guilt.[48]

3 · Entrepreneurs of the Germ

In mid-August 1881, the well-known sanitary engineer George E. Waring arrived in Washington, D.C., on a mission of national importance. The preceding July 2, the President of the United States, James A. Garfield, had been shot in the back by a would-be assassin named Charles Guiteau. Installed in a "home hospital" set up for him in the White House, the President seemed to improve briefly, then began to run a persistent fever that slowly sapped his strength. The distinguished medical men treating Garfield attributed the fever to the gunshot wound, but as he steadily weakened, the press as well as some of his advisers began to wonder if his health was being undermined by another, more subtle danger, one that required a specialist to detect. Thus George Waring was called to the White House to determine if the President was being poisoned by sewer gas arising from the mansion's antiquated plumbing.[1]

Today, Waring's mission seems ridiculous. From a modern perspective, it appears obvious that Garfield died from the consequences of his bullet wound, not from sewer gas exposure. But the White House sanitary scandal shows how differently late-nineteenth-century Americans understood the nature of infectious disease and the role of houses in spreading it. As of 1881, the sewer gas menace was so widely appreciated that both sensation-seeking journalists and respectable sanitary engineers regarded it as a plausible explanation for the President's declining health. Thus the White House story provides a good starting point for examining how the threat of house diseases was understood and acted upon in the late-nineteenth-century American home.[2]

By the early 1880s, middle-class Americans' apprehensions about their homes had become fertile ground for a new species of entrepreneurship. Like George Waring, inventors, engineers, plumbers, and manufacturers had discovered that disease-proofing houses and other buildings could be a lucrative line of business. The evolution of these commercial services allows us to move beyond advice books to consider how affluent Americans began to act upon their concerns about disease. Given that few Victorian householders confided details of plumbing renovations or disinfectant purchases to their diaries and letters, evidence about such matters found in patent applications, advertising brochures, trade journals, and business records is particularly helpful in tracking hygienic practice. The evolution of sanitary entrepreneurship opens a useful window into the search for security that transformed the landscape of the American home, from the President's house to the tenement house, in the late nineteenth century.[3]

Of course, marketplace trends do not reflect perfectly popular beliefs and behaviors. Entrepreneurs invoked only those lessons of sanitary science and the germ theory that justified the sale of their particular product. Many businessmen sought not only to appeal to what public health experts might regard as reasonable fears, but also to exaggerate and intensify those fears in unscrupulous ways. Yet allowing for this distortion and exaggeration, tracking the rise of domestic sanitary services illuminates how concerns about infectious disease transformed the customs of everyday life.

The growth of sanitary entrepreneurship confirms that well before the germ theory gained widespread acceptance, householders had already started to update their plumbing in response to the zymotic theory of disease. In the 1880s, the sanitary trades began to invoke the discovery of microorganisms chiefly to promote new versions of old sanitary standbys, such as sewer traps and disinfecting devices. In so doing, entrepreneurs shaped their conceptions of the germ to conform to prevailing fears of sewer gas. The initial efforts to "germ-proof" the American home focused primarily on its portals to the sewer system—namely, the modern conveniences of toilet and wash-basin.

The White House Plumbers

The White House sanitary scandal provides a dramatic example of the house disease concept in action just at the point when the germ theory of disease began to figure in both public debates and entrepreneurial initiatives. The 1881 controversy capped off decades of concern about the executive mansion's healthfulness. As did the city of Washington itself, the mansion at 1600 Pennsylvania Avenue had a long reputation for breeding fevers due to its proximity to the Potomac marshes. Built in 1817 to replace the original building that the British had burned during the War of 1812, the White House's water closets and sewer lines had been updated in only a piecemeal fashion. By the late 1870s, its plumbing defects had come to seem not only undignified but also positively dangerous.[4]

Several years before the Garfield controversy, the officer in charge of the White House building and grounds had recognized that the President's home fell far short of the sanitary standards observed in the better homes of the era and had begun the work of updating it. Because Congress had to approve any appropriations for its improvement, the officer had to make a strong case for every penny spent. In 1879, he managed to convince a stingy Congress to finance an overhaul of the White House plumbing. Standard devices were added to exclude sewer air, old water closets were replaced with more sanitary models, and an "upright side sash ventilator" was installed above the winding staircase to increase the flow of air to the upper floors.[5]

These repairs did not suffice to give the house a clean bill of health, however. Soon after Garfield was inaugurated in the spring of 1881, the new First Lady, Lucretia Garfield, fell ill of a prolonged and serious fever, which some observers immediately blamed on the house. An indignant *Baltimore American* reported, "The Potomac Flats have . . . filled up so much lately that they back the sewer gas right into the President's House, and Mrs. Garfield is suffering from that form of poisoning." The paper urged that "that old and shabby mansion [be] completely renovated or a new one built." Mrs. Garfield's illness prompted the new officer in charge, Colonel Almon F. Rockwell, who had a degree in medicine as well as engineering, to order more plumbing renovations—including better traps and ventilation of the house's connection with the main sewer line—and to modernize the west-wing bathroom.[6]

Sensing that profits were to be made in "doctoring" houses, sanitary entrepreneurs immediately tried to cash in on Rockwell's renovations by using their Capitol Hill connections. E. E. Rice, the inventor of "Rice's Patented Ventilation System," called in favors from an Iowa Republican, Senator William B. Allison, who wrote Rockwell testifying to what Rice's system had done in his own home; another Republican, former Senator Eppa Hunton, wrote Garfield directly, urging him to make the executive mansion a "healthy habitation" by installing Rice's device. With an unctuous tone, Hunton wrote, "I hope in the multitude of your official cares you will take time to consider the question," and he assured the President that the system would not cost more than $4,000. "I don't think the country quite willing to see [Vice-President] Arthur President and always willing to spend money to secure the health of its Chief Magistrate and his family," he concluded.[7]

But before Rockwell could decide among the many claimants for his sanitary patronage, President Garfield was shot on July 2, 1881, as he prepared to leave the capital for a vacation. He was taken immediately to the White House, where his physicians found that the bullet was lodged too deeply in his back to be removed safely. A lower chamber in the mansion was transformed into a "home hospital," and for a brief time Garfield appeared to rally. Then, in mid-July, he began to run a chronic, wasting fever that worried his medical attendants. In testimony to the widespread appreciation of Lister's work by 1881, critics immediately suggested that antiseptic procedures had not been followed carefully enough in handling the wound, a charge that Garfield's doctors vigorously denied.[8]

Harkening back to Mrs. Garfield's illness, some commentators began to wonder if the President was being harmed by the "malarious influence" exerted by the Potomac Flats. In late July, the *New York Herald* took up the sewer gas issue in its customary sensation-seeking style. For one story, *Herald* reporters interviewed a "well-known plumber" who confidently asserted of Garfield's case that "the real trouble is sewer gas, which is ten times as bad and even more poisonous" than simple foul air and claimed that "there is not a perfect working [sewer] trap in the Executive Mansion." A few days later, another intrepid *Herald* reporter described touring the odoriferous Potomac Flats with a "well known scientific gentleman of this city" and finding "masses of fecal matter and filth" floating in the water.

"Your nose is a good and safe guide," the "scientific gentleman" reportedly said, "and it has already told you that the air we are breathing is charged with the germs of disease." That same air, the "doctor" pointed out dramatically, was blowing in the direction of the White House where the feverish President lay in his sick chamber.[9]

Sanitarian-minded well-wishers were quick to respond with suggestions about protecting the President against sewer gas. After reading in the *New York Herald* about the sad state of the White House plumbing, the Sanitary Committee of the Master Plumbers of New York sent Colonel Rockwell a long letter filled with criticisms of the White House sewer traps and offering to help out in the crisis for free, "having the wellfare [sic] of the President and his family at heart." In a similar spirit, various inventors offered their devices for protecting against sewer gas, such as the ornamental street lamp, which was recommended by the firm of Ogilvie and Bennem as "a simple, effectual and economical method of destroying the poisonous gases emanating from sewers, and preventing the transmission of the same into buildings." In a sharper tone, E. E. Rice, whose overtures Rockwell had neglected the previous spring, offered to install his device, observing, "It was unfortunate that my improvements had not been made before the President was shot—it would have saved expense, unpleasant feelings, & accomplished all & more that the present arrangement is claimed to," a reference to the patent air-conditioning device Rockwell had installed to make the President more comfortable.[10]

In public, the President's physicians dismissed the notion that his fever had anything to do with sewer gas. But in private, Garfield's inner circle of advisers worried about the growing controversy. Hoping to put the matter to rest, the Attorney General of the United States, Wayne MacVeigh, finally suggested that George Waring be called in to inspect the White House plumbing. Rockwell agreed to the plan.[11]

Colonel Waring, as he was called in deference to his Civil War service, was a logical choice for this delicate mission. His rapid rise to national prominence attested to the growing salience of the "house disease" problem in the 1870s. Originally a specialist in agricultural drainage, Waring became interested in home sanitation in the late 1860s and gained widespread fame with his writings on the subject in the *Atlantic Monthly* (1875) and *Scribner's* (1876)—articles that were later reissued in book form. When a devastating yellow fever epidemic hit Memphis, Tennessee, in 1878, President Rutherford B. Hayes sent him to oversee improvements in the city's drainage. Even more relevant to the crisis at hand, Waring was familiar with the capital city's drainage problems. In an 1880 lecture delivered at the Smithsonian Institute, he had warned that sewer gas arising from the city's homes posed a far greater danger to health than did the notorious Potomac Flats, not only because of the gas's "own direct action" but also because "it so often acts as a vehicle for the germs, or causative particles of specific diseases."[12]

On August 23, 1881, Waring delivered a preliminary report on the White House to Rockwell. Without taking up the floors or breaking into the walls, which might have disturbed the ailing President, he stated that he could only guess at the quality of the "hidden work" in its interior. But based on what he could see, the house was in "a very much less unsafe condition than current reports would lead one to expect." Yet Waring confirmed that "the plumbing appliances of the Executive Mansion do not conform to what are now accepted as the necessary sanitary requirements of a safe dwelling." For example, the large basement kitchen had a sink with such a large trap "as to constitute a permanent cesspool." Upstairs, all sorts of improperly trapped pipes prevented the "free discharge of filth." The water tank that supplied the house sat in an upper room that was "practically a water closet" due to contamination from the improperly ventilated bathroom directly below it. Most worrisome, in the light of the President's illness, Waring found that the room next to his sick chamber had a bathroom with an old-fashioned "pan closet" served by an circuitous and unventilated soil pipe. The report concluded with a list of emergency repairs to be done immediately.[13]

The attorney general wrote Rockwell in early September, urging that Waring's report, if "favorable," be summarized and released to the Associated Press. "It would allay, I am sure, much apprehension about the present unhealthy surroundings of the President," he reasoned. But not surprisingly, Rockwell did not comply; although Waring meant his report to be reassuring, his frequent allusions to cesspools, filth, and foul decomposition hardly made the White House seem like a healthy place for the President's recovery.[14]

Although Garfield's medical attendants never gave any public credence to the sewer gas theory, Waring's inability to give the White

House a clean bill of health may have entered into their decision to move the President to his summer home in New Jersey, despite fears he would not survive the trip. Two weeks after Waring submitted his preliminary report, on September 6, Garfield was taken from the White House, but the change did not improve his condition. He died in New Jersey on September 19. The autopsy revealed that the immediate cause of death was a ruptured aneurysm of his splenic artery, damage probably traceable to either the bullet or the probes used to locate it.[15]

But the controversy over the sanitary state of the White House did not end with the President's death. Waring's report was released posthumously, and even the sober *New York Times*, which had steered clear of the *Herald*'s sewer gas crusade, carried long excerpts from the piece. The *Sanitary Engineer* published it in full, along with a scorching editorial lamenting that it had taken "some great calamity," such as "Mrs. Garfield's illness and the President's lingering sufferings," to bring the White House's defects to light. "The blame rests with Congress, who, by their niggardly appropriations, have made it impossible to correct the blunders that for several years have been apparent to those who have been familiar with the White House, but who could not secure the necessary funds to defray the expense of remedying them." The editorial concluded that until the mansion's plumbing "is pulled out and entirely reconstructed, the White House will be behind our better class of tenement-houses in this important particular."[16]

The new President, Chester Arthur, was so convinced of the White House's derelict condition that he refused to live there until the plumbing was redone. Under his watchful eye, Rockwell proceeded with the renovations that Waring had recommended in his preliminary report. But high anxieties and professional rivalries typical of the business of treating house diseases hampered progress on the project. Already committed to a consulting job for the French government, Waring had to oversee the White House renovations from overseas. In his absence, Rockwell decided to use the local plumbing contractor who usually did the White House work rather than the expensive Boston firm Waring had recommended. When Waring learned of the change, he wrote indignantly from Paris that he could not have these all-important repairs done by a "plumber of whom I know nothing" because "my reputation is now involved in the matter and I shall be held accountable for work that I doubt."[17]

No doubt with a nervous Chester Arthur looking over his shoulder, Rockwell compromised with Waring; he agreed to purchase the custom-made materials Waring had ordered and to hire one of Waring's assistants, William Paul Gerhard, to supervise the local plumbers. From then on, the renovations proceeded smoothly, and the White House's antiquated plumbing was replaced with new airtight pipes and state-of-the-art sewer traps and venting systems. The old water closets were replaced with white earthenware toilets that Waring had designed himself. He described his new "Dececo" model as flushing so completely that no "foul matter" could remain. It also stood free from the wall and was surrounded by white tiles instead of being encased in a wooden cabinet, so that "the whole apparatus is in full view" and no filth could accumulate in its hidden work. Seizing the entrepreneurial moment, the companies that manufactured the Dececo toilet and patented sewer trap that Waring had chosen began to advertise their goods with the slogan "recently used at the White House."[18]

In his final report, Warning certified that the White House was now safe to be lived in, but he insisted that more renovations were needed if the first family continued to use the mansion as a residence. Because the very ground on which the house sat had become saturated with sewage, Waring recommended "as a most important sanitary measure" that the whole building be raised on "piers and groined arches . . . to secure a complete separation between the ground and the building." President Arthur used Waring's report to seek Congressional support for a more radical solution—namely, to pull down the White House and rebuild a more sanitary replica, complete with modern plumbing, to the south of its present site.[19]

So powerful was the "house disease" concept that in June 1882, the Senate actually approved $300,000 for the construction of a "good, healthy, and convenient dwelling place" for the President and his family adjacent to the old building. But sentiment in favor of preserving the historic mansion eventually prevailed, and the House of Representatives never passed the bill. Chester Arthur had to be content with a thorough sanitary overhaul of the existing building. The sanitary engineers and plumbers once again took over the White House and replaced the old sewer lines underneath it with sturdy cast-iron pipe. During the four-year Garfield-Arthur administration over $110,000 was spent on the mansion, the largest outlay on its maintenance since its destruction by the British almost seventy years earlier.[20]

Entrepreneurs Discover the Germ

As the White House controversy reveals, by the time the germ theory of disease had begun to gain widespread public notice, both anxieties about sewer gas and commercial sanitary services were firmly established on the American scene. Much as converts to the germ theory commandeered the truths of sanitary science to craft their own preventive gospel, sanitary "experts" sought to expand their profits by incorporating the germ menace into the already booming business of domestic disease prevention.[21]

To this end, entrepreneurs of the germ sought to convince homeowners who thought their homes were sufficiently disinfected, trapped, and ventilated that they were wrong, that "true" protection required adding a new germicidal element to the home's sanitary defenses. In making this argument, entrepreneurs of the germ employed many of the same arguments found in the popular advice literature of the day. Their appeals illustrate not only how easily the germ sell was incorporated into existing rationales for household disease control, but also how the germ's commercial champions sought to unsettle homeowners' confidence in their existing state of safety from house diseases.

The marketing of disinfection aids illustrates both of these trends. Under the aegis of the zymotic theory, manufacturers had introduced a wide range of new chemical disinfectants that supposedly neutralized disease "ferments" more effectively than did the old lime and sulfur standbys. By the early 1880s, consumers could choose among solutions of carbolic acid, potassium permanganate, sulphate of iron (also known as copperas), and chloride of zinc, as well as patent disinfectants that combined a variety of agents in a unique, "secret" formula. As the germ theory gained recognition, disinfectant manufacturers were quick to use it as a way to gain a competitive advantage in a crowded marketplace. The protective powers of disinfectants had to be redefined to include the capacity to kill germs as well as neutralize chemical impurities in the air. Some companies played the germicidal angle by naming their products "Listerine" or "Pasteur's Marvellous Disinfectant," in homage to prominent figures

associated with the germ theory. Others sprinkled their advertisements with terms such as "germ-destroyer" and "germicide" and
claimed that the growing authority of the laboratory supported the
merits of their product.[22]

By 1884, the claims for protection were so many and so outrageous
that the American Public Health Association appointed a committee,
headed by the surgeon George Sternberg, to test the germicidal
properties of the disinfectants in common use. Carefully distinguishing a product's effectiveness in killing germs from its efficacy in
removing foul odors from the air and in arresting the processes of
decay, the committee took as the true test of a disinfectant its ability
to kill pathogenic bacteria. By this test, most disinfectants came up
sadly wanting.[23]

Sternberg's committee railed against the claims being made for
the germicidal properties of disinfectants. For example, Sternberg
wrote, "Pasteur's Marvellous Disinfectant" was virtually useless, "yet
this fluid is by some contrivance, to be thrown into the water-closet
of every germ-fearing citizen when he pulls the handle, so that it may
catch the germs on the fly, and extinguish their power for mischief
before they reach the sewers." Sternberg stressed the importance of
a more modern disinfectant practice, stating, "The time has passed
when *pater familias* can complacently congratulate himself upon having disinfected his house with a bottle of carbolic acid, which he has
brought in his vest pocket from the corner drug store." But for all his
pleas for rigoi in evaluating these compounds, Sternberg's committee inadvertently contributed to the growing appeals to "science" in
marketing disinfectants. Almost as soon as he published his first study
of commercial disinfectants in the 1885 *Medical News,* the makers of
Wither's Antizymotic Solution began (falsely, as it turned out) claiming that he had endorsed it as "the best" disinfectant.[24]

Given the complete lack of regulation in the late 1800s, disinfectant manufacturers were free to make any claim that they liked,
regardless of its basis in fact. More to the point here is that entrepreneurs moved swiftly and surely to add the promise of germ protection
to their list of a product's virtues. At the same time, they did not
abandon older claims to neutralize foul smells or arrest the processes
of decay. Although public health experts continually denigrated the
connection between foul smells and disease, consumers evidently still
preferred disinfectants to be powerful deodorants. Thyme, eucalyp-

tus, and pine were particularly favored as antidotes to foul odors and the germs that they supposedly carried. As an advertisement for the "Botsford Automatic Fountain" explained in 1893, "Thymol and eucalyptus are perfect germ destroyers and preventatives of disease from noxious vapors."[25]

Given the prevalent fears of sewer gas, household plumbing devices were another prominent site for germicidal improvements. Entrepreneurs had already discovered that water closets, sewer traps, and the like were a profitable line of trade. As the germ theory gained acceptance, they simply took existing devices for dissipating sewer gas and added chemicals or intense heat to kill disease microbes. Patent applications and advertising copy stressed that these new designs went beyond the mere "mechanical" protection offered by old-style sewer traps to offer a more complete degree of sanitary security. As the brochure for Tayman's Disinfectant and Fumigating Company of Philadelphia explained, "Experience abundantly proves that mechanical devices are insufficient, that we must seek the aid of chemistry and obtain some agent that will antagonize and destroy the seeds of diseases."[26]

In making claims for germ protection, entrepreneurs sought to put germs on an equal footing with the noxious gases that were already considered dangerous. For example, in his 1883 patent for an "apparatus for eradicating sewer-gas and destroying germs," Frederick C. Hubbard explained, "Heretofore all efforts have been directed toward counteracting the odors of sewer-gas, urinals, etc., and no combinations of disinfectants have been combined with agents for destroying the germ that arises from fungus growth in the sewer-pipes." His invention solved that deficiency by continuously dripping a combination of three disinfectants into the water closet. Similarly, George W. Beard's sewer trap and ventilator, patented in January 1886, supposedly neutralized all three dangerous elements in sewer gas—carbonic acid, sulfurated hydrogen, and germs—with a complex system that combined a current of electricity with a gas flame.[27]

No doubt because of the ease of piggybacking the germ danger onto the established threat of sewer gas, the germ sell appeared first and most forcefully in the promotion of toilets, sewer traps, and the like. Entrepreneurs were somewhat slower to develop the commercial potential of household water filtration systems. From a modern perspective, this delay seems odd, because water filtration ultimately offered greater sanitary protection than did traps for sewer gas. But the dangers of aerial infection had so long dominated both scientific and popular thinking about infection that fears of sewer gas initially overshadowed concerns about waterborne germs. Only gradually, as the bacteriological certainty about waterborne disease increased, did protecting the domestic water supply begin to take on the same importance as neutralizing the dangers of the toilet.[28]

Starting in the mid-1880s, companies in large cities, where concern about public water supplies traditionally was great, began to market domestic filtration systems that saved the trouble of having to boil drinking water. In Philadelphia, for example, where the water gained an especially poor reputation after many visitors to the 1876 centennial celebration developed diarrhea, the Hyatt Pure Water Company marketed filters that could clean from one-and-a-half to eight gallons of water per minute. A list of patrons given in its 1890 brochure included such prominent Philadelphians as the publisher Alexander McClure and the banker Anthony J. Drexel. For less affluent customers, the Sub-Merged Filter Company of Philadelphia sold a simple charcoal and sand filter that could be attached to a water cooler or home water reservoir. The company brochure assured the reader that its filters could remove all the filth from Schuylkill River water, along with "the innumerable minute worms" that throve there. As did purveyors of sewer gas devices, water companies often promoted their services by disparaging the protection offered by competitors. For example, the Hyatt Company, whose filters incorporated a special coagulant to remove bacteria, dismissed filters that lacked that capacity, explaining that "the smallest micro-organisms found in water are about 1/25,000 part of an inch in size, and for these companies to assert that their filters remove them without coagulation is unjustified."[29]

Selling the Germicide

The history of the Germicide Company allows us to see in more detail how one group of entrepreneurs tried to cash in on the revelations of the germ theory. The Germicide was a disinfectant dispenser that attached to a toilet. It was invented and patented in 1880 by Edward J. Mallett, Jr., of New York City, who was described in credit reports as a "g[oo]d square man of considerable ability" making a comfortable income from patents on his various inventions. His invention attracted the interest of a New York City collection agent, Leopold Cohn, who invested about $8,000 to manufacture and market the device. Leopold's son Casper L. Cohn became general manager of the new Germicide Company, and after it proved successful in New York City, he set up similar companies in Boston, Chicago, Cincinnati, Baltimore, Philadelphia, and Washington, D.C.[30]

The company's first brochure, published in 1882, described the Germicide as a "simple and neat contrivance made of Black Walnut" that dispensed two disinfectants: chloride of zinc, which dripped into the toilet basin and supposedly created a germicidal water-barrier against sewer gas; and thymol, which sprayed into the air to kill bacteria every time the toilet lid opened or closed. The device was rented rather than sold, and thus "requires no attention whatever from the inmates of the house, as it is always under the supervision of the Company's uniformed, experienced inspectors." The cost of installation and monthly service was fifteen dollars a year, a sum the brochure claimed was low enough "to bring it within the means of the most humble householder."[31]

The description of the device was followed by a series of medical testimonies to its efficacy. Appropriately, C. L. Cohn started with a quote from John Tyndall on the germ theory of disease and noted that "although the reception of the germ theory of disease has been gradual, it may now be said to be clearly established in the public mind." Fusing the dangers of sewer gas and the germ, he made the usual arguments about the need for chemical as well as mechanical protections. He assured readers that *"there is no safety or security in any other method"* and that only the Germicide offered real protection "from the insidious foe that stealthily enters our homes and destroys our happiness."[32]

To buttress his claims, Cohn provided lengthy scientific testimonials from various public health authorities. For example, R. Ogden Doremus, a physician and professor of chemistry and toxicology at the Bellevue Hospital Medical College, reported on a series of experiments that he had conducted in 1882 at the New York Academy of Medicine. The experiments demonstrated that the gases commonly found in sewers could pass through brick and brownstone walls, as well as through the water seals used in conventional sewer traps. Doremus stressed the difficulty of detecting this danger, "as I know and feel most keenly in the recent loss of a beloved member of my family" (a reference to his young son's recent death from typhoid fever). To stop "vile gases and disease-breeding germs" from entering the home, Doremus heartily recommended the rental of a Germicide.[33]

The technical details included in this and subsequent editions of the Germicide brochures indicate that its promoters aimed primarily at a sanitarily sophisticated clientele, that is, the sort of audience who could understand the arguments for chemical over mechanical means of protection. In turn, the endorsements of eminent doctors and scientists were used to impress a less learned public. To this end, the brochures included ever-longer lists of physician patrons in various cities, as well as institutions that had installed the device. The list for Philadelphia included the microscopist Joseph Richardson, an early advocate of the germ theory, as well as the redoubtable hygiene expert Henry Hartshorne; in New York City, the *Herald,* that journalistic scourge of sewer gas, reported that its Germicides gave satisfaction. Humble homeowners who had installed the device could be gratified to know that they were in the same sanitary company as New York's Union Club, Philadelphia's Continental Hotel, and Chicago's Public Library.[34]

A credit report on the Pennsylvania Germicide Company filed in 1884 by an agent of R. G. Dun and Company stated, "The Direction is a very respectable one, composed of leading Lawyers and Physicians[,] and has good prospects of success." An 1886 follow-up concurred that "the officers are men of g[oo]d standing & push this enterprise with zeal," and said of the device, "It is a sanitary remedy that appears to have given consid[erable] satisfaction & is becoming generally extended in its use." Starting in 1884 with $10,000 in cash to cover the manufacturing and servicing costs, the Philadelphia branch installed two thousand Germicides within the first year, doing brisk business among private residences, hotels, and other public institutions. The Pennsylvania Germicide Company continued to prosper until 1888, when problems with the apparatus and the rental arrangement led to customer dissatisfaction. The company was reorganized that year, but no further credit reports were filed so its subsequent history remains unknown.[35]

Whatever its eventual fate, the long-term success of companies such as the Germicide was ultimately limited by the ability of the established sanitary trades, particularly plumbing manufacturers, to provide similar protective services. A Dun report on a contemporary of Cohn, H. P. Clement—whose company, the New York Scientific Sanitary System, installed a patent device to exclude sewer gas from the home—pointed to this element of competition. "He seems to have uphill work to get his system introduced, although it is said by those having them in their houses to be the only perfect system known for the total exclusion of sewer gas from dwellings," the agent wrote in 1883. He concluded that Clement was "determined to fight it out and get his machines known, although he has the plumbers vs. him."[36]

Practical Sanitarians and the Germ Theory

Small entrepreneurs such as Cohn and Clement faced increasing competition as sanitary engineers, plumbing contractors, and plumbing manufacturers all realized the profitability of the germ sell, especially when piggybacked with the fear of sewer gas. The coverage of the germ theory in the *Sanitary Engineer,* the premier journal of the sanitary trades, suggests how easily the microbial menace was added to their approaches to the prevention of filth diseases. Founded in 1878 by Henry C. Meyer, a successful plumbing manufacturer, the publication aimed at keeping plumbers, engineers, architects, physicians, and interested lay people abreast of the latest developments in sanitary science.[37]

To this end, the *Sanitary Engineer* started at around 1880 to include reports on the progress of the germ theory of disease. The reports were not uniformly favorable; for example, the journal expressed reservations (correctly as it turned out) about early claims to have isolated the microbial cause of malaria. Yet on the whole, the *Sanitary Engineer* took a very respectful view of the germ theory's validity. In 1881, the editor acknowledged that the "educated public opinion is demanding more exact knowledge of the causes of disease, and the demand is creating the supply." As an example of the progress of science, he cited the fact that "the announcement by Pasteur, or Koch, or Burdin-Sanderson [sic], of the discovery of a new fact in the life history of some minute and apparently insignificant organism, at once becomes a basis for means of disinfection provided by the chemist or engineer, or for legislation in preventing the spread of disease."[38]

Although it reported favorably on the latest bacteriological discoveries, the *Sanitary Engineer* made clear that engineers already had the necessary technical expertise to deal with the newly discovered threat from disease germs. In response to the kind of arguments presented by C. L. Cohn, the journal staunchly defended the standard sewer traps in use, insisting that water barriers did indeed prevent the passage of germs into the home. In a long, scornful critique of an 1882 *Popular Science Monthly* piece on sewer gas written by Dr. Frank Hamilton, Garfield's onetime surgical consultant, the editor ridiculed the "fallacies" of Doremus's experiments showing that germs could pass through water barriers—experiments that had been cited with such approval in the Germicide brochures.[39]

Plumbing contractors, or "master plumbers," also sought to appropriate the new scientific insights into disease prevention as a way to elevate their prestige and income. Many plumbing firms not only did installations and repairs but also manufactured their own lines of sanitary wares, so it was worth their while to stay up-to-date on the latest theories linking plumbing and disease. Although they had little formal education, many master plumbers prided themselves on being "practical sanitarians" whose hygienic knowledge matched that of their middle-class counterparts in engineering and medicine.[40]

The importance of being up-to-date on matters of sanitary science was stressed at the very first meeting of the National Association of Master Plumbers (NAMP) in 1883. The delegates adopted as one of the group's primary aims "the advancement of the trade in all the latest discoveries of science appertaining to sanitary laws" and set up a sanitary committee to give annual reports on progress in sanitary science. In subsequent years, the NAMP sent a representative to attend meetings of the American Public Health Association and commissioned bacteriological studies of sewer gas.[41]

Compared to the sanitary engineers, the master plumbers were slower to exploit the germ menace, but by the early 1890s, they had clearly begun to incorporate the new views of disease into the justification for their services. After a trip to the American Public Health Association meeting in 1890, Andrew Young assured his fellow delegates, "We stand nearer the health of the home and the sanitation of men's abodes than any others." He contrasted the image of the physician, who arrived at the front door "in broad cloth, silk hat and gloves" and was treated with every courtesy, with that of the humble plumber, who entered by the back door dressed in "greasy overalls and blouse" and was ignored by servants and family members alike. Yet, Young concluded, it was the plumber who got the real job of disease prevention done, for he was "in the basement, fighting with his skill and scientific knowledge the disease germs invading and threatening the life of the household."[42]

To be sure, many of the changes in plumbing practice came from outside the trade, as cities and towns passed detailed plumbing codes designed to prevent the spread of infectious diseases. Precisely because domestic plumbing became so closely linked to public health, the trade became more tightly regulated after 1880. Although it proved difficult to enforce, a New York law passed in 1881 served as a model of such legislation: it required that the boards of health in New York City and Brooklyn approve all plumbing plans for new housing and maintain a registry of licensed master and journeymen plumbers, and it made violations of the code a misdemeanor subject to fines and imprisonment.[43]

Although they frequently complained that boards of health knew nothing about sanitary plumbing, the master plumbers generally supported the movement toward more stringent codes and registration. By seizing the hygienic high ground, they gained a useful advantage over their less affluent and sanitarily savvy fellow plumbers, who were more likely to run afoul of the regulations. Fostering the image of the "practical sanitarian" allowed plumbing contractors to align themselves with the genteel middle classes whose business they hoped to cultivate, and whose ranks they aspired to, both professionally and socially.

The Triumph of the White China Toilet

Popular anxieties about germ-laden sewer gas and commercial interests bent on assuaging them converged dramatically in the transformation of the late-nineteenth-century bathroom. In its first incarnation, the gospel of germs was perhaps best symbolized by the white china toilet, which became the premier symbol of the bacteria-proof home. George Waring's installation of the Dececo model in the White House proved to be the forerunner of a revolution in design that gave rise to the modern American bathroom.

Prior to the 1880s, bathrooms were designed with the same opulence as other rooms in the home, complete with carpets and heavy drapes. The requisite cast-iron enameled pieces—the water closet, the bath tub, and the washbasin—were encased in wooden cabinetry. In contrast, the new style bathroom abolished cabinetry in favor of a free-standing white porcelain or vitreous china toilet surrounded by white tiles. Not only did this style eliminate the dangers of "hidden work" inside the cabinet areas; it also provided bathroom surfaces that could be scoured more effectively. In addition, the new Dececo model and its imitators had a stronger flushing action, ensuring that no waste materials remained to mar the shiny white interior of the toilet bowl.[44]

The growing identification of the white toilet with state-of-the-art germ protection is neatly illustrated in a tract put out in 1887 by the Sanitary Association of Philadelphia, a group of citizens organized to promote voluntary health reforms. Using a rhetorical device often employed in late-nineteenth-century hygiene literature, the author, identified only as a "Layman," began by addressing a question to the reader: "Have you lost a child, a husband, or other relative from Scarlet Fever, Diphtheria, Typhoid Fever, or other zymotic diseases?" If so, the reader was invited to consider "whether the source of trouble may not be in the water-closet, or in the other plumbing arrangements of your house."[45]

Layman went on to condemn the custom of enclosing the water closet, describing how inside such wooden structures "the organisms escaping with the faeces are carried by the gases into contact with the surfaces surrounding water-closets and may there germinate and decompose, loading the atmosphere with the deadly agents which we have been at so much pains and expense to exclude." To guard against such dangers, Layman called for "skillful mechanics" to turn their ingenuity to making an "ideal water-closet" of porcelain, whose smooth, easily cleaned surfaces would offer no "dangerous lodging places for the invisible foes that sanitary engineers must guard against." Again demonstrating how quickly plumbing contractors picked up on such hygienic suggestions, the Cooper Brass Works of Philadelphia designed just such a toilet in response to the original article, which the Sanitary Association then promoted as a just commercial reward for its civic-mindedness.[46]

In Layman's words, the white china toilet deserved to be regarded as one of the "essential conditions of sanitary arrangements to exclude the germs of typhoid fever and other zymotic diseases." From the 1880s onward, the new style toilet quickly came to embody the cutting edge of hygienic design. In both the popular advice literature and the trade brochures, the more expensive porcelain or china toilet became endowed with superior power to protect its users against both sewer gas and germs. Like the Cooper Brass Works, plumbing contractors gladly responded to the sanitarians' demand by manufacturing and marketing the more expensive new design.[47]

So popular was the new style toilet that it rapidly displaced older models and fueled the rapid growth of the sanitary pottery industry in the 1890s and early 1900s. Companies such as the Trenton Potteries Company (Tepeco) and the Standard Sanitary Manufacturing Company turned handsome profits on the new designs, which required a highly skilled process of casting and polishing the china fixtures. Thus plumbing manufacturers reaped the financial benefits of a fear of sewer gas and germs that more short-lived entrepreneurial ventures such as the Germicide Company had helped to foster.[48]

The privilege of good plumbing, whether embodied in the discrete Germicide attachment or the more obvious show of the white china toilet, became a symbol of the hygienic protections fundamental to a genteel way of life. Those who aspired to that life had to pay a high price: "first-class" sewer traps, toilets, and plumbers did not come cheap, as Colonel Rockwell had learned in renovating the White House. Sanitary goods and services were necessarily used only by those families who had sufficient disposable income to afford them. At a time when the recommended "living wage" for a working-class family, which few actually received, was less than $600, the expense of devices such as the Germicide ($15), the Hyatt water filter ($60), and the white china toilet ($40) undoubtedly confined their use to the more affluent households.[49]

To be sure, there were enough "do-it-yourself" versions of more expensive sanitary measures so that cost alone did not have to limit adherence to sanitary recommendations. For the conscientious homeowner determined to ventilate a room, disinfect a toilet, or install a water filter, a range of alternatives existed at different prices. Yet it was not true, as entrepreneurs such as C. L. Cohn frequently asserted, that sanitary protections cost so little that they were "within the means of the most humble households."[50]

The practice of the more exacting kind of cleanliness fostered first by sanitary science and then by the germ theory remained out of reach for most Americans until the 1920s and 1930s. Although new sanitary goods and services provided affluent families with a greater sense of safety, the vast majority struggled on with the old-style plumbing fixtures that sanitarians associated with certain death. In old-fashioned Calvinist terms, the sanitary "elect" remained few in the late nineteenth century. Not until the early 1900s did the gospel of germs begin to take on more far-reaching forms.

· II ·

The Gospel Triumphant, 1890–1920

4 · Disciples of the Laboratory

In 1895, an iconoclastic San Francisco physician named Albert Abrams penned a satiric portrait of a fictitious group he called the Antiseptic Club. In order to join, members were required to undergo a lengthy disinfection, wear special antiseptic clothing, and "renounce handshaking and kissing as modes of salutation." They met in a building made of zinc and asbestos, "fired monthly to render it germ-proof," and had their meals served with antiseptic gauze covers on gelatine plates. Their meetings explored topics such as why germs were called germs ("out of respect for the Germans, who had done so much toward elevating the moral tone of microbes") and why typhoid germs were so common in railroad cars and transatlantic steamers (they could live without air for many days).[1]

As Abrams's irreverent spoof of the Antiseptic Club makes evident, the widening of germ consciousness in late-nineteenth-century America inspired satire as well as praise. Although many newspapers and magazines carried admiring stories about the latest laboratory discoveries, others aired the view of skeptics, usually men, who ridiculed the new obsession with germs as fit only for hypochondriacs and health fanatics. In a regular column entitled "Science," a correspondent for the New York *Independent* lamented in 1892, "To read much that is written, one may wonder that mankind survived more than a few generations after leaving Paradise." He noted, "Municipal governments are spending immense sums in frantic efforts to preserve the community from bacterial plagues; and even children, jokers say, are taught to be good under the dread that a 'microbe may catch em.'"[2]

91

But in the early years of the twentieth century, educated and affluent Americans became increasingly dependent on laboratory knowledge as a guide through the invisible shoals of infection. Their deepening confidence in the laboratory reflected the continued progress of bacteriology as a scientific discipline. Between 1885 and 1900, experimentalists made dramatic progress in understanding the relationship between microorganisms and disease. The growing respect for bacteriology also reflected sweeping changes in the larger society, which made the lessons derived from the laboratory seem all the more relevant to the conduct of everyday affairs. In many arenas of life, from industry to entertainment, Americans experienced a growing sense of interdependence and interconnectedness among people, objects, and events far separated in space and time. What Alan Trachtenberg has termed the "incorporation" of America set the stage for a new generation of germ fears that reflected a widening involvement in a mass-consumer society.[3]

In the older, sanitarian-dominated version of the gospel of germs, the risks of interconnectedness had centered primarily on the circulation of human excrement. In the 1880s and 1890s, the sewer system had figured as the great disease leveler, linking every toilet and washbasin to the potentially lethal wastes of a city of strangers. The new bacteriology expanded that sense of interconnectedness to include mass-produced consumer goods, public transportation, and commercial services and entertainments. Beginning in the late 1890s, the educated Americans' mental map of potentially dangerous forms of intercourse with the outside world expanded dramatically to include such habits as riding on a streetcar or buying underwear. In light of the new bacteriology, the gospel of germs was revised and expanded to guard against a host of additional hygienic perils in modern life.[4]

The New Bacteriology

In the late 1800s, experimentalists in Europe, Japan, and the United States succeeded in turning the germ theory from a debatable hypothesis into an accepted scientific fact. This progress depended first and foremost upon dramatic improvements in experimental method. Starting in the late 1870s, the flasks of turnip broth and homemade incubators employed by the first apostles of the germ theory gave way to much more sophisticated techniques requiring considerable training. The new discipline of bacteriology, a term invented in the 1880s to denote the systematic study of bacteria, placed the germ theory on a much firmer scientific footing and produced a steady stream of new insights into the transmission of infectious diseases.

The central figure in this transformation was the German physician Robert Koch, whose early work on anthrax played a key role in advancing the germ theory's credibility. Koch made two critical innovations in experimental technique. First, he replaced the flask method, in which bacteria were grown in liquid solutions of beef tea and peptone, with a method whereby solid media were placed in flat plates or test tubes. Bacteria grown in this fashion formed colonies of distinctive colors and shapes, which made them far easier to identify and isolate. Second, Koch began to stain bacteria with chemical or aniline dyes; different species, it turned out, absorbed different dyes, providing another method of distinguishing between otherwise similar-looking strains. One dye in particular, the crystal violet, became a standard tool for sorting bacteria into Gram-positive and Gram-negative classes (named in honor of Hans Gram, the researcher who pioneered its use). Staining also made bacteria stand out from their surroundings, especially when a special light-enhancing condenser was added to the standard microscope. When stained and fixed on a microscope slide, images of bacteria could be photographed and published, allowing for precise comparisons of what researchers were seeing in laboratories thousands of miles apart.[5]

The power of these new methods was conclusively demonstrated by what was probably the single most dramatic discovery of the golden age of bacteriology, Robert Koch's isolation of the tubercle bacillus. Since 1865, when the French army physician Jean Antoine Villemin had first demonstrated that he could infect guinea pigs with pus taken from a tubercle, experimentalists had suspected that consumption was communicable—yet none had been able to identify the causal microorganism. By using his new culture and staining methods, Koch finally detected the culprit, now known as the *Mycobacterium tuberculosis*. In a celebrated paper to the Berlin Physiological Society on March 24, 1882, he presented his case for the tubercle bacillus's role in the disease employing rules of evidence subsequently referred to as "Koch's postulates," which are used to this day. The postulates require that the microorganism must be located in tissue from the sick individual; that a pure culture be obtained from that tissue; and that when injected into a healthy individual, the culture causes the identical disease. Following these guidelines, Koch reported that without a doubt, the tubercle bacillus was the cause of tuberculosis in all of its forms.[6]

Using the same basic methods, Koch identified the comma bacillus of cholera as well as the bacteria responsible for wound infection. In a remarkable burst of experimental work between the late 1870s and early 1890s, other researchers identified the causal organisms of diphtheria, typhoid, scarlet fever, erysipelas, pneumonia, leprosy, gangrene, tetanus, and gonorrhea. By the early 1900s, the pathogens of bubonic plague, dysentery, whooping cough, syphilis, and gangrene had been added to the list. Whereas once anthrax had stood alone as the sole example of the specific germ-disease connection, by the turn of the century experimentalists had documented a whole host of microorganisms responsible for the most fearsome killers of the time.[7]

Medical journals and lay periodicals reporting these discoveries detailed, sometimes in awestruck terms, the new laboratory techniques used to track down these killer microbes. As Edward Trudeau recalled upon first reading Koch's paper on the tubercle bacillus, he was full of wonder at the "ingenuity of the new methods," which "read like a fairy tale to me." The lore of the petri dish, the dyes with exotic names like methylene blue and gentian violet, the silver needles used to inoculate plates, and the strange shapes and colors of bacterial colonies all became part of a new laboratory mystique. Although by modern standards the bacteriological laboratories of the 1880s and 1890s were still quite modest in size and sophistication, especially in the United States, at the time they seemed impressive testimony to the power of experimental science.[8]

The evidence derived from these marvelous new methods placed the human body at the center of the infective process. By sampling its varied surfaces, tissues, and fluids, bacteriologists substantiated the early germ theorists' assertion that the sick were a prolific source of pathogenic germs. Rigorous microscopic scrutiny of the human form revealed that even seemingly healthy people harbored pathogenic germs. In the late 1890s, Robert Koch proposed the existence of "healthy carriers"—that is, people who could transmit diseases such as diphtheria and typhoid without having symptoms themselves. This hypothesis was proven in the early 1900s. Thus the new bacteriology confirmed that *all* bodies, healthy as well as sick, had the potential of passing contagion to others.[9]

Combining laboratory insights with clinical observations, experimentalists slowly began to revise the map of infectious encounters inherited from the sanitarians by widening conceptions of disease-producing "filth" beyond fecal matter. In the sanitarian-dominated gospel of germs, nose and mouth discharges had warranted little attention as contagion bearers. The proof that the tubercle bacillus was carried in consumptives' sputum ushered in a whole new era of public health practice. A heightened concern with spitting, coughing, and sneezing became the chief marker of the new, more bacteriologically oriented gospel of germs.[10]

Having affirmed the connection between bodily discharges and the spread of disease, researchers still had much to learn about how pathogenic microorganisms spread from the infected body into the larger environment: Under what conditions of temperature and moisture could they survive outside the human body? Through what intermediaries, such as sewage, milk, or ice, could they circulate from the sick to the well? In search of answers, bacteriologists applied their new, more sophisticated means of tracking bacteria to the study of air, water, and other channels of contagion.

Bacteriological studies led researchers to revise the previous generation's assumptions about the infective properties of airborne germs. Early converts to the germ theory thought microbes were so light that they might rise up from damp soil or sewers and become "a plaything for air-currents to scatter broadcast," in the words of the German Carl Fraenkel, whose influential text on bacteriology was translated into English in 1891. Remarking on Tyndall's floating matter, or "sun dust," Fraenkel observed that "every one of these particles was supposed to be either a germ or a bearer of one." Likewise, the work on the virtually indestructible anthrax spores reinforced the traditional belief that infective particles had a long and hardy life outside the sick person's body.[11]

The new bacteriology questioned these assumptions. In the first place, experimentalists demonstrated that the germs recovered from the air, as in the early experiments of Pasteur and Tyndall, were mostly harmless. The dreaded "sun dust" turned out to be mainly the innocuous hay bacillus. In short, the most prolific germs found in everyday life were quite tame. As for the few microbes that were pathogenic, they bore little resemblance to the feather-light particles blown about by the wind envisioned by the first generation of germ theorists. Observing bacterial cultures growing on petri dishes, it was obvious, Fraenkel concluded, that "they cannot be torn from a medium on which they have once become firmly rooted, even by the strongest draught of air."[12]

Bacteriologists also found that unlike the fabled anthrax spore, the vast majority of pathogenic bacteria died within minutes, at best hours, after they left the human body. Moreover, no heroic measures were needed to destroy them; even brief exposures to sunlight, dryness, or high temperatures rendered them harmless. There were important exceptions to this rule, such as the spore forms of tetanus, gas gangrene, and botulism. But most disease germs were truly menacing only while they existed in the safe harbor of the living host or in a medium such as water or milk that could convey them to the next victim. Even the dead body, once thought to be a prolific source of disease germs, offered only a short-term haven for the pathogenic germ.[13]

Deadly Dust

Although the new bacteriology revised earlier notions of the germ's indestructibility, it made other discoveries that continued to justify preventive rituals predicated on a hardier form of germ life. Chief among these was the "dust theory of disease," which identified house dust as a prolific source of bacteria. Of all the theoretical insights of the late-nineteenth-century laboratory, this wedding of dust and disease had perhaps the most profound implications for the hygiene of everyday life.

The dust theory of infection grew out of the research of Georg Cornet, one of Koch's students. In the mid-1880s, he took finely powdered dust from consumptives' sickrooms, dissolved it in liquid, and injected the fluid into the abdominal cavity of guinea pigs; within a few weeks, almost all of the guinea pigs had died from tubercular infection. Hypothesizing that dried germs were far more dangerous than moist ones, Cornet warned that the greatest risk of infection from consumptives was not their fresh sputum, but rather the dried expectorations captured in their handkerchiefs, clothing, and furnishings. As Fraenkel's influential 1891 textbook summarized, "Cornet ascertained that the tubercle bacilli are by no means scattered all about us without choice or difference (as was formerly supposed); that they are not ubiquitous; but that they are only met with in definite, narrowly-circumscribed regions, the center of which is regularly a tuberculous and phthisical person."[14]

In the late 1890s, another German, Carl Flügge, suggested an alternate mode of infection. Placing a harmless microorganism, *Bacillus prodigiosus,* in the mouths of human volunteers, Flügge showed that when they coughed, sang, or talked, they sprayed the bacteria several meters around them. From this observation, Flügge surmised that in tuberculosis direct droplet infection was probably more common than was dust transmission. This view was borne out by later studies showing that the organisms found in dust possessed little infective power.[15]

Still, the dust theory of tubercular infection remained widely accepted well into the twentieth century, perhaps because it so neatly reconciled older beliefs about the air's infective properties with newer views of the germ. The reformulation of the airborne germ into a more specific dustborne menace is evident in the work of T. Mitchell Prudden, one of the first American bacteriologists trained in Koch's laboratory. Prudden's influential 1890 treatise, *Dust and Its Dangers,* became a foundation for turn of-the-century domestic hygiene.[16]

Prudden's work recast the house disease concept in bacteriological terms. Summarizing his own and other researchers' experiments, he argued that the greatest dust danger came from indoor, not outdoor, air. In a very dirty city, he noted, one was likely "here and there to encounter veritable germ-showers," but, in general, the air volume and wind currents diluted the microbial content of the outside atmosphere. In contrast, dust and germs were much more likely to accumulate in the home. Although Prudden claimed that good ventilation could reduce the suspended microorganisms in indoor air—an observation that supported the previous generation's obsession with fresh air—he asserted that it could not remove the bacteria-laden dust that settled into carpets and upholstered furniture. Moreover, Prudden found that ordinary methods of dry sweeping and dusting simply stirred the germs up into the air, where they were breathed in by unsuspecting inhabitants.[17]

Although Prudden agreed with Fraenkel's view that few of the germs ordinarily present in the domestic environment were pathogenic, he warned, "But few as they are they have an extreme significance." The tubercle bacillus was a case in point. Underlining the passage for emphasis, he wrote, *"every person suffering from consumption of the lungs may be expectorating every day myriads of living and virulent tubercle bacilli, and the life and virulence of these bacilli are not destroyed by prolonged drying."* Like Cornet, he concluded, "The reason why consumption is so widespread . . . is simply that consumptive persons, either from ignorance or carelessness, are distributing the poison not only everywhere they go, but everywhere the dust goes which has been formed in part by the undestroyed germ-laden material expelled from their lungs."[18]

From this perspective, Prudden harshly condemned the "expectoratory prerogative" that allowed American men to spit anywhere they liked, and he conjured up vivid images of the multitude of ways germ-laden dust circulated in the everyday environment. For example, he noted the fine bacterial crops that had been cultured from the hems of ladies' skirts: "Could women, walking upon our streets, leaving cars, and descending from elevated railroad stations, but see themselves and their environment as others see them, the management of the skirts of walking-suits would, it would seem, command from them a more careful attention."[19]

Because tuberculosis was so hard to cure, Prudden concluded, the best way to reduce its toll was prevention. The first step was to teach consumptives to be more careful in destroying the discharges from their lungs. In addition, Prudden recommended that the public be schooled in "the practice . . . of more intelligent and efficient systems of cleaning, and particularly the adoption of appropriate means for getting rid of the floating or settled dust," especially in railway cars, public meeting places, and most importantly, the home. Assuming that house dust was a serious source of infection that should be handled with great care, he outlined a detailed series of cleaning rules to be followed in its removal. As we shall see in subsequent chapters, Prudden's conceptions of dust and house diseases became the template for domestic disease control at the turn of the century.[20]

Germs with Legs

The new bacteriology pointed to another previously unsuspected agent for transferring infected matter over long distances: insects, particularly houseflies. The idea that flies might transport anthrax had been suggested as early as the 1860s; as the germ theory gained acceptance, so did the hypothesis that they might well be a vector of germ diseases in general. In an 1883 article in the *Popular Science Monthly,* Dr. Thomas Taylor of Washington, D.C., wrote, "Considering the habits and habitat of the house-fly, it will appear evident that, should it prove to be a carrier of poisonous bodies, its power to distribute them in human habitations is greater than that of any other known insect." Due to improvements in transportation, Taylor noted, a single fly "may feast to-day in the markets of Washington, and to-morrow in those of New York, and in a like manner it may be transported from a hospital for contagious or infectious diseases to homes in the vicinity, or even in remote localities." In imagery that was to be repeated frequently during the next three decades, Taylor sketched vivid pictures of the fly moving ceaselessly between manure pile and milk pail, spreading untold numbers of illnesses.[21]

Bacteriological studies in the late 1880s and 1890s confirmed the role of flies as germ carriers. For example, researchers allowed flies to feed first on the bowel evacuations of a cholera patient, then on sterilized milk; when cultured, the milk yielded a rich harvest of the cholera bacilli. Research on typhoid conducted during the Spanish-American War pointed to a similar fly-path from feces to food. Flies were also implicated in the spread of tuberculosis: guinea pigs developed the disease when inoculated with the intestinal contents of flies fed on the tubercle bacillus, and the TB germ was cultured from flies caught in hospital wards where consumptive patients were treated.[22]

The fly was not the only insect indicted in the spread of animal and human diseases. In the late 1880s, Theobald Smith and Frederick L. Kilbourne conducted an elegant set of controlled experiments that demonstrated the tick's role in spreading the deadly Texas fever among Western cattle. By the early 1900s, researchers had linked tsetse flies to African sleeping sickness, mosquitoes to the spread of malaria and yellow fever, lice to typhus fever, and rat fleas to bubonic plague. Although the average American household was not plagued by cattle ticks or tsetse flies, the discovery of these exotic vectors of disease reinforced the idea that the common fly and mosquito were dangerous enemies of the public health to be guarded against with religious zeal.[23]

As a result, the expanded version of the gospel of germs that emerged at the turn of the century was far more insect-conscious than its predecessor had been. Protecting the home, especially the food supply, against the depredations of the fly became a cardinal tenet of early twentieth-century public health practice. The fly and dust theories reinforced one another: the perceived infective power of dust was amplified by the likelihood that it contained fly excreta, and the fly provided, in conjunction with the breeze, another way for germ-laden dust to move beyond the diseased person's immediate surroundings. Along with the crusade against spitting, the war against the insect, particularly the housefly, became a hallmark of the new, improved gospel of germs.

Pathogenic Water and Food

The new bacteriology confirmed that the water supply was a frequent source of fecalborne diseases such as cholera and typhoid, as had long been suspected. In the late nineteenth century, public health departments began using laboratory testing to link outbreaks of typhoid fever with fecal contamination of drinking water. At early public health laboratories such as the Lawrence Experiment Station in Massachusetts, researchers studied the best methods for removing potentially dangerous bacteria from water with mechanical filters and disinfectants such as chlorine. The persistent finding that death rates from typhoid and other diseases fell in the wake of large-scale water purification reinforced the appeal of both household and municipal filtration systems as safeguards against waterborne diseases.[24]

Bacteriological scrutiny of the water led to new precautions concerning the food supply as well. In the early 1900s, Americans relied heavily on ice to preserve and prepare many foods. But T. Mitchell Prudden and other early bacteriologists found that although cold temperatures retarded their growth, typhoid bacteria retained their infective power after thawing. Very few bacteria, it turned out, survived prolonged freezing, but because commercial manufacturers often made and sold their ice quickly, without allowing time for the bacteria to die, the ice supply posed a serious public health concern.

The dangers of contaminated ice were underlined by several well-publicized typhoid epidemics that were traced to ice cream made with tainted ice.[25]

Milk also attracted close bacteriological scrutiny. Sanitarians had long regarded milk as a substance highly receptive to disease "ferments," and bacteriological testing confirmed that it was indeed a hospitable medium for the microorganisms causing typhoid, diphtheria, and scarlet fever. Bacteriologists pinpointed many avenues by which milk acquired bacterial contaminants, starting with dirt on the cow's udder and the milker's hands and continuing with its unsanitary distribution via uncovered milk pails and unsterilized dippers. In addition, dairies sometimes diluted milk with water containing typhoid or other disease germs.[26]

Milk was also implicated in the spread of some forms of tuberculosis. The fact that cows as well as humans suffered from tuberculosis was well known in the 1880s and 1890s; only in 1898 did Theobald Smith demonstrate that the human and bovine forms were in fact two distinct diseases. Researchers assumed that most nonpulmonary forms of the disease, including tuberculosis of the glands and bones, were contracted from milk. Because these forms were more common in children than adults, the link with milk drinking seemed convincing. Testing of American dairy herds indicated that tubercular infection was widespread, and samplings of market milk in large cities revealed that it often contained tuberculosis germs. To safeguard milk, public health authorities began to recommend "pasteurization"—a process of lengthy, gentle heating that killed the microorganisms without ruining the milk's flavor and consistency and could be done at the dairy or in the home.[27]

In addition to confirming older concerns about water and milk, the new bacteriology pointed to new routes of foodborne infection. Given that laboratory culture media were often derived from familiar foodstuffs such as beef tea and gelatin, it was logical to assume that the kitchen offered microbes a similarly fertile field for multiplication. Many common food-handling customs, such as leaving uncovered bowls of cooked food on the table between meals or selling fresh fruit and vegetables from open piles, seemed to invite dangerous deposits of bacteria-laden dust and fly excreta. In addition, both cooked and uncooked food might be touched or coughed upon by food handlers, including healthy carriers whom no one suspected of being sick.

Initially, these new concerns about foodstuffs centered on the spread of diseases such as tuberculosis, typhoid, diphtheria, and scarlet fever. But researchers soon discovered that there were other species of pathogenic bacteria that had a particular affinity for food. In 1884, bacteriologists George Sternberg and Victor Vaughan studied a Michigan epidemic of acute illness caused by a batch of cheese, from which they isolated the same bacteria, the *Staphylococcus aureus,* that Louis Pasteur had found four years before in human pus. In 1888, while investigating a wave of illness associated with spoiled beef, the German bacteriologist August Gartner isolated a microbe that he named "bacillus enteritidis," now known as *Salmonella enteritidis.* Over the next two decades, both the *S. aureus* and Gartner's bacillus were linked to numerous outbreaks of illness involving spoiled food.[28]

Such foodborne diseases were often referred to as "ptomaine poisoning," in homage to the so-called ptomaine theory popular in the 1880s. According to this theory, when the proteins composing the animal body decayed, they produced a deadly substance known as ptomaine, or more gruesomely, "cadaveric" or "corpse" poison, which could spread via the air and water supply. Further research showed that certain species of bacteria did indeed poison food by producing powerful chemical toxins. A key discovery came in 1895, when the Belgian researcher Emile-Pierre-Marie Van Ermengem isolated the microbe responsible for a rare but often fatal form of food poisoning known as botulism, which had been first observed in the eighteenth century in connection with spoiled sausage. Van Ermengem discovered that the *Clostridium botulinum,* a near relation of the microbes responsible for tetanus and gas gangrene, emitted a powerful neurotoxin. Moreover, it required no oxygen to live and formed extremely resistant spores, making it perfectly suited to thrive in canned goods.[29]

Thus by the turn of the century the new bacteriology had produced convincing evidence that the food supply represented a significant source of germborne disease. Both scientific and public awareness of microbial food contamination, or ptomaine poisoning as it continued to be called in colloquial terms, rose steadily in the early 1900s. As a consequence, the need for exacting care in storing and cooking food became a major concern of the expanded gospel of germs. Slowly but surely, both the private kitchen and its commercial suppliers came under the rule of new laboratory-inspired disciplines designed to safeguard the food supply against dangerous microorganisms.

Asepsis and the Surgical Model of Cleanliness

Lessons learned in the late-nineteenth-century operating room further reinforced attention to the fine details of cleanliness in everyday life. In the late 1880s and 1890s, surgical technique was radically transformed by the discovery that seemingly insignificant forms of casual contact had the capacity to spread potentially deadly wound infections. The new bacteriologically informed standard of aseptic cleanliness that became the norm in surgery profoundly influenced the broader practice of cleanliness in modern life.[30]

As we saw in Chapter 1, Louis Pasteur's early work on fermentation gave rise to the antiseptic method in surgery, which reflected the assumption that airborne germs were the main cause of infection. In the 1880s, Robert Koch's research on wound infection demonstrated that contact with the unsterile body and instruments of the operator, not the infective properties of the air, was the chief source of postsurgical infections. As the American surgeon Carl Beck explained in his 1895 text on the new surgical techniques, Koch's insight "gave birth to *Asepsis,*" which represented a great improvement over Lister's conception of antisepsis. Instead of trying to neutralize germs already in the operating chamber, surgeons sought to exclude them through careful rituals of cleansing and sterilization.[31]

The patients themselves proved comparatively easy to render aseptic; their skin could be shaved and scrubbed, and the area to be incised could be surrounded by sterile sheets. But the active bodies of the surgeon and surgical nurse posed a more difficult challenge. As Hunter Robb wrote in his influential 1894 text *Aseptic Surgical Technique,* "No method has yet been discovered by which the skin can be rendered absolutely sterile." In search of the perfect cleanser, surgeons switched from carbolic acid to mercuric chloride, also known as corrosive sublimate, which as its name suggests, proved very hard on the skin. The disposable, sterile rubber glove, which first came into use at Johns Hopkins Hospital in the 1890s, eventually replaced bare hands in the operating room.[32]

To protect patients against the microbial life found in hair and mouth discharges, surgeons in the 1890s began using skull caps and face masks. Sacrificing beards and mustaches "on the altar of asepsis," in Beck's words, surgeons embraced the clean-shaven look for its convenience and safety. As one doctor explained, "We got tired of tying our beards up in flannel bags every time we operated." After Flügge and other researchers had demonstrated that talking expelled germs, operating teams also sought to speak as little as possible during an operation.[33]

Objects in the operating room represented another potential source of contamination. Because cloth was widely believed to act as a "fomite," or contagion bearer, up-to-date surgeons replaced their frock coats with white, sterile operating gowns. (The color was later changed to green to minimize the glare from the operating room's bright lights.) Surgical instruments were completely sterilized with steam or chemicals between every use. During the operation itself, great pains were taken so that no sterile object came into contact with an unsterilized surface.[34]

As Hunter Robb noted, aseptic technique made the operator acutely conscious of the little habits of everyday life that spread microbial contamination. "Any one who has been trained in a bacteriological laboratory will have exalted ideas of *surgical cleanliness,* and cannot fail to see the many inconsistencies that occur during the majority of operations," he wrote. As examples, he cited a surgeon who touched an unsterilized blanket and, even worse, a nurse who wiped her nose during an operation. The new surgery, like bacteriology, taught an appreciation, in Robb's words, of "the necessity of paying attention to the most minute details," for the "smallest slip may invalidate the whole procedure." The search for asepsis fostered a powerful awareness of how the touch conveyed infections not only between people, but also between people and things. Those ideas writ large came to dominate the practice of what was often called "surgical cleanliness" in everyday life.[35]

Hand, Mouth, and Fomite

The insights derived from the new aseptic surgery helped increase public awareness about the role of casual contact in spreading infection. Bacteriologists' revelations about the vibrant germ life found on the skin and in the mouth opened up troubling new suspicions about seemingly innocent forms of touching. Sexual intercourse had long been recognized as dangerous, but now even such common social gestures as handshakes and kisses figured as possible means for transmitting disease germs. As one impassioned physician-critic observed in urging an end to the handshake, "The microscope and bacteriology have opened our eyes. Now torpidity where danger lurks becomes a crime."[36]

The potential traffic in germs widened even more when one took into account the objects that people shared. Experimentalists demonstrated that contact with the mouth or the hand left germs behind on cups, pencils, and pieces of paper. Because even otherwise fastidious people routinely touched their noses and mouths with their hands and then handled objects passed on to others, this "universal trade in human saliva," as Charles Chapin dubbed it, was difficult to guard against. The hygiene conscious could easily evade kisses or handshakes, especially from people whose health status was unknown or questionable; but it was impossible to know simply by looking at a cup, a doorknob, or a paper dollar what horribly diseased individual had handled it in the past.[37]

The common drinking cup became the subject of particular bacteriological scrutiny and condemnation. By swabbing the lips of common cups used at city water fountains or culturing the dregs of wine left in communion vessels, investigators isolated all kinds of pathogenic bacteria. With the new attention to mouthborne diseases, the idea of a common drinking vessel touching the lips of countless strangers day after day became especially loathsome. As Chapin noted in his standard text on the sources and modes of infection, "Hundreds of thousands of persons must be each day in this manner exchanging the secretions of the mouth."[38]

For similar reasons, cloth and paper came under renewed scrutiny as disease carriers. If the unsterilized coat or towel could contaminate the surgical field, it was logical to assume that ordinary cloth or towels, especially if handled by a sick person, could transmit disease germs. The well-known propensity of cloth to absorb dust added to its potential disease-bearing properties. As Prudden noted in *Dust and Its Dangers,* pathogenic germs could be cultured from upholstery, carpets, handkerchiefs, and wallpaper. Researchers also recovered bacteria from the surfaces of books and money, forms of paper that enjoyed a wide circulation among the sick and the well. Hygienists warned that such commonplace habits as wetting a finger to turn a book's page or licking a stamp to mail a letter could conceivably circulate microbes among large numbers of people.[39]

In retrospect, it seems apparent that the extent of fomite infection was exaggerated in the public health teachings of the period. Experimentalists soon appreciated that the life expectancy of germs deposited on the drinking cup or a child's toy was quite short and that bacteria cultured from upholstery or paper money had relatively little infective power. But the fear of fomite transmission loomed large well into the twentieth century, in large part because it accounted so easily for cases of disease that were otherwise difficult to explain.[40]

The Dark Side of Contagion

Anxieties about casual contact and fomite infection were amplified by growing concerns about the spread of venereal diseases. Due to public reticence concerning discussion of sexual matters, the dangers of gonorrhea and syphilis were rarely acknowledged with the same directness and openness as were fears about tuberculosis and typhoid. Yet the recognition that venereal diseases were also germ diseases subtly but strongly influenced the changing discourse of disease prevention in the late nineteenth century. One of the first pathogens identified by the new experimental methods was the bacterium responsible for gonorrhea; it was isolated by the German dermatologist Albert Neisser in 1879. Although researchers did not isolate the spirochete that caused syphilis until 1905, in the meantime, it too was widely assumed to have a microbial cause.[41]

The new understanding of venereal diseases as germ diseases coincided with a dramatic increase in their incidence that was due largely to the spread of commercialized prostitution. The stigma attached to sexually transmitted diseases made their incidence hard to track, but few urban physicians doubted that cases of syphilis and gonorrhea were rising steadily in the second half of the nineteenth century among men of all classes. As early as 1859, the physician William Sanger estimated that 40 percent of New York City prostitutes had syphilis. Prince Morrow, in his influential 1904 study *Social Diseases and Marriage,* estimated that one in seven marriages were rendered childless as the result of venereal infections.[42]

As the new bacteriology heightened awareness of the microbe-ridden body, physicians and lay people alike became increasingly fearful that venereal diseases could spread through casual forms of touch. Intimate contact with the warm, moist areas of the penis and vagina had long been perceived as a disease risk; now the more "innocent" membranes of skin and mouth were implicated in disease transmission as well. As dermatologists were well aware, many syphilitics developed open sores in their mouths and on their hands. Kissing or shaking hands with them, or sharing their cups or towels, could conceivably transfer the disease.

L. Duncan Bulkley, a New York City dermatologist, developed this scenario at length in his 1894 book *Syphilis in the Innocent*. Based on his own extensive clinical practice, Bulkley asserted that nonvenereal transmission of syphilis was far more common than physicians realized. Besides the obvious dangers of kisses, handshakes, cups, and towels, he pointed to a wide range of fomite transmissions involving items as varied as bathing suits and wooden pencils. For example, he warned of the dangers of smoking cigarettes or cigars made by syphilitics. While treating a young physician of supposedly spotless character who had developed a chancre just where he held his cigar in his mouth, Bulkley reported that at "about the same time" a Connecticut cigar factory had discovered that a syphilitic operative "had for some time been rolling and moistening cigars on the lips, which were the seat of abundant mucous patches." Sure enough, when Bulkley's patient checked his brand of cigars, he discovered that they were made in Connecticut.[43]

In another case, Miss W. K., described as "an intelligent and modest girl, aged 25, a sales-lady in a dry-goods store," developed a syphilitic sore on her tongue. After questioning her about her private life, which she insisted was abstemious, Bulkley wrote, "I was led . . . to accept the probability that the poison was conveyed by the means of *pins,* which injured and inoculated the tongue, as she was in the habit of placing them in that side of the mouth, regardless of where she had obtained them; she herself remarked that she must have been poisoned in this manner, as she was conscious of having thus wounded the tongue."[44]

One can almost hear the saleswoman's sigh of relief when Bulkley arrived at this conclusion. Read a hundred years later, with the knowledge that nonvenereal syphilitic infection is rare, Bulkley's case histories suggest that both doctor and patient were eager to find less embarrassing explanations for what was considered a highly immoral disease. Still, at the time, the attribution of venereal infection to casual contact was taken quite seriously and strongly reinforced the emphasis on fomite transmission. Bulkley's chart of the "domestic and social transmission" of syphilis in his 1894 text reads like an inventory of the newfound risks involved in the everyday exchange of objects.[45]

The association of casual contact with venereal infection added a frightening element to an already grim landscape of disease hazards at the turn of the century. In numbers and heartache, the silent epidemic of venereal disease posed a serious public health problem. Yet unlike tuberculosis, it was difficult to address its prevention frankly due to the strictures on open discussion of sexual matters. Thus general warnings about the dangers of casual contact and infection had to do double duty as tacit reminders of the darker side of contagion as well. In their jeremiads against the common drinking cup, reformers frequently noted that it was responsible for the spread not only of consumption, but also of other "loathsome" diseases that could not be named. Hearers probably understood the double message quite clearly: the same modes of infection that spread more "respectable" ailments such as typhoid and tuberculosis transmitted venereal diseases as well.[46]

The Incorporation of the Germ

Disciples of the new bacteriology prided themselves on having a more precise, scientific view of the microbe than did the first apostles of the germ theory. As Charles Chapin noted approvingly in 1902, modern bacteriologists had advanced far beyond the old "filth theory" of disease and its hazy understandings of the microbe. Yet in its place, the new experimentalism promulgated a dust theory of infection that was virtually as broad as previous notions of airborne infection. It also introduced an infective trio of "flies, fingers, and food" in ways that had enormous implications for the conduct of daily life.[47]

Coincidental developments in the larger economy and society ensured that the new bacteriological critique of everyday life had even more far-reaching effects than did the previous generation's obsessions with filth and sewer gas. The deepening germ consciousness at the turn of the century mirrored a growing perception of incorporation and interconnectedness. Not only did economic progress accelerate the circulation of mass-produced consumer goods, from canned goods to underwear, that bacteriologists had identified as potential sources of disease, but also the new institutions of urban society fostered greater and more promiscuous minglings of people—in Pullman cars, amusement parks, restaurants, and hotels—that fostered potential contacts with infection.

The cultural implications of the new bacteriology were amplified by economic changes that linked the local marketplace with national and even international corporations. Due to innovations in mass production and distribution, customers from New York to San Francisco were increasingly buying the same brand-name goods produced farther and farther from their points of sale. The industrial transformations making possible this cornucopia of goods simultaneously created concerns about their hygienic status. Many companies sacrificed the purity and quality of their products for a greater profit margin, assuming that distant buyers would never be aware of such potentially health-threatening shortcuts. Company managers also cared little for the health and safety of their employees; they cut wages, expanded hours, and allowed unsanitary conditions in the workplace, all of which increased the risk of workers' illness as well as injury.[48]

From the middle-class consumers' standpoint, the health status of industrial workers mattered little, as long as the products they made did not convey disease into the home. But many of the most important new sectors of mass production were *domestic* goods, including clothing and food, that public health doctrine now emphasized as potential carriers of contagious diseases. The image of a consumptive seamstress stitching a baby's nightgown, or a tuberculous cow producing milk for the dairy, suddenly made the connections among diseased work places, sick workers, and American homes real and frightening.

The expanding transportation system presented additional opportunities for microbial Russian roulette. From the 1870s onward, Americans of all classes became increasingly dependent upon long-distance passenger trains, local commuter railroads, streetcar companies, and subways. The maturation of the transportation system greatly facilitated the circulation of people and their attendant germs. By the early 1900s, a sick person could take a train from New York to Chicago in two days and reach the West Coast in less than a week. As the "streetcar suburbs" expanded in the late 1800s, the daily commute became a potentially infective rite of passage for millions of urban Americans. Although the fares charged were sufficiently high to discourage the most destitute and presumably (from the affluent rider's viewpoint) disease-ridden sorts from their use, the new forms of transportation still represented a leveling experience, both socially and hygienically.[49]

The flourishing commercial entertainments and personal services of Progressive-era cities and towns presented their own distinctive dangers. Department stores, which had been redesigned to make shopping an indulgence for all the senses, exposed customers to salesclerks and fellow shoppers whose hygienic status was unknown; goods displayed to entice the eye also invited the stranger's touch or the fly's visitations. As more and more Americans ate out, they increased their risk of foodborne illness. Vendors offering foods to be eaten festively in the open air observed no protections against the dust and fly infestations common to the city street. Fine restaurants provided a cleaner, more homelike setting, but there was always the possibility, even at the most expensive places, that the victuals had been contaminated during storage or preparation.[50]

Going out to play in the Progressive-era city involved further risky contacts. As Victorian dictates about class- and gender-specific activities weakened, commercialized forms of entertainment drew Americans of diverse backgrounds into convivial new associations. Young people in particular flocked to plays, cabarets, amusement parks, penny arcades, and dance halls. The advent of film took them to nickelodeons and eventually to movie theaters. All of these amusements packed large numbers of bodies into spaces that were often poorly ventilated and cleaned. They also fostered forms of sexual intimacy that reformers feared would bring not only moral debauchment but also the continued spread of venereal diseases.[51]

In more subtle ways, modern living increasingly depended upon the circulation of items such as paper money and postage stamps that linked masses of people into an expanded traffic of germs. The most commonplace objects became suspect as intermediaries of infection between the self and the unknown stranger: borrowing a library book, for example, involved the risk that a previous reader had been a syphilitic or a consumptive; and receiving change at the store or opening the daily mail carried the threat of encountering a paper-loving microbe.

In short, the new hurly-burly of economic and social life in the Progressive period encouraged the very forms of casual contact and fomite exchange that the new bacteriological perspective had implicated as potentially hazardous to one's health. The Philadelphia physician Lawrence Flick summoned up this frightening vision of urban society in an 1888 tract about the dangers of contracting tuberculosis: "It is a matter of daily occurrence that people who have consumption and who are constantly expectorating infectious matter, fill positions in which they must necessarily contaminate the clothing, food, and drink of others." Ticking off the commonly used services of modern life, he warned, "There are consumptive tailors and dressmakers, consumptive cooks and waiters, consumptive candy-makers, consumptive bakers, consumptives indeed in every calling of life." As Flick suggested, the life of the modern city dweller was inextricably linked with a horde of fellow citizens, many of whom had no idea they were ill and thus took no precautions to shield others from their potentially fatal secretions.[52]

The changing ethnic and racial composition of the urban population during these same years only compounded city dwellers' anxieties about infection. Between 1870 and 1914, more than 30 million immigrants came to the United States, predominantly from southern and eastern Europe, Asia, and Central America. By 1920, almost 60 percent of the inhabitants of large American cities were either first- or second-generation immigrants. During these same decades, a small but growing number of African Americans moved to cities in both the South and the North. Highly diverse in language, religion, and ethnicity, these so-called new immigrants seemed to white Anglo-Americans to be particularly unassimilable and disease-prone. The diseased straphanger, the consumptive garment worker, or the sickly candy-maker was often perceived by the white majority as a foreign or dark-skinned face.[53]

But however much they might have wanted to, affluent white Americans could not seal themselves off from the potentially disease-ridden transactions of everyday life. Participating in the pleasures of a mass consumer society, whether by buying a new dress or eating out at a restaurant, involved a certain level of risk. Modern living brought with it new forms of physical interdependence that meant the diseased status of one could compromise the health status of all. Thus Progressive-era city dwellers felt a need to protect themselves against infection in ways that went far beyond their parents' reliance on sewer traps and white china toilets.

In the 1890s, two movements sprang up that offered to guide Americans through the microbial hazards of modern life. Although they pursued different aims and audiences, the antituberculosis crusade and the domestic science movement played similar roles in introducing diverse groups of Americans to the more expansive gospel of germs derived from the findings of the new bacteriology. Both movements took the lessons of the laboratory and shaped them into exacting codes of everyday behavior. By disseminating their gospels of correct behavior far more effectively and extensively than the previous generation of popularists had broadcast theirs, antituberculosis workers and domestic scientists gradually made the germ a household word in early twentieth-century America.

5 · Tuberculosis Religion

In September 1893, the Philadelphia physician Lawrence Flick received a letter, signed only with the initials "WMP," describing a painful scene that the writer had just had with his wife, who was dying of pulmonary tuberculosis. Her husband and her physician had agreed to keep the diagnosis from her, telling her instead that she suffered from "a malignant tumor"—a deception indicative of the relative terrors of consumption and cancer in this era. But other family members had insisted that Mrs. P be told the truth because they feared that she might unwittingly infect her little granddaughter Lillian by such affectionate intimacies as feeding the baby with her spoon. "Of course it became my duty to say so to Mrs. P and I have the usual credit of the bearer of bad news," her husband wrote glumly to Flick. To make his wife understand what a danger her behavior posed to her adored granddaughter, Mr. P gave his wife a simple lesson in the germ theory, explaining "that the mycrobes [sic] or bacilli might carry the disease to a young child and that for the sake of all of us she must not kiss the child or give it a chance of inhaling her breath or feeding from her plate of food."[1]

This sad vignette reflects what was probably the single greatest challenge of turn-of-the-century germ "gospellers"—namely, to convince Americans that consumption was a communicable disease and train them out of the many commonplace habits that spread it. Of all the revelations associated with the new germ theory of disease, none was more unexpected or difficult to assimilate than Robert Koch's discovery of the tubercle bacillus in 1882. Although folk beliefs in the

113

disease's contagiousness had long existed, the educated had almost universally held consumption to be a hereditary ailment. Thus nothing in the old gospel of prevention had anticipated the realization that the "captain of all these men of death," in John Bunyan's famous phrase, was a germ disease.[2]

Conversion to this new belief was frequently described in the language of evangelism. Addressing a Pennsylvania group in 1909 on "Preventive Measures against Tuberculosis," Dr. M. V. Ball said of the "doctrine" of tuberculosis contagion: "When . . . one has this conviction strong in his heart he becomes as zealous in the promulgation of what he calls 'the light' as the most devout convert of some new religious truth. The antituberculosis campaign has in it much of the fervor of a new religion." This tuberculosis "religion" was the first and most robust fruit born of the new lessons of the laboratory described in Chapter 4. More than any other disease crusade of the Progressive period, the antituberculosis movement reflected the importance of experimentally derived hygienic truths in reshaping and expanding the gospel of germs.[3]

In 1892, Lawrence Flick, himself a cured consumptive, began the anti-TB movement by founding the Pennsylvania Society for the Prevention of Tuberculosis (PSPT), which was dedicated to educating the public about new views of the disease. Similar groups formed in other cities and states, and in 1904, the National Association for the Study and Prevention of Tuberculosis, soon renamed the National Tuberculosis Association (NTA), was formed to coordinate their activities. The NTA enjoyed the backing of prominent physicians, philanthropists, and politicians; three presidents of the United States—Grover Cleveland, Theodore Roosevelt, and Warren G. Harding—served as its honorary officers. Although its leadership was almost entirely white, male, Protestant, and middle-class, the antituberculosis movement attracted wide-ranging support across gender, class, and racial lines. At the peak of its influence in the late 1910s, the NTA and its approximately thirteen hundred affiliates enlisted thousands of American men, women, and children in the work of preventing tuberculosis. Their methods were so spectacularly successful that they were imitated by subsequent groups organized to combat mental illness, cancer, diabetes, and infantile paralysis.[4]

The anti-TB societies mounted the first truly mass health education campaign directed at a single disease. Working closely with city and state departments of health, they produced countless lectures, exhibits, posters, films, and pamphlets that preached to millions of Americans from all walks of life the same hygienic message: tuberculosis was a deadly communicable disease that could be prevented by careful hygiene. In the process, the anti-TB movement profoundly changed individual conceptions of public health morality—that is, the responsibilities that ordinary people should assume in order to guard themselves and others against infection.[5]

Ironically, the crusade against the great white plague commenced at a time when death rates for tuberculosis were declining from the heights of the early nineteenth century. The reasons for that decline, which began sometime in the mid-1800s, are still the subject of heated debate. But at the time, public health reformers were confident that their preventive labors accounted in large part for the reduction in deaths. Indeed, they cited the falling mortality rates for tuberculosis, as well as for other infectious diseases, as proof that the program of popular education and municipal reform, begun by the sanitarians and continued by the germ theorists, was working.[6]

Although already in decline, tuberculosis posed a fiercesome threat: in 1900, it still was a leading cause of mortality, accounting for roughly one in ten deaths overall, and one in four deaths among young adults. Despite numerous claims to have discovered a miracle cure, researchers had failed to find a vaccine or drug capable of halting its destructive course. The hopes for an effective therapy raised in the 1890s by Robert Koch's isolation of tuberculin, a substance derived from pure cultures of the tubercle bacillus, were soon dashed when it proved to be useful only as a diagnostic tool. The most successful treatment remained a long, expensive stay in a sanitarium, a European institution introduced to the United States by Edward Trudeau in 1884. The methods used at his "cottage sanitarium" near Saranac Lake, New York, were widely copied in the succeeding decades, but even with prolonged bed rest, a special diet, and continuous exposure to fresh air, most patients died. For all of these reasons, prevention remained the most reliable and effective way to combat the white plague.[7]

Selling the TB Gospel

Antituberculosis societies pursued many goals, such as encouraging research, sponsoring legislation, and promoting sanitarium construction, but none was more important than educating the public. Anti-TB workers fervently believed that circulating information about the contagious nature of the disease, and the preventive measures to avoid it, was one of the most effective ways to save lives. This commitment to popular education stemmed not only from a characteristically American faith in individual enlightenment as the key to social reform, but also from the practical realization that in a democratic society, change was easier to achieve through education than by coercive legislation.

To spread their version of the gospel of germs, anti-TB societies initially relied upon traditional nineteenth-century forms of persuasion such as pamphlets, popular lectures, and newspaper articles. Among its first activities, the PSPT issued a series of short pamphlets or "tracts" (a term originally applied to religious literature) for the public. But although they were written in simple language, these dense, dry texts made few concessions to readers with little education and were probably read only by consumptives and their families.[8]

TB crusaders quickly realized the desirability of developing more accessible forms of propaganda. As an NTA publication noted in 1910, "Many people do not read books or pamphlets, and are therefore most difficult to reach and yet most in need of instruction. They can only be attracted in some striking way." The national organization led the way in finding such "striking ways" to interest the public, both by initiating its own educational programs and by publicizing innovative work done by its state and local affiliates.[9]

The NTA's early staff brought to this work an eclectic mix of experience in academic social science, social work, journalism, and advertising. The first executive secretary, Livingston Farrand, had earned an M.D. from the prestigious College of Physicians and Surgeons of New York, but instead of practicing medicine had gone on to study sociology and anthropology at Columbia University with the famous cultural anthropologist, Franz Boas. The NTA's first publicity director, Philip Jacobs, was a onetime journalist with a bachelor's degree in theology and a Ph.D. in sociology; before coming to the NTA, he had worked at the Charity Organization Society of the City of New York, the most important social work agency in the city and an early proponent of TB prevention work. Evart G. Routzahn, who created the first traveling tuberculosis exhibits, had previously been employed by the Young Men's Christian Association, one of the most successful religious groups of the era. Charles De Forest, who was hired to run the Christmas seal campaign, had done promotional work for his inventor brother Lee, a pioneer in early radio.[10]

Searching for more effective methods of mass education, these early tuberculosis workers found inspiration in the burgeoning field of advertising. In the late 1800s, as ever more companies competed for consumers in a national marketplace, advertising emerged as an increasingly important area of business expertise. The traditional disdain for the bold, "carnivalesque" methods used by P. T. Barnum and other nineteenth-century entrepreneurs began to weaken as merchandisers realized the virtues of snappy copy and visual "eye appeal." Companies such as Eastman Kodak, the National Biscuit Company, and American Tobacco dramatically increased sales of their products by using jingles, trademarks, advertising cards, and other promotional gimmicks.[11]

For public health educators, these new advertising methods represented a veritable gold mine of persuasive techniques. As Philip Jacobs noted in his 1923 handbook, *The Tuberculosis Worker,* "For the last fifteen years, the tuberculosis campaign has utilized practically every educational device that has been invented by advertising and business agencies and has devised many new ideas of its own." Given the modern association of advertising with manipulation and distortion, Jacobs's use of the word "educational" may seem far-fetched. But in many ways, the new advertising approaches offered solutions to the limitations of traditional print forms of health education.[12]

Perhaps the greatest advantage that advertising offered public health education was its varied techniques for embodying the invisible. One of the most perplexing tasks that germ "gospellers" faced was making the unseen microbe and its menace seem real. Not only was the tubercle bacillus invisible to the naked eye, but even when rendered visible by the microscope, it was not particularly impressive looking. Microphotographs of individual germs or even a whole bacterial colony hardly conveyed a sense of their dangerousness. To spur compliance with the rituals of prevention, TB workers had to develop more compelling ways to represent the disease's menace and to associate it with specific objects and behaviors.

Advertisers faced a similar challenge: to invest seemingly commonplace objects and behaviors with fresh meaning. Starting in the 1890s, advertising companies pioneered a range of innovative methods for endowing the products of American agriculture and industry with seductive charm. By combining short texts with arresting pictures, advertisers sought to associate objects with the power to convey such intangible qualities as beauty, happiness, and well-being. They invited viewers to imagine how much better their lives would be if they possessed some particular fruit of American abundance.[13]

These creative ways of reimagining the world held enormous potential for popular health education. Instead of having to make do with the unassuming tubercle bacillus, advertising methods suggested bold new strategies for visualizing the germ. By combining short texts with effective illustrations, TB crusaders were able to open up whole new story lines linking the dread germ to familiar objects and situations.

At the simplest level, the new advertising taught the virtue of the well-chosen phrase, or "jingle"—for example, Schlitz's tag "the beer that made Milwaukee famous," or Kodak's slogan, "You push the button, we do the rest." Although tuberculosis did not lend itself quite so easily to jingle making, the basic principle of writing advertising copy—make the message short and memorable—worked equally well in health education. Imitating the advertisement, tuberculosis workers sought to compress their preventive gospel into brief slogans or mottos, such as "Don't spit," "Fresh air promotes health," and "Flies carry disease." The layout of educational materials was also changed to make the slogans easier to read at a glance. The slogan style was particularly well adapted for display advertising in newspapers, streetcars, and billboards; in Topeka, Kansas, even the sidewalks were put to use by being laid with bricks that read "Don't Spit."[14]

Anti-TB societies also exploited the concept of the trademark, which companies such as Procter and Gamble and Quaker Oats had developed in the late nineteenth century to distinguish their supposedly superior wares from cheaper, generic brands. In 1906, the NTA adopted the double-barred Lorraine cross as its symbol in order to distance its work from competitors such as the White Cross League, which sold toilet articles to raise money for consumptives. The Lorraine cross lent itself well to display advertising and became a focal point in all of the NTA's publicity campaigns.[15]

In more fundamental ways, TB workers appropriated the new visual culture of advertising to sell their gospel of prevention. At the turn of the century, many companies began hiring artists and photographers to embody their product's virtues in striking images, a trend that dramatically changed the look of modern advertising. Similarly, tuberculosis workers soon learned to use both artwork and photography to convey their hygienic story lines more dramatically. Frequently citing the old adage that one picture was worth a thousand words, they too recruited artists and photographers to provide compelling illustrations for their hygienic messages.[16]

Admittedly, some of their first efforts in this direction were clumsy. For example, in 1908, the tuberculosis committee of the New York City Charity Organization Society designed a poster for use in Italian neighborhoods featuring a scenic Venetian view surrounded by TB slogans. The picture had no relation to the message; it was intended solely to draw tenement dwellers to look at the poster in the hope that they would read the mottoes on its borders. As it turned out, the society had miscalculated badly: the Italians of the Lower East Side came predominantly from southern Italy and heartily disliked their northern neighbors, so the poster had little appeal for them. By learning from such mistakes, the movement's publicity experts gradually became more adept at making graphics and copy complement each other and in adapting their message for different ethnic and racial groups.[17]

The early antituberculosis message relied heavily on what advertisers referred to as the "reason why" approach, a strategy that was particularly easy to adapt for educational purposes. In the same way that copywriters produced a list of reasons why customers should buy Fleishman's yeast, TB workers presented a list of arguments for why citizens should buy Christmas seals to support the antituberculosis crusade. Public health educators also favored parental-style admonitions along the lines of "do this because it's good for you," which were common in early twentieth-century ads as well. This straightforward, indeed heavy-handed, approach is evident in a poster series that the NTA distributed in New York City with the assistance of the local bill poster's association: one pictured a drunkard with his bottle and stated in bold type, "Intemperance and other excesses lead to consumption"; another showed the consumptive in his home with wife, babe, and spittoon nearby and announced, "A Careless Consumptive Is Dangerous to His Family."[18]

Moralistic appeals were frequently combined with more direct, fear-based warnings about the consequences of carelessness. In a particularly striking poster issued during World War I, the Illinois Tuberculosis Association used death's-head imagery to dramatic effect. The poster showed a skeleton astride a giant fly, dropping bombs labeled with the names of infectious diseases on the people below. One of the accompanying slogans proclaimed, "Death lurks in the filth on a fly's feet." In the same vein, a health exhibit company marketed what it called a "death rate illusion," which consisted of a tiny doll that turned into a skeleton every thirty seconds. This exhibit illustrated the fact that every time the illusion occurred a person somewhere in the world died from tuberculosis.[19]

But TB workers soon realized that both moralistic and fear-based advertising repelled some viewers, and they began to vary the drunkard-and-giant-fly approach with appeals to other, more positive virtues. For example, during World War I, TB posters featured images of brave American soldiers and equated disease prevention with a patriotic act on their behalf. Along the same lines, mother and child representations became increasingly popular; in a 1920 poster contest, the Pennsylvania Society gave the top prize to a poster whose fresh air slogan was illustrated with a lovely drawing of a woman and child sitting on a park bench.[20]

This lighter touch was particularly evident in the health materials designed for children, as in a series of attractive posters, commissioned by the PSPT and widely reproduced in card and book form, that illustrated each preventive motto with a charming drawing of a child. In the 1910s, TB workers also began to experiment with cartoons as a way to reach both adult and child audiences. A cartoon series developed by American public health workers for use in France, where the Gallic mind was assumed to be hostile to the "do-gooder" approach, proved so popular that they were reproduced and reprinted for use in the United States. Cartooning also opened up new possibilities for picturing the tubercle bacillus as either a menacing or whimsical character, a tactic that proved especially useful with children.[21]

These same strategies for visualizing and dramatizing the gospel of germs carried over into what became another key innovation of the anti-tuberculosis movement, namely its pioneering use of motion pictures. Like advertising, the medium of film offered health reformers exciting new ways to translate their preventive message into attention-getting images and story lines. Not surprisingly, given their interest in publicity, TB workers were quick to appreciate the educational potential of movies. In 1910, only five years after the first commercial films were made in the United States, the NTA commissioned Thomas Alva Edison to produce a series of films for its use. Combining as they did information and entertainment, public health movies proved extremely popular; through them, the educational tableaux created in still photographs and posters could be made to come to life. For example, the *Temple of Moloch*, produced in 1914, presented the taming of the "careless consumptive" as high melodrama. Instructions about proper coughing, sweeping, and ventilation were woven into a rousing tale about an idealistic young public-health doctor trying to help a working-class family. This kind of celluloid propaganda quickly became a centerpiece of the anti-TB movement and of American popular health education in general.[22]

The advertising and entertainment industries influenced not only the content and style of public health propaganda, but also the ways in which that propaganda was publicized and distributed. At the turn of the century, manufacturers and merchandisers began to develop elaborate promotional campaigns and media events designed to pique consumer interest in their products. In similar ways, TB workers started to boost their educational efforts with carefully coordinated publicity events. The NTA's first attempt at a national media campaign developed in conjunction with its traveling tuberculosis exhibit, a movable museum composed of charts, drawings, posters, and models concerning the disease and its prevention. First used in Baltimore in 1904, an expanded and revised tuberculosis exhibit toured the United States between 1906 and 1912. When the exhibit opened in 1908 at the American Museum of Natural History in New York City, complete with a giant replica of the fly, tens of thousands flocked to see it.[23]

To our media-jaded eyes, the excitement generated by these dreary-looking exhibits is difficult to comprehend. But by the standards of the time, they represented a remarkable advancement in making information *visual*—that is, in turning text lessons into more accessible photographs and charts. Moreover, exhibit organizers used all the methods of an intensive advertising campaign—"somewhat similar to those followed by a big traveling circus," as one tuberculosis worker put it—to attract visitors. Weeks before the exhibit arrived, the host city was flooded with posters, handbills, and advertising cards announcing its arrival. To herald the opening, TB societies staged circus-like parades, complete with elephants, to build excitement.[24]

The Christmas seal campaign, which the NTA began in 1908 in conjunction with the American Red Cross, represented another highly successful adaptation of modern publicity methods. Once a year, the TB societies sold attractive stamps, to be placed on Christmas cards and other holiday correspondence, to finance their activities. By combining an intensive educational program with a fund-raising appeal, the Christmas seal crusade worked so well that it rapidly became the centerpiece of the TB societies' educational work. Using the same methods employed in national advertising campaigns, the NTA boosted proceeds of the seal sales from $250,000 in 1915 to almost $4 million in 1920.[25]

Even by today's standards, the scale and cleverness of the early Christmas seal campaigns is impressive. In the months before the sale began, TB societies saturated their communities with publicity materials by getting billboard companies and newspapers to post their ads for free. Churches were asked to observe "Tuberculosis Sunday" with an appropriate sermon and special collection for the work. Local associations signed up hordes of volunteers (comely young women were especially encouraged) to sell the stamps, which each year featured an attractive new design. TB workers expended considerable energy trying to think up clever "stunts" to draw attention to the sale. As the popular writer Samuel Hopkins Adams observed admiringly of the Wisconsin seal sale, "There may be commercial organizations in this country which have as complete and pervasive a system . . . but I have yet to hear of them." He concluded, "How any citizen of Wisconsin, unless he is a real badger and lives in a burrow, can escape getting his chance to purchase a seal and thus contribute his cent to the cause is difficult to imagine."[26]

Perhaps the most far-reaching educational program developed by the NTA began as an adjunct to the seal campaign. In 1915, Charles De Forest, director of the seal sale, had the clever idea of enlisting children as "crusaders" in the cause. Appealing to the medievalism then in fashion, he proposed giving the child-volunteers "rank" in the crusade based on their sales; as they sold more stamps, they moved up from squire to knight to the highest honor, a seat at the "round table of health chivalry." To give the scheme greater educational value, De Forest modified the plan in 1917 to make progress up the ranks depend on completion of eleven health chores that embodied the essentials of tuberculosis prevention and general hygiene. In its revised form, this Modern Health Crusade became enormously popular, spawning its own specialized educational materials such as chore cards, games, playlets, and songs. By 1922, over 7 million American children were enrolled as participants in the crusade.[27]

The popularity of the Modern Health Crusade reflected the growing recognition among health educators that children responded to hygienic direction more quickly and positively than did their parents. At the same time, the crusade served other, more publicity oriented needs—for example, the anti-TB societies made frequent use of photographs showing attractive little crusaders selling stamps to famous people. In their white capes and elaborate hats, they became a sort of human figure trademark; like the manufacturers of Sun Maid Raisins and Packer's Tar Soap, the anti-TB societies realized, in the words of one Brooklyn worker, that "children are a very good advertising medium."[28]

The Antituberculosis Message

Employing methods of persuasion that were well adapted to the new consumer economy, anti-TB societies put together the first mass health education campaign in American history. Thanks to the scope and sophistication of their efforts, the antituberculosis crusaders probably did more than any other single group to promulgate a new health code based on the gospel of germs. The lessons that TB workers promoted—via the poster, the advertising card, and the public health film—contained much more than a simple set of rules about how to live the healthy life. Embedded in their texts and their images were both a vision of the microscopic dimensions of everyday life and a set of assumptions about what individuals and communities owed one another to prevent the spread of disease.

Science was certainly the touchstone of the anti-TB movement's vision of how one could live safely with the microbe. In pressing their cause, TB workers continually invoked the legitimacy of the laboratory as a guide to human affairs. Louis Pasteur, Robert Koch, and Edward Trudeau were the patron saints of the new tuberculosis religion. A postcard advertising the 1909 International Tuberculosis

Exhibition in Philadelphia bore the dramatic legend, "The two emancipators: Lincoln wiped out slavery. Science can wipe out consumption."[29]

But the tuberculosis movement mixed its appeals to science with an older language of Protestant evangelicalism; advocates continued the tradition of calling pamphlets "tracts" or "catechisms" and referred to health rules as "commandments." A circular for consumptives distributed circa 1908 by the Illinois State Board of Health began, "FOLLOW THE GOLDEN RULE: Do unto others as you would that they should do unto you," and was followed by more prosaic injunctions such as "DON'T EVER SPIT ON ANY FLOOR, BE HOPEFUL AND CHEERFUL, KEEP THE WINDOW OPEN." In like fashion, the Modern Health Crusade modeled the war against tuberculosis on the Christians' crusades against the infidels; as one young crusader observed in a prize-winning essay, "The germs are the Turks."[30]

Because the NTA was an avowedly secular enterprise, the invocation of religious imagery sometimes made its leaders uneasy. In adopting the double-barred cross, for example, the NTA tried to make it less like the original Christian symbol by tinkering with the length and design of the cross-arms. Yet the double-barred cross remained a powerful reminder of the evangelical roots of health reform. For all the new authority of science, religious conversion still best approximated the kind of transformation in belief and behavior that the anti-TB movement sought to induce.[31]

In the new tuberculosis religion, spitting constituted the worst of the mortal sins. Although etiquette writers and health reformers had long complained about the "expectatory prerogative," especially when it was combined with tobacco chewing, the turn-of-the-century anti-TB campaign took the attack against spitting to unprecedented new heights. TB tracts waxed endlessly and eloquently on the perils associated with mouth discharges. As one NTA circular described the consumptive's spit, "The germ, which is a microscopic rod, is found in millions in their spit from very early in the disease, and it is through this spit almost alone that it reaches others." Slogans popularized by TB workers, such as "Spitting is dangerous, indecent and against the law" and "No spit, no consumption," continually emphasized its potentially lethal properties.[32]

The tubercle bacillus was endowed with extraordinary abilities to fly through the air for great distances, to attach itself to common objects, to mingle with dust and dirt, and to taint liquids such as water and milk. A tuberculosis "catechism" prepared for eighth graders in New York City schools described the bacteria's persistence and mobility in these terms: "The germs will live in the darkness and dampness for a long time, and are stirred up in dusting and sweeping these rooms, and float in the air and may be breathed into the lungs, or may fall upon articles of food and be taken into the body in that way." At times the infective power of germs seemed almost supernatural; one author described the disease as "an ever present demon" and an "insatiable, microscopic destroyer"—in language that echoed evangelical Christians' descriptions of Satan as the "Destroyer."[33]

The fact that many tuberculosis sufferers were able to maintain a normal routine until their disease was far advanced made vigilance especially important. TB workers often pointed out that unlike victims of cholera or smallpox, consumptives could move freely in society, spreading the seeds of disease for many years. Moreover, unlike older conceptions of a "filth" that manifested itself in foul odor and squalid appearance, the dangers of mouth and nose discharges did not always announce themselves so clearly. The new lessons of the laboratory only intensified the warning that surface healthiness and cleanliness were deceptive guides to safety. For example, in a tract warning about the dangers of "the Tuberculous Cow," the reader was presented with a series of photographs of dairy cows, each of which was virtually identical to the untrained eye but differentiated with captions such as "dangerous tuberculous cow" and "exceptionally dangerous cow." Similarly, a slide of a normal-looking mother and child sitting in a clean, comfortable bedroom was labeled with a caption that consumptive parents put their children at terrible risk.[34]

In a world of deceptive appearances, true safety lay in a "sort of moral regeneration," in the words of one TB worker, that would turn the truths of tuberculosis religion "into so many thousand worries and torments to prick the public conscience into action." Familiar ways of behaving had to be transformed by a new sense of their hygienic significance. By turning this cosmology of germ dangers into a concrete code of hygienic behavior, tuberculosis workers extended accountability for disease prevention into every nook and cranny of daily life.[35]

The scope of behaviors implicated in tuberculosis prevention was mind-boggling. The lists of dos and don'ts covered an astonishing number of potential sanitary sins; there were "right" and "wrong" ways to mop the floor, blow one's nose, prepare a baby's bottle, and set a table. Health exhibits featured "good" and "bad" rooms that contrasted the hygienic virtues of specific floor coverings, drapes, and furniture styles. The tone of admonitions accompanying these displays was fierce and unwavering: "It should be an absolute rule," intoned one author, "never to put a baby or a young child on the floor to play, as is so generally the custom." Such detailed specifications for behavior were coupled with the warning that even the tiniest infraction could lead to death. "We are each one of us in hourly danger," began a 1910 article on the "Little Dangers to Be Avoided in the Daily Fight against Tuberculosis," which followed with a dizzying array of warnings against drinking cups, paper money, wooden pencils, the prick of an old pin, and the forks supplied at oyster stands.[36]

Although the scope of tuberculosis religion was unlimited, reformers left no doubt that its practice had to begin in the home. Tuberculosis was repeatedly described as a "house disease" to be guarded against by exacting household rituals (which I will detail more fully in Chapter 6.) It is interesting to note that the White House once again came under sanitary scrutiny during this period; after First Lady Caroline Harrison died of consumption there in 1892, leaders of the PSPT advocated "the effectual disinfection and renovation of the dwelling rooms, indeed of the whole structure," as "imperatively necessary."[37]

Although they portrayed the home as the primary site of infection, antituberculosis tracts also emphasized how contagion spread outward from the home into the wider circle of human interactions. Occupying any space that might previously have housed a consumptive—such as a steamship cabin, a Pullman berth, or a hotel room—was presented as a serious health risk. To avoid the dangers of contaminated blankets in hotels, one TB tract advised, "The careful traveller will . . . insist that the blanket be covered by a fresh clean sheet the turn down of which shall cover it for a distance of two feet from the top." This practice was adopted by many hotels in this period, as I will discuss in Chapter 7.[38]

The anti-TB movement asked converts to abandon other familiar habits in the light of science. For example, following the lead of the modern surgeon, men were asked to renounce the long full beard, a traditional symbol of masculine authority and distinction. One physician who urged American men to spare their loved ones the curse of hairy, germ-laden kisses came up with the inspired slogan of "Sacrifice Whiskers and Save Children." Another tract counseled, "If we know that a friend has tuberculosis, we shall be wise if we avoid shaking hands with him, and we shall be extremely careful whom we kiss or whom we ALLOW OUR CHILDREN TO KISS." Mothers were urged not only to prevent strangers from kissing their children, but even to give up the practice themselves.[39]

At the same time that they stressed elaborate rituals of germ-avoidance, TB workers took pains to emphasize the other side of the preventive equation: maintaining the body's "resisting power." As Lawrence Flick explained, "Sufficient fresh air, sufficient food, and sufficient rest and sleep are the watch-dogs of health, and where they are on the alert consumption can never enter." The seed and soil metaphor remained a favorite for making the point that healthy individuals furnished the tubercle bacillus only a "rocky soil" for propagation. This imagery helped to reconcile the new germ theory of tuberculosis with older hygienic traditions that stressed the importance of a strong constitution and a clean environment.[40]

The emphasis on resistance also connected the anti-TB crusade with other health reforms of the period, such as the temperance and mental hygiene movements. TB tracts routinely included warnings like the one given in the New York City catechism for eighth graders: "Alcohol weakens the body so that it cannot resist the disease germs." In addition, a balanced, optimistic mental outlook was presented as a protection against infection. "Play enough to relax the working faculties, and to exercise the glad and joyous powers of the mind and body," advised a physician in a pamphlet on how to prevent tuberculosis in everyday life.[41]

The Chain of Disease

The precepts of tuberculosis religion clearly emphasized the individual's impelling moral obligation to prevent the spread of the disease. But this emphasis on personal reformation by no means precluded more collective responsibilities for prevention. As with the previous generation of public health reformers, the burden for hygiene was laid on both the individual and the community. Given the mobility and vitality of tuberculosis germs, there could be no simple dividing line between the contagious individual and the rest of society.

A particularly fine explication of this interdependency theme appeared in an 1895 article provocatively entitled "The Microbe as a Social Leveller," which was written by Cyrus Edson, the New York City health commissioner. "The microbe of disease is no respecter of persons," he wrote; "it cannot be guarded against by any bank account, however large." The "Socialistic side" of the microbe required that the wealthy be concerned for the poor, Edson warned. "The former cannot afford to sit at his well-covered table and forget the absence of food in the latter's poor room, because that absence of food means, sooner or later, that disease will break out in the room, and the microbes or their spores will in time pass the heavy curtains on the windows of the mansion to find their prey inside." He concluded, "This is the Socialism of the microbe, this is the chain of disease, which binds all the people of a community together."[42]

Turn-of-the-century health statistics made evident that poor, immigrant, and nonwhite Americans were increasingly more likely to contract the disease than were their affluent, native-born, white peers. TB workers were well aware that these groups suffered a greater risk from tuberculosis due to conditions such as low wages and poor housing that were largely outside their personal control. As the New York City Department of Health's catechism explained, "The kind of people most likely to get tuberculosis are those who are run down or ill from poor or insufficient food, from living in dark, overcrowded or ill-ventilated rooms, or from overwork." Many TB workers explicitly attributed the high rates of disease among the poor to the moral failings of builders who put up shoddy, unsafe tenements; landlords who failed to provide adequate water supplies or toilet facilities; factory owners who paid their workers poorly; and shopkeepers who sold unclean food. Walter Rauschenbusch, one of the founders of the social gospel movement, wrote a prayer for Tuberculosis Sunday that articulated this sense of collective moral responsibility: "Since we are all jointly guilty of the conditions which have bred their disease, may we stand by those who bear the burden of our common sin, and set the united will of our community against this power that slays the young and strong in the bloom of their life."[43]

In its critique of older laissez-faire notions of economic and social welfare, the rhetoric of the antituberculosis crusade shared common elements with the larger Progressive movement. For many public health reformers, scientific revelations about the incorporation of the germ seemed to demand an end to rugged individualism in public health affairs. As one observer put it, "*Laissez-faire* is the paltriest of all philosophies in sanitation—as in everything else." Akin to the ways in which earlier reformers had used the cholera menace to strengthen state support for public health, their Progressive-era counterparts pointed to the TB problem to argue for stronger tenement house laws, factory inspection, food regulation, and other collective sanitary measures.[44]

Yet by encouraging concern about "how the other half lived," to use Jacob Riis's famous phrase, the antituberculosis movement also allowed ample scope for exercising the ethnic and racial prejudices that abounded in the Progressive period. As did Progressive reformers in general, TB workers disagreed over whether individual behavior, hereditary defect, or environmental conditions were most to blame for these group differences in the disease's incidence. Some observers saw high rates of tuberculosis as reflecting poorly on the cleanliness and temperance of the new immigrants from southern and eastern Europe. In a 1905 talk on the "Home in Its Relation to the Tuberculosis Problem," the influential Johns Hopkins physician William Osler provided a chart comparing the "personal and household cleanliness" of tuberculous patients treated at a Baltimore clinic. The chart purportedly showed that 70 percent of the Russian Jewish homes, 56 percent of the "colored" homes, and 30 percent of the "white" homes were not clean. He noted, "It is exceptional to find the former [the Russian Jews] in a condition, either in person or house, that could be termed in any way cleanly." But Osler's chart also listed other factors over which the immigrants clearly had little or no control, such as "bad sanitary location," "insufficient light and ventilation," and "overcrowding." It seemed that uncleanliness alone was hardly responsible for the immigrants' plight.[45]

Other TB reformers offered less blame-laden but still stereotypical explanations for variations in TB susceptibility. A North Carolina sanatorium superintendent, Lucius Morse, writing in the *Journal of the Outdoor Life,* noted that "primitive people" in their natural state did not have tuberculosis and that once they were exposed to it by contact with Westerners, they often succumbed quickly. He attributed the higher rates of tuberculosis among American Indians and African

Americans to their relative lack of exposure to the disease. In contrast, the Jews, "a people who for two thousand years have been city dwellers," enjoyed a "well-known circumstance of racial immunity."[46]

Although the antituberculosis crusade certainly reinforced common stereotypes of the "other" as dirty and dangerous, its public rhetoric tended to emphasize the integrative mission of health education. Compared to contemporaneous efforts to limit immigration and pass sterilization laws, the anti-TB movement chose a more inclusive strategy of trying to bring groups at high risk from the disease within a protective circle of hygienic knowledge. Tuberculosis workers were fond of vignettes illustrating the "melting pot" aspects of health education: for example, in an article describing the Modern Health Crusade as a "Democratic Movement," a Red Cross nurse in the mining community of Gilman, Colorado, waxed eloquent about the good influence of its commandments on Austrian miners, Mexicans, Negroes, Indians, blind and deaf children, and even a pair of Siamese twins.[47]

No aspect of the movement better symbolized its reintegrative goals than did the antituberculosis parades and pageants of the period. These community rituals recast nineteenth-century parade forms, which had displayed the orderly ranks of society by class and occupation, into an army of citizens marching against disease. Conducted with circus-style promotional flourishes, antituberculosis parades were often led by children—an honor befitting their special role as "evangelists" of health reform—and included delegations and floats from businesses, labor unions, women's clubs, Young Men's and Women's Christian Associations, and the like.[48]

For communities deeply riven by social differences, the common fight against an invisible enemy—the tubercle bacillus—served as a positive means of redefining what it meant to be an American. In the rhetoric of the anti-TB movement, the "socialism of the microbe" pointed the way to a more democratic society in which good health, particularly freedom from the white plague, was the birthright of all of its members, regardless of their sex, ethnicity, class, or race. The yardstick of public health morality could be used to chart a path for every citizen's reclamation, both in personal and social terms.

Yet at the same time, the tuberculosis crusade undoubtedly promoted darker, more divisive emotions. Even when they tried to adapt advertising methods to "pretty up" their subject, TB workers still dealt in images of disease and death that often frightened rather than inspired compassion. This jarring contradiction was evident in such publicity gimmicks as the Cincinnati "death calendar," which listed the daily mortality rates from the disease; although the local anti-TB society had it "printed in attractive fashion," it was hard to imagine the calendar cheering up anyone's parlor or kitchen. Bombarding people with images of the invisible, inescapable world of microorganisms inevitably made many fearful of ordinary human interactions. Despite TB workers' criticism of "phthisophobia"—exaggerated fears of the consumptive—their own literature is filled with stories about people who lost jobs, homes, or sweethearts when their diagnosis was revealed. Inevitably, this prejudice fell particularly hard on those who suffered the highest rates of the disease: the working classes, immigrants, and African Americans.[49]

Moreover, when faced with intractable social problems such as poor housing or low wages, commentators often shifted the locus of blame back to the individual and family. Indeed, the harshest moral condemnations to be found in the antituberculosis literature were those leveled at the "careless consumptives" who knew they were sick yet failed to protect others from their germ-laden discharges. TB workers might rail in principle against scurrilous landlords and greedy factory owners, but when it came time to personalize the causes of tuberculosis, they almost always chose the homes and the habits of the consumptives themselves. For example, in the NTA's standard slide lecture, the prime "breeding grounds" of tuberculosis portrayed were dirty homes and yards—that is, the domestic spaces inhabited by the victims. The only human faces shown to whom blame could be readily attached were those of ignorant parents exposing their innocent children to the disease.[50]

In effect, the reformers communicated a double message about the consumptive's status as simultaneously victim and menace. On the one hand, they repeatedly insisted, as one tract explained, that "it is not dangerous to live or work with a person who has tuberculosis if he is cleanly and is very careful to destroy all the sputum which he coughs up." On the other hand, the endless warnings about the ubiquitous bacteria to be found in the consumptives' vicinity inevitably created a revulsion toward their company.[51]

No matter how often TB workers acknowledged the socialism of the microbe, their emphasis on both education and careful observance of sanitary rituals favored the more privileged strata of American society. Thus the disease became associated with not only the moral failings of individuals but also the sanitary defects of the "lower orders." The "careless" or "unteachable" consumptive was frequently assumed to be poor, uneducated, foreign-born, or nonwhite. As public health nurse Ellen La Motte wrote in 1909, "The millionaire, the professional man, and the bank clerk" might be counted on to learn, if only by rote, what they needed to know to protect others from consumption. In contrast, she wrote, "The day laborer, the shop girl, the drunken negro belong to a class which, *by reason of the very conditions which constitute it a class,* is unable to make use of what it learns."[52]

The Debate over the Common Communion Cup

Nowhere were the contradictions inherent in the new "tuberculosis religion" more painfully evident than in the turn-of-the-century debate over the common communion cup among Protestant churches. The dilemma of whether to abandon the common cup, which symbolized the community of all believers, was unique to Protestants, for in the Roman Catholic tradition, only the priest took the sacred wine. The question precipitated a long, wrenching debate among Protestant denominations precisely because it touched upon such deep issues of trust and community.

In 1887, M. O. Terry, a physician in Utica, New York, first publicly called for the abolition of the common communion cup on hygienic grounds. In 1893, physicians in Rochester and Philadelphia pressed the issue in their local churches, noting that the common cup had been implicated in the transmission of tuberculosis as well as of other "loathsome diseases" that they hesitated to name. As anti-TB societies formed in many cities, their members played a key role in agitating the communion cup issue throughout the country.[53]

The proposal to abolish the common communion cup initially met with deep resistance. For many Protestants, the fact that Jesus and his disciples used one vessel at the Last Supper was sufficient reason to forbid any change in the practice. As W. M. Parker, a Massachusetts physician, put it in 1892, "We may safely believe that He who instituted the sacred feast will be equally strong to guard His children against such dreadful danger." Parker argued that instituting the individual cup would be "obnoxious," implying as it did "distrust as to the cleanliness of the communicants." When the General Assembly of the Presbyterian Church was first queried on the issue in 1895, it agreed that the hygiene issue was insufficient reason to alter "the primitive and historic method of administering the Lord's Supper."[54]

Supporters of the reform responded to these arguments on both practical and scientific grounds. First, they noted that many large congregations had already abandoned the single cup because its use made the communion service too long. If the custom had already been breached as a matter of convenience, reformers argued, surely the use of individual cups could be justified for more compelling hygienic reasons. Citing the experimental evidence that pathogenic bacteria had been cultured from communion cups, critics of the custom reminded congregants, in the words of one, "It is not every earnest Christian who has always been a Christian, or who has not at some time strayed from the narrow path."[55]

In an impassioned plea against the common cup, the physician Ellen Wallace, state superintendent of hygiene for the Women's Christian Temperance Union, personified its dangers in a way calculated to make her audience profoundly uncomfortable. She reported the story of a fellow physician who upon kneeling at the communion table saw two people he knew to be diseased, one with tuberculosis, the other with an "odious disease" (presumably syphilis), and both of whose mouths he knew to be "in a condition dangerous to his neighbor." As a result, when the communion cup came to him, he simply let it pass. For many communicants, Wallace argued, the thought of sharing a cup with a diseased person had become so "offensive" that the rite no longer demonstrated "the spirit of unity which Christ sought to exemplify." The individual cup movement was necessary, she concluded, to restore the sacred meaning of the ritual.[56]

The communion cup issue divided individual congregations over the relative importance of religious doctrine and hygiene. At the Walnut Street Presbyterian Church in Philadelphia, the matter came to a head in 1898, when a congregant who was both a physician and a member of the PSPT raised the issue. The church board appointed a special committee to investigate the matter, which reported back in favor of preserving the common cup; the committee argued that some 13 million Protestants had long used it without obvious injury. "If the usages of the communion service, which have prevailed among Christians . . . have been attended with risk to life and health, it only seems fair to assume that experience and common observation would have long observed the danger." But the report obviously did not end the controversy, and the church decided to take a vote on the issue. Two-thirds of the congregation voted in 1898 to adopt the individual cup system, and it was instituted soon after.[57]

Evidently many Protestant congregations moved in the same direction at the turn of the century. A brochure for the Sanitary Communion Outfit Company of Rochester, published around 1900, listed hundreds of churches across the country that had purchased their patented communion sets, which allowed easy sterilization of individual glass cups and serving trays between uses. The denominations represented included Baptists, Congregationalists, Lutherans, Methodists, Presbyterians, and Universalists. Taking advantage of this trend, the Presbyterian Historical Society acquired a magnificent collection of silver and pewter by simply writing to the Presbyterian churches on the list and asking for their discarded communion sets.[58]

Subsequent generations of Protestants would take their communion wine from those little glasses, unaware of the triumph of hygiene over religious doctrine that they represented. With the advent of tuberculosis religion, spiritual community came to depend upon such discrete but firm separations from its diseased members. As Howard Anders, a vigorous proponent of the individual cup put it, even the pious felt a desire of "preserving one's self from that which is manifestly unclean, unsanitary, unnecessary, unmannerly, and unchristian." The communion controversy illustrates how the gospel of germs fostered a new sense of community that abjured the casual contacts that could spread disease, even in the most holy of ceremonies.[59]

For better and worse, the antituberculosis movement provided Americans with a remarkably expansive set of beliefs about the nature of germs and the need to avoid them. In the process of educating Americans about the white plague, anti-TB societies fostered a sweeping new vision of the "chain of disease" that linked all Americans, sick and well, into the rights and duties of public health citizenship. The "worries and torments" of tuberculosis religion became the foundation for a growing awareness of the germ that would eventually reshape every facet of daily life. And for those who had the disease, a harsh new existence had just begun.

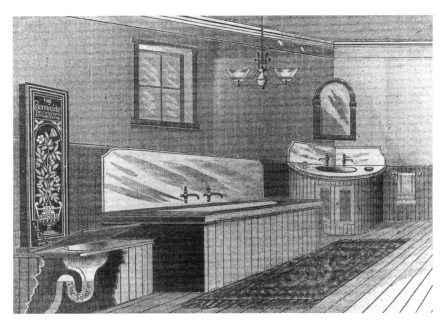

The Germicide, a disinfectant device that attached to the toilet, was one of the many devices marketed as a protection against sewer gas. This illustration from a promotional brochure shows the Germicide installed in a typical middle-class bathroom of the 1880s, which was furnished with wood paneling, wallpaper, and rugs much like other rooms in the house. (From The Germicide Company of New York, *Sewer-Gas and the Remedy* [n.d]. Courtesy of the Warshaw Collection of Business Americana, National Museum of American History, Washington, D.C.)

In response to the growing popular awareness of germs, the old, opulent style of the Victorian bath gave way to the smooth, hard look of the modern bathroom. Standing free of wooden cabinetry and surrounded by tiles, the white porcelain toilet shown here supposedly presented no place for filth-loving microbes to accumulate. (From *Guarding the Home: Essential Conditions of Sanitary Arrangements to Exclude the Germs of Typhoid Fever and Other Zymotic Diseases,* 1887. Courtesy of Hagley Museum and Library, Wilmington, Delaware.)

The pursuit of a more "antisepticonscious" style of living led to a growing preference for spare, smooth, easily cleaned surfaces, which profoundly changed the look of the twentieth-century American home. This 1906 advertisement for porcelain enameled ware suggests how the sanitary standards once advocated only for the bathroom expanded to include the kitchen, laundry room, and bedroom. (From *Good Housekeeping Magazine*, July–December 1906. Courtesy of American Standard Inc.)

At a time when municipal water supplies were still of dubious purity, many middle-class Americans installed home water filters to protect their households against waterborne disease germs. This drawing from an 1894 trade catalogue portrayed sickness as the inevitable consequence of failing to install such a filter. The fanciful microscopic creatures pictured here vaguely resemble insects, suggesting the ease with which many lay people equated "bugs" and germs. (From McConnell Filter Co., *McConnell Germ-Proof Water Filters, Illustrated Catalogue*, 1894. Courtesy of the Historical Collections of the College of Physicians of Philadelphia.)

The discovery that tuberculosis was a communicable disease spread by coughing and spitting ushered in a more aggressive era of public health education at the turn of the century. These posters, promoting the need for a healthy lifestyle, careful housecleaning methods, and an end to spitting, were designed by the National Tuberculosis Association for a 1910 publicity campaign. (From *Journal of the Outdoor Life*, January 1910. Courtesy of the Historical Collections of the College of Physicians of Philadelphia.)

Borrowing heavily from the advertising and entertainment industries, the National Tuberculosis Association and its affiliates developed many innovative methods of popular health education, including the use of posters, moving pictures, health exhibits, and parades. This photograph shows a parade from the early 1920s, complete with elephants. The publicity methods developed by the American anti-TB movement were widely copied by other health advocacy groups.

Originally introduced in 1915 as part of the Christmas Seal campaign, the National Tuberculosis Association developed the Modern Health Crusade into an ambitious program of child health education. This photograph shows a Modern Health Crusader in full regalia selling Christmas seals to the French war hero, Marshal Ferdinand Foch. (From *Journal of the Outdoor Life*, December 1921. Courtesy of the Historical Collections of the College of Physicians of Philadelphia.)

I·WASHED·MY·HANDS·
BEFORE·EACH·MEAL·TODAY·

The theory and practice of modern aseptic surgery brought new awareness of the microbial populations resident on human skin and hair. To combat casual infection, public health authorities promoted frequent hand washing. This poster promoting the hand washing habit was distributed by the Pennsylvania Society for the Prevention of Tuberculosis as part of the Modern Health Crusade. (From *Yearbook of the Pennsylvania Society for the Prevention of Tuberculosis for 1919.* Courtesy of the Historical Collections of the College of Physicians of Philadelphia.)

NOW IS THE TIME TO FIGHT!

Death Lurks
in the Filth
on a
Fly's Feet

AVOID IT!

Don't Let
That Fly
Become a
Grandfather.

KILL IT NOW!

The Descendants of ONE May Fly Will Number Millions—If You Let It Live

The recognition that insects could act as carriers of disease germs led to a vigorous crusade against the common housefly. This 1917 poster from the Illinois Tuberculosis Association likened fly control to a patriotic duty. The bombs are labeled with the names of diseases thought to be transmitted by flies: typhoid, tuberculosis, and cholera. (From *Bulletin of the National Association for the Study and Prevention of Tuberculosis,* June 1917. Courtesy of the Historical Collections of the College of Physicians of Philadelphia.)

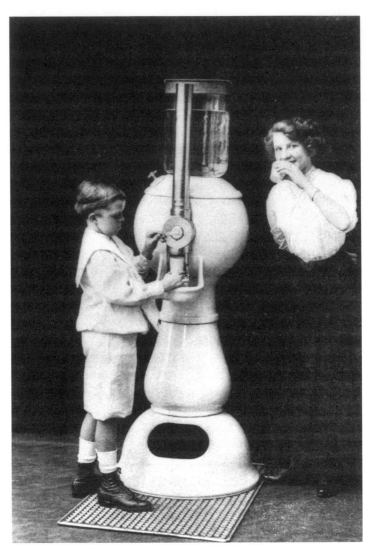

Armed with new awareness of the bacteria to be found in the human mouth, health reformers sought to remove the common cups traditionally supplied at public water fountains. In their place, businesses and public institutions began to install sanitary drinking cup vendors, such as the one pictured here, which dispensed disposable paper cups for a penny. (From *Journal of the Outdoor Life*, August 1909. Courtesy of the Historical Collections of the College of Physicians of Philadelphia.)

By linking disease prevention with specific techniques of housecleaning and food preparation, the gospel of germs made the American homemaker a central figure in the domestic "battle with bacteria." This photograph shows a visiting nurse instructing an immigrant mother about how to prepare a baby's bottle. In addition to demonstrating such specific techniques, the visiting nurse modeled a scrupulous sense of personal cleanliness, symbolized here by the clean white apron worn over her street clothes. (From *The Child in the City*, 1912.)

After bacteriologists cultured germs from the hems of ladies' dresses and cloaks, public health reformers began to warn against the hazards of the "septic skirt." In the late 1890s, women formed "Rainy Day Clubs" to encourage shorter, more hygienic hemlines. This 1900 cartoon from the illustrated New York weekly *Puck* suggested that the germs on a fashionable lady's cloak threatened not only the maid who had to clean it but also the owner's children standing nearby.

The growing emphasis on the human body as a source of infection led some hygiene reformers to criticize social customs, such as handshaking and baby kissing, that fostered the exchange of germs. Mothers in particular were warned to be leery of letting anyone, even relatives, touch or kiss their children. This toddler wore a bib bearing the words "I Don't Want to Get Sick. Do Not Kiss Me." (Courtesy of the City of Milwaukee Health Department.)

As the new bacteriology increased awareness of foodborne illness, public health workers and home economists warned against the common practice of leaving foods uncovered and unrefrigerated for long periods of time. Yet many poor families could not afford the expense of even the simplest icebox. This photograph taken by Lewis Hine, circa 1902–1904, was titled "Tenement Frigidaire and Infant Mortality," suggesting the link between improper food handling and the high death rate among the young. (Courtesy of the George Eastman House, Rochester, New York.)

Outbreaks of foodborne illness fostered new attention to the hygienic condition of both private kitchens and public eateries. This photograph shows a sanitation inspector from the Milwaukee Health Department looking for hygienic infractions in a private kitchen. The discovery that a small minority of healthy people carry the bacteria responsible for typhoid and other diseases led to growing scrutiny of food handlers as well. (Courtesy of the City of Milwaukee Health Department.)

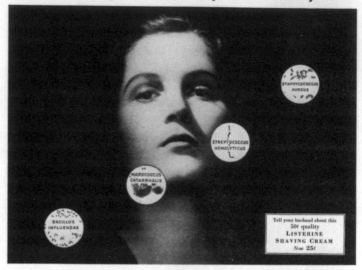

Although infectious diseases began to decline in the early 1900s, the "germ sell" remained a popular one in American advertising. The manufacturers of Listerine promoted frequent hand washing and gargling with disinfectant as a protection against disease. This striking 1931 advertisement highlighted the dangerous germs to be found in the mouth of a beautiful woman. (From J. Walter Thompson Company Archives, Special Collections Library, Duke University, Durham, North Carolina. Courtesy of the Warner-Lambert Company.)

Anxieties about foodborne illness greatly stimulated the development of new forms of packaging. To build popular demand for cellophane-wrapped goods, the Du Pont Cellophane Company ran advertisements such as this in the late 1920s and early 1930s, which repeated public health warnings about the dangers of the casual touch. Concerns about germs steadily increased consumer tolerance for the costs of disposable packaging. (From Hagley Museum and Library, Wilmington, Delaware. Courtesy of the Du Pont Company.)

None of these will give you AIDS.

Many of the specific practices advocated by early-twentieth-century health reformers, such as hand washing and protecting against coughs and sneezes, remain the foundation of modern disease prevention. Yet fears of casual contagion have impeded public understanding of the human immunodeficiency virus, which does *not* spread by casual contact, insect vectors, or food contamination. As this poster attests, AIDS educators have had to help Americans unlearn the lessons so successfully taught by previous anti-disease crusades. (Courtesy of the William H. Helfand Collection, New York, New York.)

6 · The Domestication of the Germ

During the same decades that the anti-TB movement reached its zenith, women reformers began a parallel effort to bring the gospel of germs to the American housewife. In their mission to modernize and rationalize the home, these "domestic scientists," as they styled themselves, found the new lessons of the laboratory particularly relevant. A bacteriological perspective brought significance to the most humble of the housewife's chores; whether dusting or canning, planning her parlor decor, or nursing a sick child, science suggested that she had an important contribution to make to the war against dread disease. As Ellen Richards, one of the founders of the domestic science movement, explained, "Sweeping and cleaning and laundry work are all processes of sanitation and not mere drudgery imposed by tradition, as some people seem to think." For that reason, Richards and her colleague Marion Talbot declared in 1887, "A knowledge of sanitary principles should be regarded as an essential part of every woman's education, and obedience to sanitary laws should be ranked, as it was in the Mosaic Code, as a religious duty."[1]

This conception of the housewife's "Mosaic Code," a reference to the laws of cleanliness imposed on the ancient Hebrews, reflects the growing importance of female voices in the late-nineteenth-century discourse about disease prevention. Prior to the 1890s, the field of domestic hygiene had been dominated by male experts, primarily physicians and sanitary engineers. As women's access to higher education widened in the late nineteenth century, a talented generation of college-educated women began to take up the work of applying

135

scientific principles to woman's traditional sphere of the home. Their teachings, as embodied in the new discipline of domestic science, were disseminated by the vast network of women's voluntary associations and clubs that grew up at the turn of the century, including the Association of Collegiate Alumnae (now the American Association of University Women), the General Federation of Women's Clubs, and the Women's Christian Temperance Union. Domestic science also became the foundation of new professional careers for women in the fields of social work, visiting nursing, and home economics.[2]

The domestic science movement and the anti-TB crusade shared the conviction that the home was the primary site for preventing the spread of contagious diseases, yet they approached their reform work in different, albeit complementary, ways. In contrast to the anti-TB work, the domestic science movement was founded by women, led by women, and directed at women. Its leadership conceived of the home as a breeder not just of tuberculosis but of all germ-related ailments, from typhoid and diphtheria to food poisoning. As a result, its preventive teachings covered in detail issues such as food storage and preparation, which the tuberculosis crusade passed over with only a mention.

The domestic science movement encompassed a wide range of voluntary and professional initiatives, all of which became to some degree conduits for transmitting the gospel of germs. But it was the new discipline of home economics that became most committed to both theorizing about the home as a microbial environment and systematizing popular education about germs for American women and girls. Visiting nursing, the other profession most deeply involved in teaching domestic disease prevention, concentrated more on practice than theory (I will examine this field in more detail in Chap. 8). The broader conception of housework as microbe management, which influenced the whole philosophy of domestic science at the turn of the century, can be most clearly discerned in the teachings of "household bacteriology" that were promulgated by early home economists.[3]

The "Woman Problem" in Public Health Education

The domestic science movement addressed a long-standing problem in the public health movement—namely, how to move beyond the ranks of hygiene enthusiasts to reach a broader female audience that had little if any science education. Although many middle-class women were clearly interested in hygiene issues, it was not easy to convey the fruits of the new bacteriology to them in terms that they could easily understand. As of the 1890s, even the best educated women had little exposure to experimental methods in general or to bacteriology in particular. The vast majority of public health experts were men with limited interest in teaching housewives about the mysteries of the germ. Female physicians, who often had a greater commitment to educating women on hygiene issues, still constituted only a very small minority within the medical profession.[4]

Harriette Plunkett and Ellen Richards, two of the first popular authors to interpret the germ theory of disease for their sex, acknowledged the difficulty of their work. In an 1893 letter to Lawrence Flick, Plunkett wrote that the "carefully prepared articles that are garnered up in the State Board Reports, and in those of the American Public Health Association have one drawback—they are read principally by physicians and sanitarians and do not reach the minds of the *women* of the land." To help her interpret the new scientific findings, Plunkett asked Flick to pass on "striking facts" and dramatic anecdotes about consumption and its dangers, explaining, "I try to 'sugar-coat' my facts as much as possible with personal anecdotes, for some sharp observer has said there is no use in trying to interest women in anything that isn't personal." Ellen Richards registered a similar complaint in 1898, lamenting the fact that inadequate educations often left women prey to ignorance and superstition. As she put it eloquently, "They have feared the thunder and ignored the microbe."[5]

But even as Plunkett and Richards penned those complaints, a younger, college-educated generation of women was coming of age who proved far more receptive to the lessons of the laboratory. In the decades after the Civil War, white women gained much greater access to higher education, both at the elite women's colleges, known colloquially as the "seven sisters," and in the expanding state university systems. As women students began to take more courses in both the physical and social sciences, an enthusiastic minority pressed on to take graduate training and become researchers and teachers themselves. Of those who acquired graduate degrees, some found teaching and research positions in the women's colleges; others carved out small, precarious fiefdoms in public universities. Partly because other routes were closed to them, and partly because they found domestic subjects most interesting, many in this pioneer generation took on as their particular mission the application of scientific principles to the home.[6]

Ellen Richards and Marion Talbot were among the first women to both take advantage of the new science training and see its broader potential. After earning a B.A. at Vassar in 1870 and a B.S. in chemistry from the Massachusetts Institute of Technology in 1873, Richards went to work in the laboratory of bacteriologist William Sedgwick, where she helped to develop new methodologies for measuring water pollution. While marrying and raising a family, she continued to teach, research, and write on the subject of sanitary science and domestic hygiene. Marion Talbot earned a B.A. from Boston University in 1880 and a B.S. from the Massachusetts Institute of Technology in 1888. After lecturing at La Salle Seminary and Wellesley College, she was hired by the University of Chicago to teach sanitary science in 1892. She later became its dean of women, a position she used to promote other women's academic careers.[7]

These women experts, trained in the new experimental and statistical methods prevalent in the late-nineteenth-century sciences, presented themselves as translators and guides for their less educated contemporaries. Echoing Plunkett's observations, the social scientist and reformer Florence Kelley noted that the publications of state and local public health agencies were filled with detailed information about industrial conditions, food adulterations, and other public health issues of grave consequences to the individual woman. Yet none of it was presented in a practical form that allowed the individual consumer to avoid "buying smallpox," as she put it, along with her new garments.[8]

In bringing the new sciences to bear on women's supposedly "natural" interests in home and family, domestic scientists sought to make their sex more productive and contented citizens. They repeatedly stressed that men need have no fear that better science education would divert women from their traditional sphere of home life. As Charlotte Angstman explained in an 1898 article "College Women and the New Science" published by the *Popular Science Monthly,* the new female experts were interested in improving, not destroying, the home, for "home is the magnet to which their thoughts and efforts are continually drawn."9

Domestic reformers' efforts to disseminate the lessons of the laboratory to their less educated sisters were couched in the language of disinterested female benevolence. They believed theirs to be an altruistic mission of enlightenment based on purely objective principles of science. Yet in retrospect, it seems apparent that the domestic science movement did much to advance its proponents' own personal and class interests. By setting themselves up as experts on the home, reformers carved out interesting new spheres of social and political influence for themselves. The rhetoric of home improvement and enlightenment fueled the expansion of such professional fields as social work, home economics, and nursing, and thereby increased women's avenues for economic independence and personal satisfaction. Although domestic experts portrayed their expertise as purely objective, the ideals of uplift that they promoted strongly reinforced their own sense of cultural superiority as white, educated, middle-class women. Thus the mission of carrying the gospel of germs to the American household was by no means the simple exercise of benevolence that the founders of the domestic science movement had envisioned it to be.10

Educating the Housewife

Compared to the new forms of propaganda developed in the antituberculosis movement, the channels of influence that home economists used to convey their version of the gospel of germs were comparatively traditional. The eye-catching artwork, clever slogans, and circus-style stunts that characterized anti-TB work did not carry over into the pedagogy of home economics, which continued to rely heavily on less showy forms of print literature. But the discipline did adopt one innovative form of teaching, known as the "demonstration method," that was equally if not more dramatic than the TB poster or parade.

Modeled on the tradition of public science, in which the experimenter produced his dramatic results in front of a live audience, the home economics demonstration involved performing experiments or techniques for students to observe and imitate. Demonstrations of household bacteriology sought primarily to expose the prolific germ life to be found in everyday operations such as cleaning and cooking. For example, growing a "dust garden," which involved exposing a petri dish to common house dust and showing the bacterial colonies that resulted, "will impress this whole subject much more vividly than any amount of reading," wrote S. Maria Elliott. Another popular point to demonstrate was the microbial population of the hand. During a cooking class, Elliott recalled, a student insisted that her hands were already clean and that she did not need to wash them before going to work. The teacher had the girl wash her hands in water and then introduced a tablespoon of the wash water into sterilized milk, which "proved an efficient aid in its putrefaction." Elliott reported that from then on "the cooking class never forgot to wash their hands."[11]

Home economists found a secure institutional niche for their demonstration method in the public school system, where courses on domestic science soon became a requirement for girls. Among adult women, the club movement became the chief vehicle for home economics education; in addition to literary and cultural offerings, many women's clubs in the early 1900s began offering programs on domestic science. In rural areas, the extension service played a similar role, as I will discuss more fully in Chapter 8. Through these various educational conduits, the tenets of household bacteriology reached a wide range of American women and girls.[12]

Another important educational outlet that home economists successfully exploited was the new mass-circulation women's magazines such as the *Ladies' Home Journal* and *Good Housekeeping*, which had millions of women readers by the early 1900s. In the hands of talented journalists, the often sober prose of household bacteriology took on more attention-getting forms. Magazines carried articles with arresting titles such as "Household Art and the Microbe" and "When to Be Afraid: A Common-Sense Talk about Children's Diseases." Domestic advice columns offered the housewife's equivalent of the antituberculosis slogans, such as "Know Your Ice" and "Swat the Fly."[13]

Good Housekeeping made especially effective use of the domestic science philosophy in crafting its journalistic mission. For example, the magazine established its own "experiment station," the Good Housekeeping Institute, to test consumer goods and household appliances, and the institute's reports became a regular feature. *Good*

Housekeeping also ran a column titled "Discoveries" in which house-wives were encouraged to share their experiences as "observers and experimenters." There readers traded suggestions for improving their daily lives, such as how to kill insects or to clean an ice box. One intrepid reader reported on how she avoided catching germs while using a public telephone: she talked loudly with the mouthpiece held to her chest instead of her mouth. (The editor noted that she had tried the method and it worked, although the speaker sounded far away.)[14]

The lessons of household bacteriology were also reinforced by the many pages of advertisements carried in women's magazines. In contrast to their nineteenth-century predecessors, the new mass-cir-culation journals depended upon advertising revenue, not subscrip-tions, for their success. Far from resenting this dependence, the editors of the *Ladies' Home Journal* and *Good Housekeeping*, like the TB workers, came to regard advertising as a medium of education rather than manipulation. As I will describe more fully in Chapter 7, many of the advertisements placed in women's magazines at the turn of the century reinforced the precepts of household bacteriology.[15]

Scientific Housekeeping and the Germ

At first glance, the principles of household bacteriology seem little different from the previous generation's conceptions of "house dis-eases" and "filth diseases." Much of what Ellen Richards, the founder of home economics, called the "science of controllable environment" hinged on the intelligent management of human wastes and other forms of organic dirt. But upon closer reading, home economists' understandings of which dirt was dangerous, and how science should guide household practice, clearly reflected the influence of the new bacteriology.[16]

To begin with, home economists sought to teach not only the rules for disease prevention but also the scientific rationale for why their observance was so important. To that end, they insisted that even ordinary housewives should learn the rudiments of bacteriology, an idea that was novel at the turn of the century. The Cornell home economist Martha Van Rensselaer's experience in starting a Farmers Wives' Reading-Course is a case in point. Van Rensselaer, who lacked any special training in sanitary science or bacteriology, nonetheless felt that these subjects should be an essential part of the College of Agriculture's extension work with farm women. When she approached a Cornell bacteriologist, saying, "I would like to learn about the bacteriology of the dishcloth so that I may explain to farm women the importance of its cleanliness," he responded, "Oh, they do not need to learn about bacteria. Teach them to keep the dishcloth clean because it is nicer that way."[17]

Home economists emphatically rejected this scientist's advice in their early textbooks and circulars, and they found more cooperative bacteriologists to help them. The Wesleyan College bacteriologist Herbert W. Conn, who became a staunch supporter of home economics, observed that if medical students were required to take courses in bacteriology, so should homemakers, because the microorganisms' "relation to the ordinary household, and hence to the housewife, is even more intimate than to the physician."[18]

The standard texts used in early home economics courses, such as Conn's *Bacteria, Yeasts, and Molds in the Home* (1903) and S. Maria Elliott's *Household Bacteriology* (1907), attempted to explain the fundamentals of bacteriology in simple terms. Not only did they summarize how and why scientists had come to link germs with disease by giving brief histories of the germ theory and even by explaining Koch's postulates; they also stressed the importance of conducting simple experiments and gaining access to a microscope so that the student could see this invisible world of germs for herself. Elliott encouraged the experimental frame of mind, noting, "Many of the daily occurrences in the home give rise to questions which may be readily answered if we will but turn our kitchens into laboratories and try some simple experiments." Even Van Rensselaer's short extension bulletins for farm women urged them to grow their own "dust gardens" and turn a naturalist's eye to the processes of decay and putrefaction evident in carelessly stored food.[19]

In the spirit of John Tyndall, whom the popular domestic science writer Marion Harland once called "the prince of everyday practical scientific teachers," home economists mixed the exotic with the familiar in describing the microbial world. Maria Elliott, after describing the three principal shapes of bacteria—the rod, the ball, and the spiral—likened the formation of a bacterial chain colony to "a string of sausages, often seen hanging in the windows of a market." In another passage, she compared microorganisms to human diners, noting, "Bacteria in general like the same kinds of food that man likes," in that some have a preference for milk, others for meat juices or sugary, starchy foods.[20]

Using drawings and photographs, home economics texts introduced women to the differences among yeast, molds, and bacteria and decried the popular view that all microorganisms were bad. Dairying was a popular example: the "bad" microbial species that could spoil milk or turn it into a vehicle of contagion were often contrasted with the "good" species needed to make butter and cheese. Such homely examples underlined the need for housewives to learn to discriminate among the different types of microbes and to keep the dangerous ones at bay. From the study of bacteriology, the modern woman could come to appreciate that the "science of the infinitely small has become the infinitely important," as Maria Elliott put it.[21]

Still, for all their efforts to teach an appreciation of bacteriology, the overriding goal of home economists was not to give women a deep understanding of the microbe, but rather to teach them the principles of scientific housekeeping. As Elliott explained in the preface to *Household Hygiene,* the main contribution of "the science of bacteriology" was the "correct understanding of the preventive measures which tend toward health." In this spirit, home economists directed their explications of bacteriology toward teaching a more scientific code of household hygiene.[22]

Most of the precepts embodied in the housewife's Mosaic Code were identical to the rituals promulgated in the name of tuberculosis "religion." But when the two are compared, one striking difference is apparent: domestic science texts preserved much more of the initial gospel of germs' preoccupations with toilets and sewer systems. Even as bacteriologists were seeking to dispel the fear of sewer gas at the turn of the century, home economists still perpetuated anxieties about the toilet's role as a portal of disease. Well into the 1910s, home economics texts routinely included long, detailed expositions of the pipes, drains, and traps needed to prevent germ-laden sewage vapors from fouling the interior air supply. As had her sanitarian predecessors, Maria Elliott recommended in her 1907 text, *Household Hygiene,* the periodic use of the "peppermint test" to make sure plumbing was airtight. In the domestic science credo, toilets still figured as major pieces of household technology requiring exacting daily maintenance, including soaking, brushing, and scrubbing with disinfectant soap powder.[23]

Texts written before 1920 also continued to teach a rudimentary knowledge of how to secure a safe water supply, a reflection of the fact that many poorer homes did not have access to filtered water supplies. Home economists continually warned women against trusting their senses when it came to water purity: the clear sparkling draft from creek or well could easily be laden with typhoid germs. Domestic hygiene manuals routinely included home tests for water contamination and discussed the merits of various filtering systems; they even recommended specific brands, such as the Pasteur-Chamberland filter, for their superior antibacterial action. Whenever in doubt about water safety, housewives were urged to resort to the universal germ preventive of boiling to protect the family's drinking supply.[24]

Although they perpetuated the original gospel of germs' concerns with pure air and water, the teachings of domestic science also incorporated the insights of the new bacteriology, particularly the stress on "dust dangers." Women were taught to assume that where there was dust, there were germs, and to act accordingly. Following the recommendations of T. Mitchell Prudden, whose text *Dust and Its Dangers* they frequently cited, home economists warned against dry sweeping and promoted the use of a wet mop, "dustless" broom, or oiled cloth to capture potentially germ-laden particles. Housewives were counseled to let several hours elapse between cleaning house and preparing meals to allow the dust they raised to settle back to the floor. Home economists hailed the vacuum cleaner as a great hygienic boon because it sucked up dangerous dirt so much more safely and thoroughly than did traditional carpet sweepers. As Martha Van Rensselaer responded to one housewife who asked about its value, "We do believe in the vacuum cleaner."[25]

The connection between dust and disease figured prominently in home economists' criticisms of the overstuffed furniture, thick carpets, patterned wallpaper, and extensive bric-a-brac so beloved in Victorian decorating schemes. Instead, they promoted surfaces that were smooth, washable, and free of ornamentation that might harbor dust, vermin, and germs. As Van Rensselaer's 1902 brochure on "Decoration in the Farm Home" explained, "Modern sanitary homemaking avoids 'dust-lines,'" including picture moldings on walls, carved picture frames, and heavy tasselled drapes, because, in her words, "Dust is one of the friends of disease." Germ-ridden carpets were to be replaced with finished wood floors and small area rugs, and disease-breeding wallpaper was to be abandoned in favor of washable wall coverings or plaster tinted with germicidal paint.[26]

The germ-bearing properties of cloth turned the laundry room into another site requiring special sanitary discipline. As Cornell home economist Flora Rose warned in an extension bulletin on modern laundry methods, "Whenever organic matter accumulates on clothing, even to a slight extent, there is a possibility of the presence of bacteria." For underwear, the layer of clothing worn nearest the microbe-ridden skin, she recommended ten minutes of boiling "not so much to remove visible dirt as to complete the washing process by destroying germs and thus to purify the clothing." When a family member had a cold or other communicable disease, she urged that disinfectant be added to the water used to rinse sheets and handkerchiefs. In a laundry manual published in 1912, Lydia Ray Balderston and Emma H. Gunther praised the "disinfecting power of the flatiron" when heated to 480 degrees Fahrenheit. Warning against the use of wicker clothes hampers, they noted that "numbers of bacteria . . . have found lodgement in the crevices of the reeds of the basket" and recommended using cloth bags or an enameled container that could be easily washed and "always kept white."[27]

Home economics texts also reflected the new bacteriology's teachings about insect vectors of disease. As a result of the "epoch making discoveries" in this regard, Martha Van Rensselaer wrote in 1913, "The hum of the mosquito and the buzz of the house fly have become fraught with an entirely new significance. Even the dog and the cat, with their burden of fleas, have taken on a new aspect." The experimental link between insects and germs lent a bacteriological blessing to traditional sanitarian preoccupations with standing water, outdoor privies, food garbage, stables, and manure piles, which were now vigorously condemned as insect breeding grounds. Home economists urged housewives to use any means necessary, from homemade insecticides to window screens, to keep these "germs with wings" out of their home domain. For a generation of ordinary women being introduced to the idea of the microbe, the ubiquitous "bug," a term that became a colloquial expression for the microbe around this time, served well to embody the invisible germ enemy.[28]

The sickroom was another battleground against bacteria where the housewife continued to play a leading role. Although by the early 1900s hospitals were coming into much wider use, most women still frequently had to nurse family members, especially children, through bouts of serious communicable disease; thus knowledge of home nursing procedures became an important part of the housewife's Mosaic Code. The basic principles of the "home hospital" changed little from the 1880s to the 1910s; women were still instructed to strip the sick chamber of superfluous furniture, to hang a disinfectant-soaked sheet over the door, and to disinfect the patient's bodily discharges promptly. To these traditional measures, the new understanding of dust and droplet infection added more attention to the regulation of coughing and spitting. Reflecting the concern about TB, home nursing instructions now included the sanitary care of spittoons, the virtues of disposable "crepe napkins" or paper tissues over cloth handkerchiefs, and the construction of inexpensive paper sputum cups.[29]

In general, home economists called upon women to break up what Charles Chapin once termed the "universal trade in human saliva." As guardians of babies and small children, they were assigned the special duty of modeling exemplary habits of mouth and hand cleanliness. As Martha Van Rensselaer acknowledged in 1909, "The knowledge of germs is becoming general among housekeepers, yet we are guilty of little habits which we should drop very quickly if they were brought to our attention in connection with the danger from bacteria." Among the "little sins" of carelessness that she noted were "common ways of taking to ourselves organisms," such as licking a stamp and moistening a finger to turn a page. Home economists scolded otherwise conscientious mothers who tested the temperature of the baby's bottle by taking it in their own mouths or chewed a cracker to make it easier for a baby to digest. They also warned women against the dangers of indiscriminate baby kissing, recommending that mothers put a stop to this unhygienic custom.[30]

The Bacteriology of Food

The bacteriology of food held a special place in the teachings of home economics. As Herbert W. Conn explained in his text, *Bacteria, Yeasts, and Molds,* it was largely because of woman's historic association with cookery that "the study of bacteriology and kindred subjects has in recent years come to be looked upon as a part of the necessary training of the housewife." Although some microorganisms played a positive role in cookery, "the growth of bacteria in food is nearly always undesirable," he warned, "and the housewife must always aim to prevent it." In place of old commonsense rules about how to keep food safe and "sweet," bacteriology now supplied exact times and temperatures governing the storage and preservation of food. Laboratory studies of the effect of cold on bacteria suggested that a constant temperature of 40 degrees Fahrenheit or less was necessary to retard bacterial growth on food, a finding that greatly enhanced the importance of having an icebox or refrigerator in the home.[31]

From the home economists' viewpoint, the refrigerator soon came to rank second only to the toilet as a sanitary essential for the household; yet like the toilet, it brought its own burdens of care. Not only did the refrigerator's temperature need to be carefully monitored, but its surfaces also had to be kept spotlessly clean to avoid germ contamination of the contents. As Van Rensselaer warned, "The refrigerator might be called upon to tell many tales of the life history of germs, for its recesses hide a multitude of sins." Her fellow home economist Maria Parloa concurred: "It will be far better to get along without the comfort it affords then to endanger health and life by using a contaminated article." Proper refrigerator hygiene required a thorough weekly cleaning with boiling water to get out every particle of dirt—a ritual as essential as disinfecting the toilet and boiling the laundry.[32]

Bacteriologically informed cooking required an equally fine understanding of the virtues of high and prolonged heat for killing both germs and their dangerous spores. After discussing laboratory studies of the effect of heat on germs, Conn advised his readers, "A practical lesson to be drawn from these facts is that food heated to *boiling* in its preparation is thereby, in a measure, protected from spoiling, since the bacteria are mostly killed." Home economists promulgated the basic rule of thumb that twenty to thirty minutes of vigorous boiling would make any meat, fruit, or vegetable safe to eat. For foods that would be rendered unpalatable by such prolonged cooking, a briefer exposure to heat, supplemented by natural preservatives such as sugar, vinegar, and salt, would suffice. These directives helped to turn the American middle-class table into a feast of overcooked meats and heavily sweetened, salted, or vinegared side dishes.[33]

Another common theme in the new food hygiene was the need to keep items covered against dust and flies. "A table filled with left-overs, waiting to be prepared for the next meal, is a veritable dust-garden," cautioned Van Rensselaer, "and who knows what additions it may make to our diet?" Home economists spun the lessons of the laboratory into countless revolting scenarios of daily life, such as the dusty bread covered with dried sputum bought from the careless baker or the "typhoid fly" walking on a manure pile and then lighting on the baby's bottle.[34]

The bacteriology of milk figured as a special area of concern in home economics texts. Acknowledging the common view that boiling deprived milk of both nutritional value and flavor, they encouraged women to practice home pasteurization, which required heating the milk to a temperature slightly below boiling, holding it there for twenty minutes, and then cooling it quickly. If they could afford it, women were advised to avoid this labor by buying pasteurized or "certified" milk from commercial dairies. But even then, Flora Rose cautioned, housewives had to be careful that the glass bottles used to deliver the milk were "thoroughly washed and sterilized in order not to spread disease germs which they may gather up from various homes at which they are left."[35]

In every aspect of kitchen life, home economists urged housewives to apply the gospel of stringent cleanliness. Like the laboratory or the operating room, the woman's theater of cooking had to be kept scrupulously clean. For ease of cleaning and disinfection, home economists recommended that kitchens be furnished more like bathrooms, with washable walls, enamel surfaces, and linoleum floors. The kitchen sink, where warm water and food scraps created a perfect place for germs to breed, needed to be disinfected daily. Cooking equipment, including utensils, serving dishes, silverware, and dish-cloths, were to be scalded with boiling water before every use. The cook, like the surgeon, had to keep herself clean as well by using an apron (preferably white, and frequently bleached, boiled, and ironed) to protect food from her germ-ridden clothes and by washing her hands often with disinfectant soap. She was also urged to guard against unthinking personal habits that might contaminate food, such as tasting the soup with a spoon and then using it to stir the pot.[36]

By far the most dramatic lessons about germ life in the kitchen were delivered in home economists' instructions about home canning. Canning was a favorite subject of the early discipline, especially among extension agents who advocated it as an ideal way for farm girls and women to improve their family's diets while making extra money for themselves. Canning lessons also attracted town women who liked to "put up" summer fruits and vegetables for use as winter delicacies. From a home economist's perspective, then, canning offered a splendid opportunity for illustrating the basic principles of household bacteriology to a broad range of women. As Ola Powell wrote in a 1917 text, canning beautifully illustrated the application of practical science to the home: "A worker who follows scientific principles and is watchful of sanitary conditions will have results that are uniform and satisfying." Although Powell acknowledged that many women canned successfully "with little knowledge of germs," she insisted that "to know something about these minute forms of life, which are so abundant everywhere, will make the work more interesting."[37]

As part of their education in the womanly art of food preservation, home canners were treated to remarkably detailed accounts of the natural history of bacteria, molds, and yeast. The standard canning texts used in home economics courses described the life cycles of the microorganisms most commonly found in fruits and vegetables, such as *Aspergillus fumigatus* (a mold whose favorite food was tomato sauces and preserves) and *Bacillus butyricus* (a bacteria particularly partial to canned corn). These descriptions were often accompanied by illustrations of the organisms that noted interesting features such as spores or flagella. Of particular importance to canners was the distinction between aerobic bacteria, which required oxygen to live, and anaerobic bacteria, which did not. This basic principle of bacteriology, rarely mentioned in the anti-TB literature, was of obvious relevance in canning, which provided anaerobic bacteria a perfect environment in which to thrive.[38]

The elaborate protocols of canning mimicked the world of the laboratory. In the spirit of Pasteur and Lister, the home canner learned to appreciate the need for minute care. "The principal weapon of defence against bacterial action is the practice of most scrupulous cleanliness, just as modern surgery depends upon absolute cleanliness," Ola Powell explained. Likewise, canning taught the importance of precise temperatures because killing the spores of dangerous microorganisms required much higher temperatures at longer periods than did regular cooking. As Mary Hughes reminded her readers in the *Everywoman's Canning Book,* "It is *absolutely necessary* to follow accurately the time given in the tables for processing, if success is to be assured." The penalty for failure was obvious and potentially deadly: not only would the food spoil and be wasted, but it also might pose a threat to the family's health.[39]

The specter of food poisoning hovered over all of the canning proceedings. Defective canned goods had long been associated with illness, and the bacteriological revelations about *Clostridium botulinum* only underlined their potential dangers. Home economists tended to avoid extended discussions of botulism, which might frighten away potential canners; they emphasized instead that the risks of home canning could be safely overcome by scrupulous care. Ola Powell noted delicately, "People have gradually acquainted themselves with the ways in which bacteria work for our good or ill, and it is no longer necessary to whisper when discussing their effect on canned goods." Yet the often sensationalistic news coverage of botulism outbreaks associated with canned foods served as a potent reminder that careless technique could have fatal consequences.[40]

Private Woman, Public Hygiene

The fine sensibilities to germ life that home economists sought to cultivate in the American woman were never meant to be confined within the boundaries of the home. Like many Progressive-era women reformers, home economists believed that the principles of domestic science should also be applied to the public spheres of commerce and politics. Moreover, they realized that if knowledge of scientific housekeeping were limited to their own elite ranks, they would have very little success in upgrading sanitary standards. To extend their influence, home economists sought to recruit an army of informed homemakers who would uphold those standards through wise purchases and political actions. Thus domestic scientists in general and home economists in particular sought to give American women a vision of the individual household's relations to the wider world. For a woman to be truly modern, they emphasized, she had to comprehend how her "peculiar kingdom" of the home related to the highly interdependent, complex world of industrial society. Domestic science encouraged the housewife to comprehend those connections and to take an interest in every facet of modern life, from labor relations to food regulation, that touched upon the health and safety of her family.[41]

Nowhere was this vision more powerfully realized than in conceptions of the private woman's role in the public health movement. As Maria Elliott stressed, there were many "points of contact between the housekeeper and the public in sanitary matters." The home did not exist in hygienic isolation, but rather was involved in continual and potentially hazardous exchanges with the outside world. Thus along with teaching the minutia of toilet hygiene, dusting, and canning, home economists sought to remind the homemaker that the sanitary lapses of others could have serious consequences for her. True hygienic vigilance required that she acquaint herself not only with the microbial environment of her own home, but also with the germ hazards lying in wait outside of its walls.[42]

This externalization of germ threats to the home led many domestic scientists to condemn the same problems identified by tuberculosis workers, such as unreliable municipal services, dilapidated housing, and workplace abuses. Where the domestic science movement made a more distinctive contribution to the broader project of sanitary discipline was in its attention to women's roles as *consumers*. For women reformers, the universality of consumption—that is, women's dependence on goods produced outside the home—played a similar function to the "chain of disease" concept in the antituberculosis movement. As Florence Kelley, the first president of the National Consumers' League, explained in 1899, "Since the exodus of manufacture from the home, the one great industrial function of women has been that of the purchaser." Because the vast majority of consumer goods—including furniture, books, and clothing—were "prepared with the direct object in view of being sold to women," as Kelley noted, women shoppers constituted a potentially formidable lobby. In an era when women did not yet have the right to vote in national elections, their role as consumers gave them a powerful source of influence over the public world of commerce and politics.[43]

In the rapidly expanding consumer economy of the turn of the century, this conception of woman power as purchasing power opened a whole new vista of female influence over hygiene. By educating the housewife to be more microbe conscious in her purchases and to demand from manufacturers and service providers a higher standard of hygienic observance, the domestic science movement sought to unleash a force more potent than all the public health departments in the nation combined. To this end, home economists continually invited their housewife-pupils to contemplate the home's connection with the outside, disease-ridden world. Like their counterparts in the TB movement, home economists sought to make the germ's threat manifest in commonplace objects and behaviors: they pointed out that the housewife's pristine white toilet and immaculately kept kitchen sink were hooked up to public sewer and water services that might carry the germs of disease; her windows, left open to allow in the healthful sunshine and air, facilitated entry of the germ-ridden dust of the street; her guests entering the carefully dusted parlor might carry the dangerous filth of rarely cleaned streets on their hemlines and shoes; her immaculate kitchen received meat, vegetables, and milk potentially contaminated with pathogenic germs; and her spotless laundry might be infected by tenement-made clothing carrying the germs of its worker-inhabitants. In other words, no matter how safe a homemaker tried to make her private domain, it was never impermeable to the microbial dangers of the outside world.

By sketching such scenarios, home economists continually worked to awaken the housewife's imagination to a reality that she could not perceive with her own senses. By envisioning where her milk bottle or loaf of bread had come from, the housewife learned to think of the wider context of modern home life and to appreciate the intangible, often mysterious ways that her family was exposed to the influence of other people and places far removed from direct contact with her household. The Cornell home economist Martha Van Rensselaer was particularly gifted at crafting these sorts of hygienic story lines. In a long section on milk that appeared in her pamphlet *Household Bacteriology,* she asked her readers to imagine two scenarios concerning the source of the family's milk supply. "*Look first on this picture,*" she asked, describing the good dairy where the milkman dressed in clean clothes and washed his hands frequently, the clean, well-curried cows deposited their milk in spotless pails, and the milk was kept properly covered and cool until it was delivered to the consumer. Then she asked the reader to imagine the contrast: an unkempt dairyman with soiled clothing and hands, "the cow lying in her own dirt [that is, manure] overnight, her udder soiled," the milking done with "dirt flying, flies bothering the cow." The moral of the story was simple: the woman consumer should gladly pay the extra pennies charged by the first dairy to acquire safe milk for her children.[44]

At the most concrete level, this conception of the traffic of germs into the home translated into practical suggestions about shopping. For example, women were urged to wear washable gloves when making their purchases, as a hygienic precaution both for themselves and for others, and to choose stores that guarded products against the infective trio of "flies, fingers, and food." Van Rensselaer advised women to "patronize a covered delivery wagon and a grocery in which provisions are kept under cover in preference to those in which the provisions are exposed to the air."[45]

The directives about hygienic shopping also addressed the more difficult issue of cost. Although goods packaged to be dust- and fly-proof cost more than items bought in bulk, home economists emphasized that savings should never come at the expense of the family's health. As an editorial titled "Cheap Food or Clean Food" suggested, the housewife who prided herself on saving a few cents might change her mind after "she has seen the grocer plunge a strong—but not precisely immaculate—hand into the cracker-box and bring out the crackers in generous fistfuls," or "had the pleasure of seeing the grocery cat sleeping off the excitement of its last mouse-hunt in the top of the cracker-barrel or the sugar barrel, as the case may be." The message was clear: old-fashioned notions of housewifely thrift had to give way to a more modern appreciation of the importance of germ-free food.[46]

Home economists believed that once the housewife was imbued with higher hygienic standards, she could become a sanitary investigator far more effective than the overworked staff of the public health department. By the simple act of shunning the baker who allowed vermin in his shop—to let him know "we do not like to eat the bread the mouse ran in," as Van Rensselaer put it drolly—she could exercise enormous influence. In this fashion, lessons about germs and disease mastered in the home could be applied to all of a woman's dealings with store keepers, milk dealers, and even local politicians.[47]

In all of these ways, home economists assumed that a basic education in household bacteriology would make American women not only better wives and mothers, but also more effective citizens in a modern industrial society. The linking of domestic purchasing power with disease prevention provided women sanitary reformers with a powerful form of influence over the many industries serving the American home, from garment manufacturers to canned food producers. During the Progressive era, the private side of public health widened dramatically as hygienic concerns were translated into consumer preferences. The growing appeal of "selling the germ" testified to the transformative power of those hygienic anxieties when they were multiplied many times over.

· III ·

The Gospel in Practice, 1900–1930

7 · Antisepticonscious America

In the late 1890s, a small group of New York City women formed an association to promote the cause of hygienic dress reform, in particular the wearing of short skirts on wet days. Although not all members of this "Rainy Day Club" were brave enough to venture out in public wearing their club regalia, which consisted of an ankle-length skirt and high boots, those who did expressed great satisfaction at being liberated from a highly unsanitary mode of dress. "When I see women dragging around long skirts I can't help having an 'I am not such a one as thou' feeling," said one member, who was described in the *New York Times* as wearing a particularly attractive short-skirted outfit. Rainy day clubs soon formed in other American large cities to encourage this new style of women's dress.[1]

For some contemporaries, the crusade for the shorter hemline was almost as heretical as was the assault on the common communion cup. In a "plea for the long skirt" published in 1900 in *Harper's Bazaar,* an "old-fashioned woman" protested that "modern science" was turning women's fashion into "a grim demonstration of hygiene." She urged her fellow women not to "sacrifice the grace of our mothers' skirts" on the altar of asepsis. Perhaps doctors had "figured to a dot the number of deadly bacilli possible to be gathered to a square inch of a woman's train," but "what of woman's mission to be lovely?" she asked. Instead of wearing shorter skirts, the editorial suggested that women should simply be more careful about keeping them out of the dust.[2]

The battle of the hemlines was but one skirmish in a larger controversy over hygienic standards that escalated after 1900. Thanks to the tireless efforts of the antituberculosis and domestic science movements, growing numbers of American men and women were coming to fear more acutely the hazards of the germ. Exposed to the continual refrain that "little things were no trifles," they began to look at familiar habits with new eyes. The long skirt, the Old Testament beard, and the overstuffed parlor all became symbols of an unsanitary way of life that had to be resolutely left behind in the name of science. Although some observers like the "old-fashioned" woman in *Harper's Bazaar* regretted the passing of the old Victorian order, by the early twentieth century the forces of hygienic reform were clearly winning the battle.

This rising consciousness of the germ coincided with the emergence of a new style of cultural modernism. In the glossy mass-circulation periodicals of the day, Americans were urged to cast off their parents' old-fashioned ways and to embrace a more modern way of living. The gospel of germs offered a vision of hygienic modernism that was perfectly suited to this "cult of the new" and its "perfectionist project," as historians William Leach and Jackson Lears have respectively termed it. There was no better way to demonstrate an up-to-date outlook than by rejecting the unscientific and potentially deadly customs of the Victorian era. The cult of the new and the gospel of germs were but two facets of the same enthusiastic pursuit of modernity.[3]

Under the tutelage of the first apostles of the germ theory, the Victorian generation had come to think of sanitary correctness in terms of sewer traps, Germicides, and white china toilets. During the Progressive period, their children and grandchildren broadened their understanding of germ protections to defend against the additional dangers highlighted by the new bacteriology. As the conception of the "house disease" evolved to include dust and fomites, "flies, fingers, and food," the practice of domestic hygiene expanded to accommodate a host of new precautions against infectious disease. This widening of germ consciousness radiated outward from the individual and the home to many aspects of public and commercial life. In response to rising consumer anxieties about the microbe, manufacturers and service industries that provided "home-like" services, from paint manufacturers and food processors to hoteliers and Pullman car porters, began to incorporate the lessons of the laboratory into their philosophies of doing business. In the words of William W. Bauer, writing in the *American Mercury* magazine, all of America became more "antisepticonscious" during the early years of the twentieth century.[4]

Stripping Down

One of most striking features of this "cult of the new" was a marked preference for a stripped-down aesthetic. Victorian style was notorious for its delight in surface ornamentation and complexity, both in personal attire and home decoration. To the Progressive eye, these traits represented not merely defective taste, but also a reckless disregard for health because they attracted germs and dust. Among "antisepticonscious" Americans, modernity was manifest in a new urge toward minimalism in body decoration and house design.

For middle-class men, the modern look was symbolized by the closely shaven face. Influenced by reformers' warnings about germ-ridden facial hair, American men began to abandon the full beards and long moustaches that had been popular since the 1850s. Commenting on the "passing of the beard" in 1903, an editorial in *Harper's Weekly* noted, "Now that consumption is no longer consumption, but tuberculosis, and is not hereditary but infectious, . . . the theory of science is that the beard is infected with the germs of tuberculosis." As a result, the clean-shaven face became a visible symbol of masculine allegiance to the gospel of germs.[5]

This trend was strikingly illustrated in the visages of the NTA leadership. Of the forty-eight male officers pictured in its 1922 history, only five had full beards, all of them closely cropped; the rest had either clean faces or at most slim mustaches. The younger the man, the more likely he was to be completely smooth shaven. The preference for the hairless face was manifest not only among TB crusaders, but also among American men in general. The "revolt against the whisker," commented William Inglis in *Harper's Weekly* in 1907, "has run like wild-fire over the land." Photographs and advertisements from the early 1900s confirm the new look's growing popularity, especially among younger men. Riding the crest of this trend, the visionary King Gillette made a fortune by inventing a safety razor that gave men "the security from infection of shaving yourself," as its ad copy proclaimed.[6]

For middle- and upper-class women, the new stripped-down body aesthetic manifested itself in shorter and narrower skirtlines. Since the mid-1800s, dress reformers had unsuccessfully complained of the fatigue and discomfort involved in managing the full, trailing skirts sanctioned by both fashion and modesty. In the 1890s, as the new college-educated women began taking up sports, entertainments, and careers outside the home, the inconvenience of these skirts became even more acute. At just this point in time, bacteriologists handed dress reformers a powerful weapon by indicting the so-called septic skirt and its load of disease germs. As a woman physician wrote in 1894, the trailing hemline was a sanitary abomination that "picks up all sorts of evil things from the street and elsewhere, carrying them home to be distributed to all the family without their knowledge or consent." Similarly, in 1900, the illustrated New York magazine *Puck* featured a dramatic cartoon showing the germs of disease carried by a lady's cloak. The accompanying editorial condemned the "dirty fashion" of trailing skirts, noting that "the germs of influenza, consumption and typhoid fever are the least of the evils which mothers bring home to their defenceless children on their skirts." To those who insisted that beauty take precedence over health, the author concluded, in a classic statement of the new modernism, that "nothing unhygienic can be beautiful."[7]

The hygienic argument undoubtedly contributed to the steady rise in women's hemlines that started in the early 1900s. The change occurred first in sports and walking outfits, then in daytime dresses. Formal evening gowns continued to have long trains—presumably when one went to a ball, concerns about germs could be suspended—but in other women's garments the upward trend prevailed. By the early 1910s, women's skirts had risen to a safe, hygienic distance above the ankles and had become more narrow, which reduced the surface for dirt collection. Well before the "flapper generation" made the short skirt a symbol of new sexual freedom, the rising hemline signaled American women's revulsion against the septic skirt.[8]

The same impulse to streamline was even more dramatically displayed in the turn-of-the-century home. As with dress reform, earlier warnings against home decor and disease had little effect before the late 1890s. Although a few zealots such as Ellen Richards began stripping down their parlors in the 1880s, the preference for the "artistic house," with its complex surfaces, heavy upholstery, and exotic "Turkish" corners, remained strong until the last years of the nineteenth century. But at around the turn of the century, the hygienic criticism of the American home finally began to bear fruit. Architects and home designers promoted new looks—including the colonial revival, modernist, and arts and crafts styles—that eliminated dust lines and facilitated a more bacteriologically informed cleanliness.[9]

In place of the old overstuffed look, early twentieth-century home fashion featured lighter, more easily cleaned materials, including wicker, metals, and glass, that would have been rejected as too cold and sterile in the Victorian home. Likewise, popular house plans, such as the bungalow style, demonstrated a growing appreciation for smooth surfaces and clean lines. A comparison of Victorian and Progressive interiors reveals that the inside of the middle-class American home became noticeably more "germ proof" between 1890 and 1920. The white tiled bathroom and enameled kitchen; the living room furnished with parquet floor, area rugs, scanty curtains, painted, molding-free walls, and pared-down furniture; and the spare use of decorative items all paid tribute to the new sanitary standards.[10]

To be sure, hygienic concerns were not the sole reason for these striking changes in domestic architecture and interior decorating. The styles favored in the Progressive period also represented a conscious reassertion of Anglo-American tastes, a kind of cultural eugenics embraced in response to the "new immigration." As American society became much more ethnically and racially diverse, the Victorian preference for the exotic seemed less appealing. But the allure of the new decorating fashions was also intimately related to their sanitary correctness. The "purer" look of the middle-class home distanced the old Americans from the new on both aesthetic and hygienic grounds.[11]

Pitching the Germ

Within this broad shift in preference toward a stripped-down aesthetic, the marks of a growing consciousness of the germ can be clearly discerned in the advertising images of the period. In a highly competitive marketplace, manufacturers and advertising agencies searched diligently for the best merchandizing strategies to sell their wares. That they so often invoked the new and improved gospel of germs to woo consumers attests to how deeply the teachings of tuberculosis religion and the housewife's Mosaic Code had infiltrated the affluent classes by the early twentieth century.[12]

The germ sell changed in some important ways from 1880 to 1920. During the 1880s and 1890s, the hygiene pitch based on the first, sanitarian-dominated gospel of germs had been aimed primarily at middle-class men, as the marketing of the Germicide had made clear. Commercial invocations of the microbe focused chiefly on the virtues of disinfectants, sanitary toilets, and sewer gas preventives. By the early 1900s, manufacturers were drawing heavily on the second, more bacteriologically informed version of the gospel of germs to promote their goods. They also became much more attentive to women's concerns about the microbe, a shift that reflected the success of the domestic science movement in making women more hygiene conscious in their shopping.

These commercial messages were by no means limited only to affluent Americans. Although the cult of the new originated in the Anglo-American middle class, it did not remain that group's exclusive preserve. For example, as national manufacturers searched for new markets, they began to place advertisements for brand-name goods in immigrant newspapers. As a result, newcomers to the United States were quickly exposed to the same messages equating certain products and services with germ protection that were being disseminated to their native-born contemporaries. Conforming to the gospel of germs thus became an integral part of the "Americanization" process encouraged by these exposures to mass culture.[13]

The business community's rush to sell hygienic safety to "Mr. and Mrs. Consumer," particularly Mrs. Consumer, testifies to the success of public health crusades in awakening a greater germ consciousness. In turn, the reproduction of public health claims in advertising and commerce powerfully validated the gospel of germs. Every time an advertisement for a vacuum cleaner or a household disinfectant repeated the claims of "medical experts" about the dangers of dust, that advertising subtly reaffirmed the teachings of tuberculosis religion and the housewife's Mosaic Code.

Thus turn-of-the-century advertising became a powerful medium for diffusing images about the invisible world of microbes and for promoting the daily practices—and products—needed to keep them at bay. Given the considerable resources that private enterprise had at its command, these commercial renderings of the microbe had as much, if not more, influence in shaping consumer perceptions as did the more disinterested efforts of hygiene reformers. In a series entitled "A Course in Scientific Shopping," which appeared in *Good Housekeeping*, the advertising pioneer Earnest Elmo Calkins observed that articles on proper household hygiene appeared only sporadically in its pages, whereas advertising continually carried on the work of hygiene education "like the constant dropping of water." The sanitary message of advertising, he concluded, "has the persistency and effect of a well-preached sermon."[14]

Yet for all their appeals to science, manufacturers and advertisers did not promote the lessons of the laboratory for their own sake, but rather to sell products. When the two goals came into conflict, profit clearly won out. The persistence of advertising appeals that invoked the fear of sewer gas is a case in point. By the early 1900s, public health experts no longer believed that airborne bacteria from sewer and toilet represented a real threat. Yet because so many American consumers continued to believe that these bacteria were dangerous, manufacturers played to and sustained their anxieties about toilets for decades after public health experts had repudiated this holdover from the old gospel of germs.

For example, in 1915, when nineteen New Jersey toilet manufacturers formed the "Potteries Selling Company" to promote a new trademark for the "Sy-Clo" toilet, they emphasized the direct connection between toilets and disease. A promotional brochure for the Sy-Clo model featured a line drawing of the "typhoid microbe magnified one thousand times" and informed the reader that "these microbes get into houses through the pipes of imperfect closet bowls." Harkening back to the old arguments about sewer traps, they reassured customers that "it is absolutely impossible for [germs] to work their way through the water seal of a Sy-Clo closet."[15]

At the same time, plumbing manufacturers proved more than willing to update their advertising slogans with newer views of infection as well. A 1906 ad for the Sy-Clo equated the toilet's role in the home with "what disinfection means to the surgeon—what vaccination means to the public health." The public's preference for smooth, easily cleaned surfaces associated with asepsis meshed well with manufacturers' promotion of porcelain bathroom fixtures. "Your house is not modern without a bathroom equipped with

'Standard' Ware," read a 1905 advertisement; "It is sanitary because its snowy surface is non-porous without crack or crevice for dirt to lodge."[16]

Manufacturers of home water filters had less need to revise their marketing strategies in light of the new bacteriology, given science's verification of the link between impure water and disease. But their representations of the germ were not necessarily more accurate than those used by earlier advertisers. For example, the McConnell Filter Company provided a drawing of a magnified drop of water that pictured the germs of cholera, "diptheria" (misspelled in the original), and typhoid fever as fanciful insects and crustaceans. Still, the belief in waterborne pathogens was apparently so widely accepted by the 1890s that filter manufacturers could move on to other merchandising points, such as how attractively their various designs fit into modern home decor. The Pasteur-Chamberland Filter Company, named after the two prominent bacteriologists who designed the filter, obligingly provided sketches of the decorative pedestals available with its most popular models.[17]

For manufacturers of household disinfectants, the new bacteriology's emphasis on aseptic cleanliness was a tremendous boon, especially in its equation of germs with such ubiquitous household presences as dust and houseflies. "The bacteriologist's point of view is always a purely scientific and critical one," explained the makers of one portable fumigator. That perspective seemed to indicate that the best strategy was still to disinfect everywhere and often. At the turn of the century, the range of items that the commercial gospel of disinfection represented as requiring purification was remarkably diverse. In addition to the traditional sickroom uses, the makers of Sanitas recommended its virtues as a mouthwash, dandruff preventive, and moth destroyer; it was also touted as an all-purpose disinfectant for floors and walls, furniture, ash barrels, garbage cans, and even cabs and ferry boats. In equally global terms, the manufacturers of Pratt's "Germ-a-thol" asserted its virtues as both a personal and household disinfectant, one to be used not only in the traditional sanitarian danger points of cellar, drains, and toilet, but also in the newer bacteriological hotspots of ice chest, laundry room, and carpet. Germ-a-thol was even recommended as a good wash for horses and dogs, suggesting a heightened concern about the germ-bearing properties of domestic animals.[18]

Repeating the hygienists' warning that "mere" cleanliness was not enough, advertisers often used the same moralistic tone so often found in tuberculosis catechisms. The choice of a household soap was endowed with monumental consequences for the family's health. The C. N. West Company began its "Message to Good Housewives" with the warning that the house might look clean, "but don't get the idea that you can judge simply by the appearance of things." Cleaning with ordinary soap, the company claimed, left behind a scum that attracted dirt, and "wherever there is dirt, germs can breed; and flies and vermin will come."[19]

The revelations that skin, hair, and body cavities harbored millions of germs provided abundant material for heightening anxieties about bodily hygiene. The substances that doctors used to achieve a surgical level of cleanliness, such as corrosive sublimate, were too harsh and potentially toxic to be marketed to the general public. In their place, pharmaceutical companies promoted a number of less disagreeable but still powerful disinfectants. At a time when doctors still routinely wrote prescriptions for disinfectants, manufacturers sought to market products that were supposedly strong enough for a doctor's use while being so safe that even the housewife could wield their power with confidence.

One such substance was hydrogen dioxide, better known as peroxide. An 1897 brochure from the Oakland Chemical Company cited scientific heroes such as Louis Pasteur and Joseph Lister in touting peroxide's virtues as "a synonym for personal prophylaxis." Although it was not recommended for use on dogs or ferryboats, peroxide's household applications were still quite extensive; for example, one could add it to the baby's bathwater or use it to sterilize milk. Similarly, the makers of Listerine, a solution of boric acid, sought to promote its virtues as a mouthwash and a skin cleanser. An 1899 Listerine brochure assured purchasers that the wash was sufficiently strong yet gentle enough to be used "with freedom, by injection, lotion, or spray, in the natural cavities of the body."[20]

The identification of the microorganisms responsible for pimples and boils opened a particularly fertile line of appeal for antiseptic soaps. Using the same "imagine where the product comes from" ploy beloved of home economists, an 1899 advertisement for Hyomei Antiseptic Skin Soap explained that "ordinary soap" was from made from fats and grease collected by "street scavengers." Invoking the lessons of the laboratory, the ad continued, "It is claimed that the heat used destroys all the germs of disease; but the medical profession assert the contrary." Because "the thought of using such products is not a pleasant one," the ad urged the reader to patronize the germ-free Hyomei brand, which, in further testimony to its medicinal properties, was to be found only at drugstores.[21]

Dust-Proofing the Home

The dust theory of infection offered manufacturers another made-to-order domestic menace. Household dust was a ubiquitous problem whose solution beautifully illustrated the need to embrace the modern by making wise purchases of new "scientific" cleaning tools. As Mildred Maddocks announced in a 1917 article in *Good Housekeeping,* "'House cleaning' by the old broom and dustpan method, which meant the raising of a great dust with its attendant germs, is more and more coming to be considered no cleaning at all." The portable vacuum cleaner, which used suction to remove dust and dirt from carpets and upholstery, was represented as the only truly modern way to clean the house. After testing a variety of new models in 1919, Charles J. Clarke of the Good Housekeeping Institute declared, "Relatively the broom occupies to the cleaner the same position that the ox-cart does to the modern, high-powered automobile."[22]

More than any other new appliance of the era, the vacuum cleaner derived its hygienic legitimacy from the lessons of the laboratory. For example, the manufacturers of the "Duntley pneumatic cleaner" began their ad copy with the image of particles of dust under a microscope, where they were revealed as "a mass of hard, sharp edges or a soft disintegrating substance—in either case probably covered with germs." The ad stressed that only a vacuum cleaner could remove the "real dirt—the old, ground-in, dangerous, germ-laden dirt that other methods never touch." In yet another variation on the "whited sepulcher" theme, the Ideal Vacuum Cleaner stressed the need to remove "hidden dirt" in order to be really safe: "It is just as absurd to think that a house is clean because it gives no *visible* signs of dirt as it would be to think that a person must be clean because *he* gives no visible signs of dirt."[23]

Manufacturers of area rugs and floor polishes likewise tried to cash in on the association between carpets and germs. In a 1902 brochure titled the "Passing of the Carpet," the Sanitary Manufacturing Company exalted the virtues of their Brusellete Art rugs by assuring the housewife, "They will not hold or accumulate disease germs." Similarly, the makers of wood varnishes and floor polishes promised to render wood surfaces smooth and free of dust-catching, insect-harboring cracks. The Glidden Company struck a democratic note in promoting its colored wood varnish Jap-A-Lac, stating that "SANITARY FLOORS but recently were considered luxuries which only the well-to-do could afford," whereas now, "In the light of our growing knowledge of what sanitation prevents, they have become a necessary safeguard, and by virtue of JAP-A-LAC, possible in every home."[24]

The same preference for smooth, hard surfaces extended to walls. Advertisements for Sanitas, a washable wall covering, proclaimed, "The surface of SANITAS is smooth and non-porous; there is no place for germs to lodge, hence none of the dangers that lurk in old-fashioned wall coverings." Originally promoted for bathrooms and kitchens, the manufacturer of Sanitas soon broadened its uses to include all rooms in the house. Paint manufacturers developed their own germ-based selling points. For example, the Carbola Chemical Company of New York came up with the idea of a germicidal whitewash, or a "disinfectant that paints," and the Du Pont Company developed "Saniflat," an oil paint containing supposedly antiseptic ingredients. In a series of advertisements in the *Saturday Evening Post,* the Paint Manufacturers' Association promoted the idea of annual paintings as a means to disinfect walls. "Health has few allies, disease few enemies more powerful than the paint brush," the association proclaimed.[25]

The assumption that a smooth, impervious surface was resistant to germ and insect infestation reinforced the allure of enameled porcelain, tile, and linoleum in those areas of the house most closely associated with germs—the bathroom, kitchen, and laundry. As a 1906 advertisement for the Standard Sanitary Manufacturing Company put it, "Every room in the House comes under the sanitary influence of 'Standard' Porcelain Enameled Ware." Catalogues and display ads for kitchen and bathroom furnishings attest to the vogue for a white, hospital-style look.[26]

At the center of the new sanitary kitchen was the refrigerator, an appliance second only to the vacuum cleaner in the hygienic claims made for its importance to the modern home. The Bohn Syphon Refrigerator Company announced boldly, "INFANT MORTALITY would be

GREATLY REDUCED" by the purchase of its product, and explained that its "white enameled lining" was easily wiped clean, ensuring "no typhoid germs." In a nod to continued consumer anxieties about foul smells, companies also emphasized that their iceboxes gave off no unpleasant odors.[27]

Playing up the hazards of "flies, fingers, and food," manufacturers slapped the word "sanitary" on every conceivable kitchen item from roasting pans to dish drainers: the sanitary bread maker, for example, assured that "the hands do not touch the dough," and the sanitary garbage can incorporated a fly trap. The Sanitary Crystal Glass Ice Cream Freezer was represented as "absolutely sanitary and ptomaine-proof." The Wilmot Castle Company claimed that "Baby's Health and Baby's Comfort Depend upon the use of the Arnold Pasteurizer." The Paragon Manufacturing Company's "Sanitary Washer" promised "No shrinking, smell, or germs, But Sanitary, Clean and Safe."[28]

Makers of window screens and fly repellents had a particularly good disease angle to work. An advertisement for the cleverly named "Tanglefoot" fly paper pictured a mother with child in arms, delivering a curse that would have gladdened Martha Van Rensselaer's heart: "Wretched pest, you have probably come direct from some hospital, garbage pail, or stable, laden with filth and possibly disease germs." In almost sadistic terms, the ad envisioned the fly's encounter with the paper, which "will catch and hold and cover you all over, and the germs and dirt that you are carrying, with a varnish from which you can ever escape to trouble me *either living or dead.*"[29]

Sanitary Packaging

Dramatic changes in food packaging between 1890 and 1920 further attest to hygiene reformers' success in cultivating anxiety about germs' menace to the national food supply. That more and more American consumers, particularly women, really believed that food-stuffs could bring illness and death into their homes was evidenced by their willingness to pay higher prices for special sanitary packaging. In a highly competitive marketplace, food manufacturers found such added assurances of their products' "purity" well worth the effort.

The increasing use of brand names, trademarks, and sanitary packaging at the turn of the century were all part of a general effort to reassure housewives that their food was safe from both chemical and bacterial contamination. Claiming to observe higher sanitary standards in their factories than did their bulk-goods competitors, national manufacturers employed a variety of packaging materials, including cardboard boxes (also referred to as "sanitary boxes"), waxed paper, glass bottles, glassine, tin cans, and tinfoil, to guarantee the cleanliness of their goods after they had left the factory. Using adjectives such as "pure," "sanitary," "air-tight," and "dust-proof," their advertisements suggested that for a few cents more, housewives could purchase the certainty of a product uncontaminated by dust, flies, or human hands.[30]

In pressing the merits of sanitary packaged goods, manufacturers often conjured up the bad old days of food processing and transport. As one commentator put it, modern science had brought an end to the "'crackerbarrel era' of merchandizing," with its unhygienic ways. The new ideal was a product unsullied by human contact from the beginning to the end of the manufacturing process. As the makers of Gold Medal Flour assured the consumer, their product was "absolutely pure and clean . . . the hands of the miller have not come in contact with the food at any stage in its manufacture."[31]

With the image of the "consumptive baker and candymaker" to combat, bread and candy manufacturers were among the earliest food manufacturers to offer such assurances. Instead of selling bread and cakes from open pans, bakers began wrapping them in oiled or waxed paper, which had the added benefit of keeping the product moist. Although such wrappings represented a significant expense, bakers found that the increased sales more than offset it. Candy manufacturers also found that sanitary packaging helped sales, especially of the more expensive confections. As Hildreth's Velvet Candy bragged, it "is now put up in Triple Sealed Packages. Moisture, Germ and Dust Proof." After investigations of candy manufacturers revealed that workers often used their fingers to dip chocolates, companies also started to promote "fork-dipped" candies that were untouched by human hands.[32]

To offset milk's long association with disease, commercial dairies experimented with new forms of sanitary packaging, such as glass bottles, which were introduced in the late 1880s. At around the same time, Pet Milk began marketing canned condensed milk "made scientifically clean by sterilization." But the cost of both the processing and distribution of bacteria-free milk remained extremely high, making "pure" milk one of the most expensive items on the germ-conscious housewife's shopping list. Not until World War I did the glass bottle finally supplant the milk can and dipper in most parts of the United States.[33]

The canning industry tried various ways to counter fears that its products could harbor disease germs. In 1910, the National Canners Association commissioned the popular domestic author Marion Harland to write the "Story of Canning" in order to help convince women that canned goods were as safe and "dainty" as homemade food. Harland admitted that years before she herself had been an opponent of "tinned foods," which to her mind were synonymous with both cheapness and disease. But after she had visited some modern plants and learned the "methods of the conscientious Canner," Harland recanted her former opinions. The lengths to which manufacturers went to safeguard their product "sound to us like fairy tales and exaggerations ludicrously incredible," Harland wrote. Yet she had seen how even items as small as peas were washed thrice over and then processed without the touch of the human hand, except "when they pass along an endless white rubber belt where the cleanest young women pick out a broken pod or discolored pea, should there happen to be one."[34]

Recognizing that under the dictates of the new bacteriology home canning had grown so complex that the housewife was more easily tempted to let someone else do it, commercial canneries represented their plants as giant laboratories in which clean workers wearing spotless white uniforms labored as fastidiously as bacteriologists. Promoting its canned meat products, the Libby Company underlined that its "careful methods of cooking and packing are as cleanly as in your own Kitchen." By stressing their adherence to the lessons of the laboratory, food manufacturers sought to reassure housewives that she could safely leave the worry of canning to them; there would be no ptomaines in the green beans.[35]

Slowly but surely, the same markers of cleanliness came to be expected in all aspects of food processing and selling. The tradition of the white coat and aseptic technique came to characterize the "modern" supermarket and butcher's shop as well. The more that women shoppers looked for these emblems of cleanliness, the more pressure the grocer and the butcher felt to observe the tenets of household bacteriology themselves. "DO YOU WANT INCREASED SALES?" asked C. V. Hill and Company in an advertisement for its refrigerator display case, which appeared in a trade journal for butchers; if so, the ad warned, shop owners needed to display their wares in "modern sanitary show cases away from flies and dust and where they cannot be handled by the public." Similarly, the Sanitary Scale Company advertised its product with a sketch of the germs found on a meat scale, paired with a drawing of a woman consumer telling her butcher, "I do like porcelain, it looks so clean."[36]

Homes Away from Home

As hygiene reformers continually emphasized, a properly developed consciousness of disease did not stop at the door of one's own home. As soon as they had been schooled in the hygienic catechism that spit carried death and dust spelled doom, those imbued with the gospel of germs took that awareness with them wherever they journeyed in the outside world. As a result, the domestic disciplines so central to tuberculosis religion and the housewife's Mosaic Code gradually radiated outward into the public worlds of travel and entertainment, where affluent Americans went to enjoy "homelike" comforts away from their own households.

Middle-class women's growing participation in activities outside the home was an important element in this process. Beginning in the late 1800s, more and more women ventured out of the cloistered Victorian world of parlor entertainments and private dinner parties to pursue new interests and pleasures in the Progressive-era city. The domestic scientists' emphasis on the "home-to-world" theme was expressly intended to convince these women to take their finely tuned sensitivity to cleanliness into the public sphere. Those sanitary sins that a microbe-conscious homemaker would not tolerate in her own parlor and kitchen she would force the hotelier and the restauranteur to abjure as well.

Although women were often portrayed as the more hygiene-conscious sex, male converts to the gospel of germs also formed an important lobby for sanitary uplift in the Progressive period. Beginning in the late 1800s, large corporations began to employ growing numbers of middle-management executives and salesmen, many of whom developed a self-consciously modernist outlook. The new-style corporate salesmen were particularly keen on distinguishing themselves from their uncouth, tobacco-spitting predecessors. The anti-TB societies worked hard, and apparently quite successfully, to court these new white-collar male workers by portraying germ consciousness as simultaneously manly and progressive. Thus the modern businessman became a powerful force for higher standards of cleanliness as he ate out on working lunches and traveled on business.[37]

Wherever they went, these microbe-minded ladies and gentlemen took their sanitary preoccupations with them, and those who wanted their patronage faced the same pressures to improve hygiene standards as did the baker, the butcher, and the grocer. To lure patrons into public spaces where disease germs might lurk, service providers of all sorts had to assure customers that their hygienic safety was being looked after as carefully as it was at home. Thus as soon as any emblem of domestic safety became widely accepted, whether it be the white china toilet or the individual drinking glass, keepers of these "homes away from home" rapidly followed suit.

The ripple effect from domestic to public spaces appeared first and most intensely in services patronized by upper- and middle-class urban dwellers, yet it exerted a powerful downward influence as well. Industries devoted to mass transportation, lodging, and dining employed great numbers of working-class men and women who had also to be indoctrinated in the new hygienic disciplines. In service to antisepticonscious patrons, the Irish maid, the Italian waiter, and the black Pullman-car porter learned to observe the gospel of germs carefully or risk being fired.[38]

The transformation of three institutions—the hotel, the Pullman car, and the restaurant—illustrate the new sanitary consciousness that affected many service industries at the turn of the century. Not surprisingly, hotels were among the first commercial institutions to show the marks of the growing concern about infectious disease. Hotels presented all the problems of domestic disease prevention writ large; night after night, an endless stream of travelers slept, bathed, excreted, and even fornicated in hotel rooms, all of which presented formidable problems of waste disposal and purification. Due to the efforts of hygiene reformers, travelers were increasingly aware that the risks of some former occupant having tuberculosis or syphilis were uncomfortably high. Thus hotels had an enormous stake in convincing patrons that they observed stringent rules of cleaning rooms between each use. These concerns arose not only in hotels catering to business travelers, but also at resorts where many consumptives traveled to regain their health.[39]

Hoteliers first had to deal with persistent anxieties about domestic plumbing and the sewer gas "menace." From the 1870s onward, fine hotels continually remodeled their plumbing to meet the higher standards being demanded by hygiene reformers. To allay fears of sewer gas, many removed the in-room washbasins installed for the convenience of the previous generation of guests and purchased disinfectant devices such as the Germicide to attach to their toilets. Hotels also became an early, strong market for water filtration systems. The pressure for sanitary improvement only increased in the 1890s and early 1900s, as the antituberculosis crusaders portrayed hotels as hotbeds of consumption. The Pennsylvania Society for the Prevention of Tuberculosis directed one of its first tracts specifically to "Hotel Keepers." In the early 1900s, hotels became a major market for such hygienic innovations as vacuum cleaners, sanitary dishwashers, washing machines, and refrigerators.[40]

To reassure the traveling public that they were on guard against germs, the better class of hotels developed a number of cleanliness rituals, many of which persist to this day. To begin with, they strove to project an air of scrupulous yet subtle cleanliness. The authors of a regular column titled the "Practical Hotel Housekeeper," which appeared in the *Hotel Monthly,* noted that "Soap and sapolio are, after food and light, two of the most important and necessary articles in constant use in hotels." At the same time, they cautioned that disinfection must be practiced carefully. Although patrons wanted the assurance that a thorough cleaning had removed any trace of contagion, the mere suggestion, such as a harsh chemical odor, that a hotel room had recently been fumigated might alarm the nervous, particularly female, traveler. To this end, the authors warned the hotel housekeeper to be "very particular" about avoiding the use of "any kind of loud smelling disinfectant in sanitary bowls or private rooms—especially chloride of lime—in any room that is occupied." Instead, they recommended that the maid wait until the room was vacated and then clean it thoroughly with soap and lye, followed by the disinfectant copperas.[41]

Hoteliers sought to convey an ambience of hygienic correctness in other subtle ways. The *Hotel Monthly* observed in 1900 that many "first class hotels" had taken to furnishing a new cake of soap in its original wrapper for each new guest. "It may seem a small matter, but soap is one of those things that, like a tooth brush or a hair comb, a person prefers to use his own," the article noted. In lobby washrooms, hotel keepers installed individual liquid soap dispensers and sanitary roller towels in place of the old common soap dish and towel. The demand for roller towels, soap dispensers, and room cleansers resulted in whole specialized lines of sanitary goods developed just for hotels and similar establishments.[42]

The care of mattresses and bed linen was another important area of hygienic hotel practice. Hotels had to contend with a deep-seated suspicion of pillows and mattresses as contagion bearers, particularly because their beds sheltered so many strange bodies. The *Hotel Monthly's* advertising section carried ads for mattresses covered in "antiseptic pure white cotton felt" and built to be "self-ventilating." Between each use of the room, maids were instructed to air the mattress and change the sheets, pillowcases, and linens, which were then laundered so as to destroy germs.[43]

TB tracts warned that the heavy wool blankets used on hotel beds posed a special problem because they could not be easily washed between guests. To reassure their customers, many hotels began to use extra long sheets folded down over the blanket, a practice that evidently originated in western states, where many consumptives went to recuperate. A correspondent to the *New York Times* in 1910 noted that eastern newspapermen had made considerable fun of the western "long sheet movement" without understanding its hygienic rationale. "It is not a plan to keep cold feet warm, or to enable the timid to hide their heads," he explained. "The extra length of sheet is intended to be turned down over the other bed covering so that when the sheet is washed the possible exhalation of germs from tuberculous persons will be destroyed, where otherwise they would remain on the outside quilts indefinitely, a menace to health."[44]

The law made headway in the East, largely due to the pressure of businessmen's associations. In 1912, the Commercial Travelers Association of New York convinced a state assemblyman from Rochester to introduce "the nine foot sheet law" in the state legislature. As reported in the *New York Times,* the bill "provides at some detail how the upper sheet in a hotel bed must be folded forward at the head 'so that the inhalation by the occupant of bacteria &c, may be prevented and minimized.'" The legislature passed the bill in 1913, although it decided that an eight-foot sheet was sufficient protection against the germ menace.[45]

The hygienic concerns transforming the hotel industry also affected the railroads between the 1880s and the 1910s, particularly the design and maintenance of the Pullman sleeper car. A fixture of railroad travel in the late 1860s, Pullman had a virtual monopoly on sleeping car production as of the mid-1910s; in 1916 alone, 22 million passengers availed themselves of a night's sleep in one of the company's cars. Pullman's celebrated "palace cars" attracted the ire of hygiene reformers for precisely the reasons that made them seem so comfortable, namely their thick carpets, plush upholstery, and elaborate drapes. The sleeping cars on lines serving well-known resort areas for tuberculosis sufferers, such as the Smoky Mountains, the Adirondacks, and the Rockies, were particularly suspect.[46]

To solve its hygienic image problems, in 1905 the Pullman Company hired a sanitary expert, Thomas Crowder of Chicago, to improve the car's design. The wood undercarriage and flooring were replaced with steel and concrete, materials less hospitable to germ life. As Joseph Husband wrote in his 1917 book *The Story of the Pullman Car,* "Instead of insanitary woodwork, the smooth surfaces of steel which form the interior of the car offer no lurking place for germs, and soap and water at frequent and regular intervals maintain a high degree of cleanliness." The Pullman Company also adopted rigorous cleaning and fumigating rituals modeled on those used by public health departments after cases of infectious disease. Sleepers were scrubbed down after every run, and the bedding "hung in the open air for the action of that greatest of all purifiers, the sun," Husband explained. "The slight, acrid odor sometimes noticeable in a Pullman car at the beginning of a run is caused by the disinfectants which are liberally employed," he informed his readers.[47]

The Pullman Company made sure the public knew that its conductors and porters had been carefully trained to observe hygienic niceties. In a 1914 article on "sleeping cars and microbes," Edward Hungerford reported that "every Pullman conductor is under strict instruction never to re-sell a berth which has been occupied, for even a small part of the journey, by a consumptive." He also noted that the company immediately took any car discovered to have carried a contagious passenger out of service and had it fumigated by the local board of health. Last but not least, Hungerford observed that "the Pullman company exercises a remarkable discipline over its negro porters," each of whom received a portable vacuum cleaner and a jug of "a disinfectant against emergencies" as part of their basic equipment. Their training "strictly forbids the use of feather dusters or of dry sweeping while the car is underway," or the use of their hands to put ice in the car's cold-water dispensers.[48]

The new hygienic emphasis on "flies, fingers, and food," as well as the growing understanding of microbial food poisoning, wrought no less of a transformation on the restaurant industry. Dining out became increasingly popular between the end of the Civil War in 1865 and the enactment of Prohibition in 1920, attracting many women as well as men customers. As Progressive-era diners became more germ conscious, restaurants felt increasing pressure to guarantee the safety of their meals, for one serious outbreak of food poisoning could spell disaster. To assuage patrons' anxieties about cleanliness, the better eating establishments invested heavily in the accoutrements of sanitary dining, including sufficient linen to change tablecloths and napkins between each diner and sanitary dishwashers to sterilize china, glassware, and silverware.[49]

Because even the most innocent slip could cause a diner to become apprehensive, waiters and waitresses were drilled in the niceties of sanitary presentation, starting with careful attention to personal cleanliness. For male waiters, this requirement often meant getting rid of their facial hair. The *Hotel Monthly* noted approvingly in 1901 that one Fifth Avenue hotel had ordered all of its male waiters to be clean shaven. Applauding the disappearance of the bearded waiter, an editorialist observed in a 1906 *North American Review* that aesthetics aside, "The modern germ theory alone probably would suffice to deprive him of that privilege."[50]

For the same reasons, food servers had to take special care to preserve the sanitary order of the dinner table. In their column "The Practical Hotel Housekeeper," the authors warned that drinking glasses should always be dried with a clean, dry towel, for a wet one would leave a "dull, dirty, cloudy look" and "any guest will object to using that glass and he or she has a perfect right to do so." John B. Goins, who ran a school for black waiters, provided careful instructions on refilling water glasses, which he warned should always be done at the table. "Should a waiter remove several glasses at the same time from the table he is quite likely to mix them up in returning them, the mere suggestion of which is very distasteful to the guest," he explained. The Chicago hotel and restaurant impresario Albert Pick was so repelled by the idea of waiters leaving their fingerprints on the part of the utensil that went in the mouth that he invented a special "Handle-Rite Silver" to discourage such carelessness.[51]

The growing awareness of the healthy carrier problem posed a special challenge to the food service industry. The first verified case of a healthy carrier in the United States was a cook named Mary Mallon, who was arrested in 1907 after health authorities determined that she had infected people in seven households where she was employed. "Typhoid Mary," as she came to be called, was released after agreeing not to seek employment as a cook. It was discovered in 1915, however, that she had been cooking in a maternity hospital where a small typhoid epidemic had broken out. For this infraction, Mallon spent the rest of her life in confinement.[52]

The publicity attracted by Mallon's case increased the scrutiny of all food handlers. In 1910, a New York hotel announced that it was voluntarily conducting physical examinations of its staff and that "those who are adjudged likely to spread infection of any sort will be removed." In 1916, New York City amended its sanitary code to require bacteriological examinations of all food handlers and specified that no one could work in a restaurant without a certificate testifying to their good health. Other cities adopted similar codes in the 1910s and 1920s in the wake of growing concern about healthy carriers in the food industry.[53]

Toward a New Public Hygiene

As the transformation of both domestic spaces and commercial "home-like" services makes evident, consumer preferences for cleanliness made the market a powerful force for sanitary reform in the early twentieth century. Manufacturers and service providers of all sorts gave hygiene-conscious Americans what they were willing to pay for—namely, assurances of greater protection from germs. This process was by no means straightforward or even honest; companies were willing to use doubtful health claims to promote their products, and to claim to be models of antiseptic cleanliness, without necessarily living up to established sanitary standards. Likewise, hoteliers and restauranteurs aimed as much for a *show* of cleanliness as for its actual implementation; although they kept up appearances in the hotel room and the dining room, many forced their employees to live and work under intolerable hygienic conditions. In short, a great deal of chicanery and pretense accompanied American businesses' claims to be safeguarding consumers against germs.[54]

Still, no matter how uneven and duplicitous the results, the hygienic fears and sanitary expectations of affluent consumers did discipline the American marketplace between 1890 and 1920. These sanitary transformations occurred first and most completely at the high end of the market, which served those patrons most willing and able to pay the price of greater protection. The costs of cleanliness were passed on directly to the consumers—for example, in the higher prices paid for packaged goods and the substantial bills received at the end of a hotel stay or restaurant meal.

This process worked much more unevenly at the lower end of the market, where there were fewer consumer demands for cleanliness and less profit motive for providers to meet these demands. Small grocers and food producers could ill afford to make the changes necessary to guarantee the germ-free status of their products. The owners of the inexpensive boarding houses, bars, and restaurants patronized by working-class Americans lacked the resources to change sheets or tablecloths between users. Not surprisingly, these small tradesmen resented and resisted hygiene reformers' efforts to impose a new and expensive sanitary regimen on their businesses.[55]

Recognizing the limits of marketplace forces, public health departments and other governmental agencies increasingly sought to establish a sanitary bottom line through regulatory laws and inspections. For example, cities and states began to pass laws regulating the hygienic standards in establishments that offered food or sleeping accommodations to the public. Health departments passed codes for sanitary food handling and hired inspectors to make sure that all restaurants followed them. In the name of public health, hotels were gradually required to change the sheets between each patron's use of a room. Enforcing such regulations was difficult (and remains so today), but these codes did establish a minimum of sanitary services that must be provided to protect the health of consumers, no matter how poor.[56]

This interplay between private, for-profit sanitary reform and state-sponsored, compulsory initiatives was a central dynamic of the Progressive-era public health movement. It was no coincidence that the peak of popular germ consciousness among middle-class voters and taxpayers coincided with the "golden age" of American public health. During precisely the same decades that commercial invocations of the germ theory reached their greatest intensity, public health authorities enjoyed a rapid expansion of their policing powers. Private and public interest in disease prevention converged to discipline a remarkable number of institutions during the Progressive period.

Yet negotiating the boundaries between private and public hygiene was fraught with difficulties concerning issues of access, responsibility, and cost. A case in point was the elimination of the common drinking cup. Although some commercial enterprises such as restaurants and private offices absorbed with little complaint the cost of providing extra glasses or paper cups, others, such as the railroads, balked at the added expense. The cup problem was especially problematic in large cities. Traditionally, towns and cities had provided free drinking water in the form of public fountains or wells, accompanied by a cheap dipper or cup for communal use. But hygiene reformers, who regarded these cups as dangerous contagion-bearers, demanded that they be removed. Their resistance raised the thorny question of who would pay the costs of providing a sanitary alternative.

The easiest solution was to make the drinker absorb the costs of the individual cup. Indeed, many urbanites began carrying collapsible metal cups for themselves and their children so that they could avoid going without a drink of water. Companies and government agencies also began to install cup-vending devices, which sold paper cups for a penny each, in waiting rooms, municipal buildings, and other public spaces. But as the hygiene reformers realized, even the cost of a penny a cup was out of the reach of many people's means, which meant that they were tempted to use the common cup instead.[57]

Despite its loathsome hygienic image, the common cup continued to find defenders. In 1912, an editorial in *Railway and Locomotive Engineering* lamented the passing of what it saw as a humane and democratic custom, complaining, "The cranks whose senseless agitation has eliminated the public drinking cup, even in Pullman cars, have inflicted much discomfort upon ordinary people and have largely increased the business of saloon keepers." Citing the most pitiful victims as "the poor immigrant children who have need to make long journeys in hot dusty cars," the author concluded, "When this pernicious movement against free water drinking has passed, people will wonder at the blindness which enabled cranks and saloon keepers to spread so much discomfort among honest people."[58]

As the reference to saloon keepers suggests, the crusade against the common cup was complicated by the imperatives of the temperance movement, whose influence in this period easily matched that of the antituberculosis societies. Temperance advocates worried that the demise of the common cup would lead more men to patronize saloons for a drink. On these grounds, groups such as the Women's Christian Temperance Union, the Young Men's and Young Women's Christian Associations, and local charity groups began to install public drinking fountains. But as a *Survey* article on the cup controversy noted wryly, these "charity fountains" often required the patron to pay a penny for the paper cup needed to use them.[59]

The sanitary drinking fountain emerged as an attractive, economical solution to the cup problem. The so-called bubble fountains, which were introduced in 1908, allowed the user to drink without using any cup at all, and they were immediately hailed as an ideal way to provide a safe water supply in public settings. With sanitary fountains available as an alternative, many cities and states began to pass legislation outlawing the common cup. Kansas passed the first such law in 1909, and by 1912, twenty-six states and the territory of Hawaii had passed similar laws limiting or forbidding its use. At the federal level, the U.S. Public Health Service's Interstate Quarantine Regulations and the Standard Railway Sanitary Code also mandated the removal of the common cup.[60]

Although city governments accepted the idea that providing sanitary water fountains was a civic duty, many were slow to do so. New York, widely acknowledged as one of the most progressive cities in the country when it came to public health matters, still had only thirty-five "bubblers" as of 1912, and of these, fifteen were out of service. Two hundred old-style fountains continued in use, but the common cup had been removed, leaving thirsty people to get a drink only by using their hands or "turning themselves 'wrong side up' to drink from the faucet," as the *Survey* reported. The same article noted that the fountains with "Cent-a-Cup" dispensers were doing a brisk business because free fountains in places such as City Hall or Battery Park were either nonexistent or broken.[61]

Many urbanites resented paying for what once was free, and they sometimes expressed their hostilities by breaking into the cup-vending machines. The New York *Independent* reported one such incident during Woodrow Wilson's inauguration in 1912, when a crowd of soldiers waiting at Union Station "grew impatient at the delay in securing the individual drinking cups provided by the vending machines, and perhaps a little vindictive over the necessity for paying each a penny for one, smashed the glass covers of the apparatus with their rifle butts and distributed the cups." The article implied that this sort of lawlessness was to be expected when the public's right to a free drink of water had been abridged.[62]

To make matters worse, studies conducted in 1914 and 1915 showed that the supposedly sanitary bubbler fountains posed a new health threat. The bubbling action allowed the users' mouth bacteria to fall back into the basin and be recycled for the next drinker. In addition, many patrons did not understand how to use the fountains and put either their fingers or mouths on the water spout, further contaminating it. As a consequence, "bubblers" had been implicated in spreading several epidemics of streptococcal tonsillitis. Cities were predictably slow to replace them with angled-jet models that eliminated these dangers. Well into the 1920s, the vast majority of public fountains available only *appeared* to be sanitary. As one investigator put it, "We discover that instead of being a protection, our supposedly sanitary substitute for the common drinking cup is distinctly a menace to public health fully as great as the cup itself."[63]

The chief beneficiaries of the great cup crusade turned out to be paper cup manufacturers. If the sanitary fountain was not really sanitary, the only true protection lay in using a clean paper cup. Railroad companies gave in and started providing the cups for free in the 1910s, after they discovered that when riders were asked to pay for them they simply retrieved used ones from the trash. By the 1920s, progressive-minded managers dispensed free paper cups at work as a matter of course. Referring to the "condemned cup," a 1919 advertisement for the Royal Cup Company put it bluntly, "Both the criminal law and the moral law demand that you use clean sanitary *individual cups* in your office and factory *all year round.*"[64]

The controversy over the common cup suggests that as frail as consumer pressure turned out to be, it forced the changes required by the gospel of germs far more quickly and efficiently than did appeals to business or government to act on behalf of a more disinterested notion of public health. Put simply, people with the pennies to pay were far more likely to get clean cups for drinking, as well as all the other accoutrements of sanitary living, than were their impecunious counterparts. Individual forms of protection continued to provide greater safety than did reliance on the slower, more uneven interventions of the state.

For many American workers, the revolution in behavior associated with the gospel of germs came in the form of an authoritarian regime imposed by either employers or public health officials. In factories and behind shop counters, while cleaning hotel bathrooms or disinfecting Pullman cars, the vast majority of poor, immigrant, and nonwhite Americans learned their lessons about hygiene involuntarily, when forced to serve their upper-class "betters." Moreover, although business leaders showed particular sensitivity about hygiene in a few select industries, such as commercial canneries or elite hotels, most turn-of-the-century employers paid scant attention to their workers' health; they simply sought to pay the lowest wages for the longest hours. When those workers went home, they experienced deplorable sanitary conditions in tenement houses owned by urban landlords who were determined to get the maximum rent for the minimum of services.

Only affluent America, then, was really under the sway of a new "antisepticonsciousness" during the first two decades of the twentieth century. Yet popular beliefs about how contagion moved from the hovels of the poor to the mansions of the affluent made the uneven spread of hygienic enlightenment a question of increasing importance. The precepts of tuberculosis religion and the housewife's Mosaic Code worked in surprising ways to widen imperatives to improve the homes and working conditions of the poor. Beyond the facile claims of vacuum cleaner and disinfectant advertisements, some reformers tried to turn germ protection into a more expansive weapon in the crusade for sanitary uplift via the "chain of disease" concept. In the process, they opened up some unexpected common ground for those designated as particularly dangerous to the public health to speak on their own behalf.

8 · The Wages of Dirt Were Death

In 1912, Marion Harland, the popular domestic writer, breezily asserted in *Good Housekeeping* that "the least literate of housemothers has heard of germs." By way of proof, she cited her "excellent chambermaid," who upon seeing her mistress take camphor to ward off the influenza, said, "It will kill bacilli, I suppose?" As Harland observed, the germ had become a household word for many Americans by the 1910s. Yet this popular awareness was limited primarily to affluent, native-born families and those who tended most closely to their personal needs. For the millions of families living in poor urban and rural districts, the microbe remained an unknown quantity. At a time when the "antisepticonscious" middle class was purchasing Sy-Clo toilets and buying vacuum cleaners, many Americans still lived in households lacking running water and flush toilets. By the eve of World War I, the awareness of germ dangers and the practice of antiseptic cleanliness had come sharply to differentiate rich from poor, literate from illiterate, native from foreign-born, and urbanite from rural-dweller.[1]

Changing patterns of disease incidence, particularly of tuberculosis, made these differences in sanitary protection especially troubling. When the antituberculosis crusade began in the 1890s, tuberculosis was still a disease that seemed to afflict the "millionaire as well as shillingaire," in Harriette Plunkett's colorful phrase. Many of the movement's early leaders, including Lawrence Flick and Edward Trudeau, were themselves consumptives, and their cause found considerable sympathy among middle-and upper-class families who knew firsthand the ravages of the disease.[2]

But by the 1910s, the epidemiological havoc wrought by tuberculosis and other communicable diseases had become much more class and race specific. From the late 1800s onward, the ailments that the gospel of germs sought to prevent increasingly became identified with the poor and sanitarily disadvantaged. Tuberculosis slowly lost its identity as a "house disease" to be confronted by every housewife and metamorphized into a "*tenement* house disease." As death rates from typhoid declined, the disease became more closely identified with the unscreened outdoor privies frequently found in poor neighborhoods and with working-class cooks and servants such as Typhoid Mary.[3]

The narrowing scope of the disease threat by no means diminished reformers' zeal to spread the gospel of germs. On the contrary, the perception that the sanitary measures adopted by the urban upper classes had contributed to their improving health only increased the urgency of taking the same gospel to the hygienic "heathen." The most intense years of public health education among poor Americans came in the 1910s and 1920s, after the value of sanitary uplift had seemingly been proven by the experience of their affluent contemporaries. During those decades, a wide range of private and public initiatives sought to democratize the hygienic teachings that had already transformed the homes of middle-class urban Americans.[4]

Although this work was strongly supported by the male public health establishment, the gospel of germs was preached primarily by female professionals. During the first three decades of the century, countless women visiting nurses, social workers, and home economists went as sanitary missionaries to working-class and farm households. For them, the democratization of the gospel of germs was an unambiguous good: not only did it help save lives, but it also opened up new vistas of influence and independence for their own professions.[5]

For the women they sought to help, however, the preventive lessons linking "dirt, disease, and death" were not always so comforting. The advice about preventing "house diseases" came intertwined with ethnic and racial assumptions that they did not share, and it was exceedingly difficult to follow in their desperately poor housing. At a time when death rates, especially of infants and small children, remained cruelly high in poor neighborhoods, the "helping" professions created a greater sense of women's responsibility for stopping the spread of germ diseases without necessarily supplying the resources needed to do so.[6]

Models of Cleanliness

By the early 1900s, large American cities were so sharply segregated by class and race that the discrepancies between neighborhoods in both mortality rates and sanitary conditions were stark and alarming. Although middle- and working-class reformers agitated to extend to poor neighborhoods municipal boons such as filtered water, adequate sewer connections, and garbage collection, the process was slow, largely because poor residents paid fewer taxes and had less political clout than did their affluent counterparts. Public health departments tried to use their growing regulatory power, in the form of housing and plumbing codes, to bring pressure to bear on urban landlords and homeowners. But the new, stricter sanitary codes usually exempted older buildings, and cities rarely had enough inspectors to enforce the requirements among the newer structures.[7]

So although municipal authorities were beginning to force the sanitary reconstruction of poor neighborhoods, the results were decidedly slow and uneven. A 1918 survey of urban housing in the United States found that nearly one-fifth of all apartments and two-fifths of all houses had outside privies. In Pittsburgh, one of the most class-segregated cities in the country, the filtered water supply did not extend to all neighborhoods until 1909, and the municipal sewer system did not include the poor, heavily populated South Side until 1914. In the very poorest areas of the city, including those where most African Americans lived, outdoor privies were still in use in the 1950s.[8]

In light of such persistent sanitary inequities, the private side of public health continued to figure importantly in work with the urban poor. Hygiene reformers fervently believed that by teaching the rudiments of household disease prevention to immigrant mothers and their more Americanized daughters they could reduce disease and death rates in spite of nonexistent or unreliable public services. And in comparison to their sanitarian predecessors, they had many more opportunities to take that preventive message into the home.

By the early 1900s, municipal public health departments had considerable authority to impose compulsory measures such as quarantines and fumigations in homes where a case of typhoid, smallpox, diphtheria, or scarlet fever was reported. Many poorer families probably learned their first lessons about germs when someone in the family fell ill of a contagious disease and the yellow "quarantine" placard of the public health department went up on their door. Under less intimidating circumstances, the same lessons were offered by the vast network of voluntary reform groups that grew up in Progressive-era cities. Organized charities, settlement houses, Young Men's and Young Women's Christian Associations, infant welfare societies, visiting nurse associations, immigrant aid societies, and public schools all made teaching the gospel of germs to poor women and girls, especially recent immigrants, a major objective. Insurance companies such as Metropolitan Life, which had many working-class subscribers, also distributed health tracts and hired visiting nurses to teach disease prevention to their policyholders.[9]

These outreach efforts differed in important respects from the educational initiatives aimed at white, middle-class audiences. A basic rule of thumb was that the more humble the home, the simpler the message it needed about the ways of the germ. Whereas popular health tracts intended for middle-class housewives often contained short explanations of the germ theory, complete with references to Louis Pasteur and Robert Koch, those written for immigrant women consisted almost entirely of health rules and were unaccompanied by suggestions about the desirability of growing dust gardens or gaining access to a microscope. Unlike their middle-class counterparts, whose desire for modernity was taken for granted, immigrant women were also much more aggressively counseled to replace traditional wisdom with that of the professional expert.

Ten Commandments for Keeping Baby Well, a pamphlet distributed by the New York City Bureau of Child Hygiene, exemplifies how the gospel of germs was simplified for less educated audiences. The tract consisted primarily of slogans, shorn of any supporting detail, that asserted rather than explained the link between cleanliness and disease prevention. "You have heard of the Three Deadly D's—Dirt, Disease, Death. We want to teach you about the Three Healthful C's—Cleanliness, Comfort, Contentment," it read. The *Ten Commandments for Keeping Baby Well* warned that mothers should listen to doctors and nurses rather than to their relatives or friends. "If as neighbors, you want to advise one another, let that advice come through the practical example of keeping your homes and your surroundings clean," the tract concluded.[10]

In teaching poorer, less educated women, professional health workers also relied heavily on demonstrations of sanitary practices. Assuming that their pupils did not need to understand the *why* of sanitary protection so much as the *how* of its practice, they believed that hygiene lessons were best taught directly by an authoritative figure such as a teacher or nurse. The demonstration method had the added virtue of reaching immigrant audiences who could neither read nor speak English.

Here the visiting nurse, rather than the home economist, played the central role in taking the gospel of germs to poor urban households. At the turn of the century, a wide variety of organizations, including hospital clinics, charity groups, and insurance companies, began to hire nurses to care for the poor. In many cities, visiting nurse associations were founded to supply this growing demand for home health care workers. Besides caring for the sick, visiting nurses were also expected to teach basic principles of hygiene. Because many of the patients they attended were ill from infectious diseases, much of that teaching revolved around preventing the spread of infection. As a brochure for the Henry Street Nurses' Settlement explained, the presence of a dangerous disease in the home made it "comparatively easy . . . to make a school of the sick-room, with mothers and family eager students."[11]

From the beginning to the end of her home visit, the visiting nurse modeled the precepts of sanitary discipline. A set of directions prepared for Boston nurses in the late 1910s demonstrates the care taken to orchestrate the performance of cleanliness. The directions specified that upon entering, the nurse was to "remove coat, folding right side up and place on chair away from wall" and "place bag on chair or on table with newspaper underneath," thereby minimizing contamination from the potentially germ-ridden surfaces of the home. Before touching her patient, the nurse performed an elaborate cleansing of herself, which required having a family member boil water and then washing her hands surgical-style with a nail brush and antiseptic soap. After donning a clean apron to cover her clothes, she set out her shiny instruments on a clean white towel. Only after enacting these rituals of scrupulous cleanliness did the nurse finally begin to take care of the patient. In this way, she modeled the care needed to prevent the spread of disease to other members of the family.[12]

Sanitary Americanization

No matter how clear their demonstrations, visiting nurses and other welfare workers by no means found a docile audience prepared to comply with their every instruction. Immigrant housewives did not passively accept direction, but rather picked and chose among the advice they received from domestic "experts." Although they were often interested in adopting American ways of cleanliness, these women also wanted to preserve valued aspects of their own culture. Their patterns of resistance and accommodation offer a deeper understanding of how immigrant housewives translated the gospel of germs to suit the realities of their own lives.[13]

The firsthand accounts we have of tenement life, in case records and retrospective oral histories, underline the extraordinary structural obstacles poor women faced in the practice of cleanliness. Virtually every maxim of the gospel of germs was countermanded by the houses themselves: for example, tenements were so crowded together that the sunshine and fresh air deemed essential to killing bacteria and maintaining resistance were excluded, and callous landlords let roofs leak, walls crack, floors buckle, privies overflow, and garbage pile up in the halls and alleys. Implementing the gospel of germs under such circumstances was indeed daunting.[14]

To illustrate this point, a brochure for the Henry Street Nurses' Settlement recounted the experiences of a visiting nurse sent to care for Angelina, a six-year-old girl living in a Lower East Side tenement house who had contracted scarlet fever. Her grandmother seemed quite willing to master the proper isolation and nursing techniques, yet whenever the nurse visited she always found the sickroom window shut, despite her instructions that the girl have plenty of fresh air. Angelina finally confided that she was terrified of the rats outside her window. The grandmother, trying to follow the nurse's advice, had taken to throwing scalding water out of the window to scare the rats away to allow the window to be opened briefly. But the rats soon came back, so it had to be closed again.[15]

For other reasons unrelated to poor housing conditions, hygiene reformers found the gospel of fresh air to be a hard sell among immigrants. Many of the immigrants regarded the closed window as a protection *against* disease, as an Italian immigrant woman explained to a public health inspector in 1914. The woman, whose mother and child had both contracted typhoid, lived in a tiny room whose only window stood opposite a noisome privy. Asked why she kept the window securely shut even when the summer heat turned the room into a oven, the woman explained that "she kept the window shuttered in order to 'keep out disease.'" The contrary view, that air currents were useful in dispersing disease germs inside the room, was difficult to convey. A Pittsburgh charity worker reported that an Italian mother only understood the concept after seeing a play about tuberculosis germs on a visit to the fresh-air camp where her daughter had been sent to build up her strength. In an awkward rendering of her halting English, the worker reported her comment, "Now I see why you all the time open window; I go home, I take nail out window, get much air."[16]

Hygiene reformers also privileged the separate bed in ways that poor families found hard to understand. Often the visiting nurses' first objective was to isolate the person sick with a contagious disease in a single bed, yet time and time again the poor found this to be a difficult task, usually because of limited bed space. As a visiting nurse remarked of a household where the husband had typhoid fever and the wife had influenza, "It seemed a very wrong thing to have them both occupy the same bed at that time but one bed was all they had." In addition, some families regarded the single bed dictate as cruel, especially for children used to sleeping together. They "find it strange and dismal to sleep alone if it is their lot," commented a social worker.[17]

Lessons about the dangers of impure drinking water were also hard to get across, especially to recent arrivals who had come from rural areas where typhoid was rare. "You cannot make the foreigner believe that Pittsburgh water is unwholesome," observed one physician, who noted that some men, when they heard the warnings, made a show of bravado by "deliberately going to the Allegheny to quench their thirst." Accordingly, he reported, roughly half of all foreign-born men got typhoid within two years of arriving in the city. Immigrant women may have been quicker to get the message, for charity workers often commented on the extraordinary lengths to which they went to avoid tainted water supplies. As a Pittsburgh social worker reported, women living in one poor area bypassed a convenient but polluted water source, climbing up and down a very steep stairway to get drinking water from a pump farther away. Remembering the city's high typhoid rates, a Jewish immigrant woman recalled being impressed by the large water filter that her wealthy relations had installed in their home to make the water safe.[18]

Possession of an indoor toilet was deemed another extraordinary privilege. A social worker in "Skunk Hollow," a dilapidated Pittsburgh neighborhood of poor shanties, noted that "a very few have 'sanitary' toilets and shout the fact on black and white rental signs." In oral history interviews, first- and second-generation immigrant women often mentioned the acquisition of an indoor toilet as a milestone in the family's life. An immigrant woman who grew up in a home with no privy at all remarked, "To us it was . . . like luxuries" even to have a decent outhouse.[19]

Yet for all their allure, modern toilets seem to have been a puzzle to many recent immigrants. Mabel Kittredge, a home economist who made a particular study of tenement households, noted that many poor women had no understanding that the toilet had to be flushed after each visit and could not be used to dispose of garbage. Given the rural backgrounds from which many of them came, Kittredge observed, "Why should Mrs. Milewsky know what to do with modern plumbing?" She criticized charity workers for overlooking "the fact that these dazed people have no knowledge of how to use the comforts we are giving them."[20]

Immigrant housewives found particularly unappealing the new sanitary principles of house decoration. The spare styles so popular among middle-class Anglo-Americans ran directly contrary to their definitions of comfort and beauty. Ignoring domestic experts' complaints about dust and germs, immigrant families continued to spend what little money they had on plush upholstery, carved furniture, and elaborate bed linen, which suggests that they found the older Victorian style far more appealing than the antisepticonscious aesthetics of the Progressive period.[21]

Yet although immigrant women resisted some aspects of sanitary housekeeping, many of them showed an extraordinary willingness to expend back-breaking labor in pursuit of a clean home. Meticulous housekeeping was one way to maintain self-respect and to disprove the charges of being dirty and un-American that were so often leveled against them. Another compelling motive was undoubtedly the conviction, encouraged by visiting nurses and other welfare workers, that scrupulous cleaning would help defend their families against disease. In the often unpredictable world of the tenement neighborhood, what was inside the home was under a woman's control, what was outside was not. Within this gendered division of space, high standards of housekeeping served as one way to ward off the specter of dangerous disease.[22]

The importance of cleaning is abundantly clear from the high priority it occupied in many immigrant households. At a time when they often needed every penny to survive, including the contributions of mothers and daughters, the latter still expended long hours of unpaid labor cleaning house. Given the poor quality of their homes and the scarcity of hot water, the labor expended was prodigious: in oral histories, women described scrubbing floors with near boiling water and lye, dusting furniture, washing down walls, and boiling clothes and bed linen. Daughters soon learned that house cleaning was not a quick chore but had to be done "just so"; "everything had to shine," in the words of one Polish woman. Another Polish-American daughter recalled having to scrub the floor four times over because she did not get it clean enough for her mother's satisfaction. An Italian woman described the weekly custom of taking apart the refrigerator and the stove to scrub them, which she recalled "was like a ritual." Among the highest compliments an immigrant housewife could receive from relatives and neighbors was the recognition that she kept a spotless house. As one Jewish woman recalled proudly of her immigrant mother, "She was so clean, that she kept covers on the covers."[23]

These rituals of frequent scrubbing and dusting of interior surfaces were not customs that immigrant women brought from the Old World. In the southern and eastern Europe communities from which most of the so-called new immigrants came, the floors were dirt or stone and were swept rather than scrubbed. The social worker Sophinisba Breckinridge reported that many immigrant women had never seen a scrub brush before coming to the United States. So the connection between a clean house and a well-scrubbed floor and dustless furniture was largely an American-born obsession, one that was no doubt reinforced by both the extraordinary dirtiness of the early twentieth century city as well as the reformers' continual repetition of the "dirt, disease, and death" equation.[24]

Moreover, despite their often straitened circumstances, immigrant women were exposed to the advertising culture, which equated certain products with safety from disease. Recognizing that immigrants were a potentially lucrative market, American companies began to place foreign-language ads for products such as Fels-Naptha and Quaker Oats in immigrant newspapers and on radio shows. Even if they could not afford such products, women reading or hearing such advertisements still heard the lessons about disinfection, dusting, and the like that accompanied these commercial messages. Significantly, when asked to list what modern conveniences their immigrant mothers had lacked, second-generation daughters often mentioned linoleum, the washable floor covering marketed as a sanitary boon to health-conscious American housewives.[25]

Perhaps also from this exposure to consumer culture, immigrant women quickly picked up on the American obsession with whiteness. Mrs. K, a Slavic-American woman born in 1924, reported that her mother and grandmother insisted not only that the wood floors be scrubbed, but also that they be bleached white by strong lye soap. Even the commode in their cellar, which had a wooden seat, had to be bleached white. "In those days, them people believed in white wood," she recalled; "if it wasn't white, Grandma would hollar [sic] or my mother." Mrs. V, from an Italian family, recalled the same fixation with whiteness. When she scrubbed the wood floor, "If it didn't come white, I had to go back again." Unable to afford the white tile, enamel, and linoleum popular in "antisepticonscious" middle-class homes, less privileged housewives still strove to emulate their color.[26]

At a time when rates of infant and child mortality were still extremely high, keeping their children "just so" was another way that immigrant women tried to ward off the evil eye of disease. Some mothers venerated the childcare advice given by visiting nurses and local physicians as a kind of superior knowledge distilled from wealthy and successful Americans. A Russian-Orthodox woman who followed carefully her doctor's suggestions recalled with pride, "People and neighbors used to say, 'Oh your children, you take care of them like the rich people.'" One daughter recalled her immigrant mother giving her a "don't list" that might have come straight from a domestic hygiene text: "Don't put anything in your mouth and expect the child [to take it] . . . don't take a bite of the baby's food or . . . test the bottle and put it in your mouth." Yet the rituals of sanitary child-rearing sometimes became a source of tension between the generations. Young mothers recalled being called "too finicky" when they followed the newfangled American standards of child care. When her Italian-born mother came to help with her grandchildren, remembered one woman, it was a "disaster" because "she didn't think it was necessary to sterilize the bottles and to be so fussy about the formula."[27]

Tenement housewives not only prided themselves on being able to keep their homes and children clean, they also applied those standards to neighbors and relatives. Household cleanliness became an important criteria for social intimacy: a Slavic-American woman recalled that despite the racial prejudices common during her childhood, her mother allowed her to exchange visits with a "colored" school friend because the girl's mother "was just as spotless as we were." Conversely, domestic dirt was perceived as so repugnant, indeed dangerous, that some immigrant women would literally not step across the doorsill of a poorly kept house. A second-generation daughter recalled that her meticulously clean mother had only one relative in the United States, a brother who lived nearby. The mother was so appalled by how his wife kept the house that "she wouldn't go there, that's how dirty it was," and as a result the two families rarely saw one another.[28]

Children soon learned to use the language of cleanliness to express their likes and dislikes of other people; one of the worse insults one could give or receive was "dirty." A Russian-born Jewish woman who had lived in a gentile neighborhood recalled suffering from endless "name calling" there. She was relieved when her family moved to a different neighborhood where she was less afraid that "someone was going to call me a dirty Jew." A woman who grew up as the only Jew in Houston, Pennsylvania, recalled the same taunt, and remembered that it hurt "not because they called me a Jew but because they said I was dirty."[29]

Although mothers were usually the ones most concerned with instilling hygienic principles in children, fathers sometimes became involved as well. Asked in an oral history interview if her parents had ever disciplined her physically, a woman recalled that her Italian-born father, who was a chiropractor, had slapped her only once, when she was about ten or eleven. "I was fascinated at the time with cigarettes and I saw one on the ground," she remembered. Her father saw her put it in her mouth, and in her words, "He was so outraged, I am sure, because of the germ theory and who had that in their mouth first that he smacked me." Having gotten her full attention, he then gave her a lesson about the dangers of tuberculosis. Remarking that she would remember that slap until the day she died, the woman commented that she not only never became a smoker, but also remained "ultra-germ conscious," especially while raising her own children.[30]

The fears of infection fostered by the anti-TB crusade created hostility toward neighbors as well. As fear of phthisis became common among the poor, neighbors started to report consumptives to landlords or public health agencies in order to have them removed. The director of the Charity Organization Society of the City of New York recounted one instance in 1916, when the family of Mr. A, a thirty-year-old man with pulmonary tuberculosis, had been forced to ask for assistance. His wife had been supporting the family by working as a janitor in an apartment building, but Mr. A had been so "careless" in his habits that the other tenants, expressing a fear of infection, had insisted that the family be thrown out.[31]

With the revelations about spitting as a source of communicable disease, this common practice took on new and sinister significance. Describing her early married life, one woman remarked about how much she disliked her landlady, who she feared had cancer, a disease many people thought was catching like tuberculosis. The landlady routinely spat on the steps of the house, and the woman recalled, "I used to scrub those steps because I thought the baby would get sick." Spitting also became a useful tactic for men seeking to annoy hygiene-conscious ladies; a sociological study conducted in the 1920s of residents of Muncie, Indiana, noted that in spite of all the TB educational work, women "still say they walk on the inside of public sidewalks because of the constant spitting."[32]

Likewise, the well-publicized crusade against the common cup made the sharing of glasses and dishes a powerful marker of familial and group loyalty. For example, Mr. A's mother shared her son's utensils to express her contempt for the theories of disease transmission that had caused him such trouble. As the New York Charity Organization Society agent reported, she "deliberately drinks from the cup her son uses, and will not keep the dishes separate," despite warnings that her grandchildren might become infected. In another instance, a visiting nurse recalled the dilemma she faced when she refused an Italian man's offer of a glass of wine out of allegiance to her temperance principles; he immediately assumed that she did not think the glass was clean enough for her use and was deeply offended.[33]

In even more dramatic testimony to the symbolic meaning of drinking glasses, an Italian woman recalled a practice from her childhood days in Pittsburgh. Her parents ran a restaurant and bar in an integrated neighborhood and tried to remain on friendly terms with both races. African-American customers usually came to the back door of the restaurant to get their food. If a black customer "happened to insist on a drink at the bar," the woman recalled, "my mother broke the glass right in front of the whole bar after the man had a drink, to show the other people you're not going to drink out of that."[34]

Of course, far more than the fear of disease was expressed in gestures such as these. The rituals of cleanliness and separation that immigrant families incorporated into their daily lives served many purposes in the complex process of becoming more "American." Such behaviors were markers of upward social mobility and of a "white" racial superiority formed at the expense of black Americans.[35] Yet immigrant women's preoccupations with spotless homes and clean neighbors were surely related to the fear of contagion, a fear kept fresh by the high rates of illness and death common among the poor. The persistent efforts of hygiene reformers to portray death as the ultimate punishment for sanitary sin affirmed immigrant women's beliefs that fervent housecleaning and a wary stance toward the outside world were useful in warding off sickness. For all the anxiety and frustration it brought in its wake, the gospel of germs offered these women at least some control over their fates in a dangerous new world.[36]

Cleaning Up the Farm

During these same decades, the gospel of germs came to the farm as part of a larger Progressive-era effort to revitalize rural America.

Turn-of-the-century reformers, concerned that the exodus of young people to the city was weakening the agrarian virtues central to American democracy, sought to make country life more appealing by bringing the fruits of modern progress to the farm. State university "extension" programs, which began in the late 1800s, became the chief vehicle for bringing about this transformation. In 1916 and 1917, the extension movement received a great boost when the U.S. Congress passed the Smith-Lever and Smith-Hughes acts, which allocated federal funds for adult programs and public school courses in agricultural science and home economics.[37]

For rural families, it was the extension, or "demonstration," agents who took the leading role as sanitary missionaries. Their educational work was clearly divided along gender lines: the male agents, who usually had degrees in agricultural science, instructed farmers in matters such as dairy sanitation and outhouse construction, as well as crop selection, fertilization, and the use of new farm technologies. The female agents, who were trained in home economics, taught the farmer's wife the principles of household bacteriology, including sanitation, food preparation, canning, and child care. Roughly the same gendered division of labor was followed by 4-H clubs and other youth programs aimed at farm boys and girls.[38]

The extension programs developed at Cornell University under the guidance of Liberty Hyde Bailey, a professor of horticulture and later dean of the New York State College of Agriculture, were among the most successful and influential in the United States. Bailey was an early supporter of the home economics movement, which he saw as an ideal way to enlist women in the cause of rural reform. In 1900, he hired Martha Van Rensselaer, formerly the school commissioner of Cattaraugus County, New York, to develop a reading program for farmwives similar to one he had started for their husbands.[39]

Like the popular Chautauqua reading circles of the time, the Cornell Farmers' Wives Reading-Course sent out free bulletins on subjects of interest to the farm wife. Each bulletin had a sheet of study questions at the end that readers could answer and return for comments. Women enrolled in the program were also urged to form "study clubs" to work through the bulletins together. By 1914, the Cornell extension program had enlisted 36,200 women and eighty-two study clubs in the work. Van Rensselaer and her colleague Flora Rose, who had been hired in 1907 to help with the extension work, corresponded frequently with club members, answering questions and soliciting their advice about the course content. In one year alone, 1914, they received some 3,200 letters from farm women living all over New York State.[40]

Like visiting nurses, Cornell economists made use of the demonstration method and the home visit to model the principles of sanitary cleanliness. In the beginning, Van Rensselaer and Rose had limited resources to do more than give occasional talks to local clubs, so they put their energy into demonstration programs at the annual Cornell "farmers' institutes," which usually drew several thousand farmers and their wives. During "Farm Week," for example, an agent might give women a canning lesson, thereby demonstrating all the precise steps of sanitary food preparation. Or farmwives would be given a tour of the college's "cafeteria laboratory," which opened in 1913, to take in its smooth, germ-resistant surfaces, modern appliances, and general air of cleanliness. In the mid-1920s, as the extension department expanded, agents began to work more in the field, conducting so-called kitchen conferences. On these visits, the home economist met with groups of farm women in their own homes in order to give simple demonstrations and practical advice about improving household conditions.[41]

Unlike visiting nurses and their immigrant patients, who were usually quite dissimilar in cultural background, demonstration agents were often "Yankee" women from farm backgrounds who could readily identify with the problems of their farmwife-pupils. Having grown up in western New York, Martha Van Rensselaer once wrote that she had "a very high estimate of the farming population in our state," for they stood out "prominently in good citizenship and the work they are doing." Home economists assumed that farm women had common sense and sagacity, if not schoolbook learning. In a letter to Van Rensselaer, a domestic science teacher in Auburn, New York, remarked of her students, "The women are intelligent housekeepers who wish to improve their own methods." In return, farmers' wives treated extension agents with the deference that country people often accorded teachers, especially those connected with the state university. A housewife from Canton, New York, observed that if she ever tried to tell her friends or relatives how to do something new, "they would laugh and think I was too 'finicky' and too fine, but those same women will believe almost anything that is in print and particularly anything from a college that is printed."[42]

Still, farm women were hardly an easy audience to teach. Although some wanted to become better housekeepers, an agent observed in 1923 that others had "no eagerness to learn new or better ways of doing things" and at best picked up from the extension program only a few tips concerning meal planning and canning. Not surprisingly, younger women tended to be more receptive to the home economists' efforts. Writing in 1909 about the progress of her study club, Mrs. E. P. Ellinwood said of the younger members, "They are more eager to accept anything to make work easier and give them more time to enjoy life. The older women oftentimes get into a rut and are unwilling to believe that any new fashioned method of accomplishing a certain task can be better than the old way which they and their mothers and grandmothers before them did."[43]

The home economists' task was complicated by the fact that for the most part, farm women "know very little of science applied to the Home," in the words of one teacher. Some found even the simple language of the reading club bulletins difficult to follow. A woman trying to get the "ladies" in her neighborhood interested in the Cornell program reported in 1910 that "one of them said at church yesterday—she had read very word of the Bulletin but could not understand it, and didn't see that it would be of any use to her if she did."[44]

Domestic sanitation and household bacteriology were among the most difficult topics that extension agents had to cover. In the first place, they had to unsettle traditional assumptions that the country was an inherently healthy place to live. In comparison to city dwellers, farm families lived in comparative isolation, safe from faceless strangers' coughs and sneezes, and they worked outdoors in the sunshine and fresh air so often hailed as the essentials of disease prevention. Perhaps as a result, "country people are very careless about sanitation," one of Van Rensselaer's correspondents observed. Moreover, few of them had ever heard of the germ theory of disease. A farmer's wife from Brookfield, New York, urged the Cornell staff to teach more on "how to be clean about all branches of work because few enough farm wives know about germs and their danger." She claimed that there was a special need to change unsanitary dairy practices that spread sickness into so many homes.[45]

Even after they had become convinced of the need for improved sanitation, farmwives still faced formidable obstacles in complying with the gospel of germs. As late as the mid-1920s, roughly half of the homes surveyed by the Cornell extension department had no running water or indoor toilets. "Most of the houses in our neighborhood are old and naturally are lacking in many modern conveniences," a club member observed in a 1909 letter, and thus "many of the ways suggested of improving our homes to make them convenient and work lighter entail considerable expense." Another correspondent, lamenting her lack of running water and an icebox near the kitchen, said they "add more to the burden of a farmer's wife that any other two [deficiencies] you can mention."[46]

The extension staff tried to nurture the farmwife's determination to overcome such obstacles by portraying the properly maintained privy and the screened window as matters of life and death. Van Rensselaer's 1904 tract *Suggestions on Home Sanitation* used the "white sepulcher" theme to underline the urgency of these matters. The lessons were introduced with a story about a nearly grown son who came home from school to visit his parents. Having "learned the importance of pure soil, pure air and pure water" while he was away, the son alerted his parents to the dangers all around them: the "much beloved" brook carrying fecal wastes from upstream, the flies flitting from manure piles to milk pails, and so forth. *"Their eyes were opened to need of sanitation,"* and the next time the son returned home, he found everything in good order.[47]

As extension agents recognized, farm families had advantages over tenement dwellers in their ability to improve household sanitation. Because most farmers owned rather than rented their homes, they could more easily correct sanitary defects. For example, a fly-proof sanitary privy could be built at relatively little expense using simple instructions provided by the extension service. Families with more substantial means could have a cesspool dug and install an indoor flush toilet just like the kind that middle-class households had in the city or suburbs. But for this optimistic scenario to come to pass, the farmer had to cooperate; in the gendered division of farm responsibilities, external improvements such as sanitary privies and cesspools fell in his domain. It was no accident that in her 1904 morality play about household sanitation, Van Rensselaer gave men the starring roles. It was the father who, in her words, was "haunted" by the expression "a damp cellar weaves shrouds for the upper chamber" and took the initiative to put an end to the mysterious fevers afflicting the family by cleaning up the old farmstead.[48]

As Van Rensselaer well knew, men sometimes neglected the home in favor of farm improvements that paid off more directly in dollars and cents. One of the most poignant letters preserved in the Cornell correspondence expressed a woman's frustration over her husband's lack of concern about the cleanliness of their home. "Our farm is neat and clean, all but the house, yard, and cellar, and my husband keeps that for refuse," she wrote. "Even if I were capable of arranging it like a palace, so to speak, my hands are tied." She lamented, "Men, men, mud, mud, and my cellar. I wonder we are alive. Poor me, I know if everything had been kept properly my children would be alive and well."[49]

Van Rensselaer faced a difficult task in handling the "man problem" hinted at in such letters. Responding to a male correspondent who had accused her of being "too hard on the men," she replied that "constant acquaintance with conditions which are indeed deplorable" had made her "prone to think of the darker side of this question." Yet in pressing for sanitary improvements, Van Rensselaer had to be careful not to alienate the husband upon whose cooperation such improvements depended. The problem was compounded by the general hostility some male farmers felt toward home economists and their "female foolishness" regarding sanitation and diet. At the 1909 Farmers' Week, one of Van Rensselaer's correspondents overheard a man announce that he would not bring his wife next year; in his opinion, the home economics program was a "sideshow for women" run by "fools" who were "putting them up to all sorts of nonsense about oatmeal etc." Van Rensselaer agreed with her correspondent's suggestions about the "need of *getting* the *men* interested in hygienic living," which she proposed to do by getting the male agents to work up a program on the subject for the next year.[50]

Despite their primitive homes, heavy schedules of work, and sometimes unsympathetic husbands, many farmwives demonstrated no less a zeal for household cleanliness than did their immigrant contemporaries living in poor tenement houses. Their letters often expressed a sense of pride and accomplishment in following through on the home economists' suggestions. Using a biblical analogy, one wrote to Van Rensselaer, "The 'leaven hidden in the measure,' I have had of the *Reading for the Farm Home* is working very successfully in my home." In the year since she had been in the club, she proudly counted the following accomplishments: "a large warm bathroom with modern equipment, sides and ceiling painted white, good ventilation . . . also a pair of stationary wash tubs . . . and so, the good work goes on. Your labor is not in vain, you see." In a similar tone, another woman wrote that she had been inspired by the bulletins to make her farmhouse more sanitary. She vowed that she would first tackle the kitchen, where she planned to install running water and "good drains"; then she would "eliminate the carpets and finish floors for rugs in the rest of the house, making it more sanitary and easier to keep so."[51]

The proximity of farm animals to human habitations, and the variety of insects they encouraged, posed special burdens on the sanitarily minded farmwife. One woman wrote anxiously to Van Rensselaer, "Can disease be communicated by chicken manure being tracked into rooms?" After another correspondent poured out her frustrations over an insect infestation, Van Rensselaer wrote soothingly in 1909, "You have had one of the most difficult fights that a woman has to have in that line. It is, as you say, eternal vigilance which finally accomplishes something and, of course, this will in the end win out." Besides dispensing practical advice about filling cracks in wood floors and treating the edges of rugs with gasoline, Van Rensselaer reassured farm women that the battle against the bug was essential to the family's health. Moreover, she reported, insects were nothing to be ashamed of, for "the best of housekeepers seem to have them."[52]

Houseflies were by far the worst of the insect enemies. Even the farm woman who wrote rather smugly to Van Rensselaer about her "sanitary" home, bragging that thanks to screened windows and doors "we [know] nothing about the long list of household pests you speak of," admitted to having flies, who were "most troublesome as some will get in then they must be driven out of the window with flippers." During a kitchen tour in the 1920s, an extension agent named Ella Cushman reported a fly-related dilemma: after proudly showing off her kitchen renovations to her guests, a woman offered her guests apples that the flies had been freely exploring. Like the visiting nurse offered the glass of wine, Cushman found it hard to refuse the apple and noted that the woman's "improvement is encouraging and one cannot expect too much at one time."[53]

The farmwives' queries pointed up some obvious difficulties in observing the gospel of germs—for example, "how to have sunshine and air, but no flies" and "how to keep clean without cleaning constantly." Without the money to afford Sanitas or antiseptic paint, farmwives struggled with cleaning the old-fashioned wallpaper and furniture common in many farm homes; home economics agents often received questions about how to rid such household surfaces of dangerous dust and dirt.[54] One lesson of household bacteriology farm women found particularly appealing, no doubt because it was among the easier to follow, was the value of boiling water. A survey of farm homes done in the early 1930s revealed that the great majority of women scalded all the dishes and silver and regularly boiled their dishcloths. On a "kitchen tour" in Chenango County at about the same time, an elderly woman named Mrs. Meyer proudly "explained her unusual dish drainer which she invented" and showed her neighbors charts "picturing the lack of bacteria on plates scalded and air dried."[55]

Home economists' lessons about being hygiene-conscious consumers had less relevance for rural families, who grew most of their own food. In a work paper sent to the agents, one Long Island farmwife associated the issue of careful shopping not with her own circumstances, but with a trip she once made to the Lower East Side of New York, where she saw a fruit vendor spit on his handkerchief to polish an apple he wanted to sell. In general, the obsession with spitting so common in urban settings was less pronounced in the country, perhaps because the dangers of tuberculosis seemed more remote. On the other hand, the directives about home nursing were probably more useful to farm women than to their urban counterparts, given that the former usually lived so much farther from "a physician and nurse" and "probably . . . care for the sick in a house more than town women do," as Van Rensselaer noted.[56]

Even more so than their immigrant contemporaries, farm families had access to newspapers, magazines, and mail order catalogues, which promoted the new consumer culture of cleanliness. Of all the new innovations featured in their pages, farm women seem to have been most taken with the vacuum cleaner. In 1909, a study club in Barbers Corner, New York, wrote the Cornell agents asking if hand-operated vacuums would be "desirable or useful" in homes like theirs that lacked electricity. A woman fortunate to have electricity in her home wrote to Van Rensselaer at about the same time to announce, "I am just on the eve of battle—housecleaning—vacuum cleaner rented for the day as an experiment." She concluded humorously, "Shall hope for good results but the proof of the pudding is in the eating although I do not propose to eat this one. It is flavored too strongly of bacteria."[57]

The acquisition of sanitary privileges sometimes brought about new anxieties. In testimony to the persistence of fears about sewer gas, one woman wrote anxiously to Flora Rose, "There is something wrong with my septic tank and I am troubled and scared." When the sink water discharged into the tank, there was "a terrible smell," which she feared was dangerous. "Sometimes when I am miles away I imagine I smell it," the woman confided. In addition, the escalating demands for sanitary housekeeping could arouse feelings of insecurity and competition. During a kitchen conference, Ella Cushman noted that attendance at her morning talks had fallen off because "several were afraid they would have to offer their kitchens" for inspection later in the day and so had stayed home to clean instead. During one home visit, a woman in the midst of "demonstrating" to neighbors how she had located items conveniently in her kitchen "found dirt on her cupboard shelves and one of the women said something about it," casting a pall on the whole discussion, Cushman wrote regretfully.[58]

For all the home economists' promises that the new domestic science would lighten their load, many women still found that housecleaning was "the hardest work which women have to do," as members of one study club informed Cushman. Certainly the new emphasis on dust and disease did little to lessen homemakers' cares, yet it did promise a greater sense of purpose. "There is so much to be done in the house and out to rid it of dust and dirt," Van Rensselaer wrote to a correspondent in 1904, and although "it is hard work, there are returns which make it pay to undertake the effort." In another letter, Van Rensselaer advised a tired woman to take more time for herself: "You have your battles to fight every day and I believe when a woman does this well, she is as much of a heroine as the soldier is a hero on the battlefield."[59]

This vision of the farmwife's higher sanitary calling did not always find fertile soil, to use the parable of the sower so beloved by the early germ theorists. Some farm women never developed much interest in household reform; others probably tried the new ways for a short time and then abandoned them as too hard. But for many of the home economists' correspondents, their newfound sense of hygienic usefulness clearly meant a great deal. Mrs. A. C. Abbuhl wrote to Van Rensselaer in 1920, calling her "a blessing and help to us farm women" who had given them "a new idea of the importance of their work." She concluded, "You have created a new heaven and a new earth for us." Such letters help us to understand how housewives could come to see their battles against household dirt as heroic and be inspired to live up to the sanitary ideals of domestic science even under the most trying circumstances.[60]

Whatever their sense of being fortunate in receiving such knowledge, the gospel of germs surely was a mixed blessing for both immigrant housewives and farm women. On the one hand, enlisting the housewife's help in preventing fecal contamination of water supplies or in minimizing the spread of infectious diseases restored the individual's sense of control over a dangerous environment. Moreover, at least some of the housekeeping rules promoted in the early twentieth century had real utility in protecting against health hazards such as droplet infection or food poisoning. Although we will never be able to gauge precisely how much such campaigns contributed to the continuing fall of mortality rates from infectious disease from 1880 to 1940, they were not the meaningless makework some contemporary critics have made them out to be.[61]

Unfortunately, the heightened sense of responsibility for preventing disease came at a time when very poor women were still severely constrained in their ability to implement the ideals of bacteriological cleanliness being urged upon them. Without running water or sanitary toilets, even superficial cleanliness could be obtained only with backbreaking labor. In their zeal to promote the private side of public health, professional workers undoubtedly overemphasized the efficacy of individual hygienic measures. Thus when in spite of their best efforts family members, especially children, sickened or died, poor women were vulnerable to shame and grief. Like their middle-class counterparts before them, they found that believing in the gospel of germs exacted its own heavy burdens.

9 · The Two-Edged Sword

In her annual report for 1898, Maud Nathan, president of the Consumers' League of New York City, used the following curious story to underline the menace of sweatshop labor. The daughter of British prime minister Robert Peel ordered an expensive riding habit from a Regent Street tailor, so the story went. Unbeknownst to her, the garment was sent out to be finished in the home of a desperately poor worker who used the skirt to cover his feverish children against the cold. Unaware that her handsome new garment had ever left the elegant shop, much less been "germ-infected," as Nathan put it, the lovely young lady wore it proudly and soon after died of the same fever. Assuring her readers that similar "unknown tragedies" occurred frequently in New York City, Nathan concluded that the danger of the sweatshop was a "two-edged sword: it cut deep into the lives of the consumer as well as the worker."[1]

As Nathan's image of the "two-edged sword" attests, Progressive-era reformers firmly believed that the incorporation of the marketplace bound the health of workers and consumers together in significant new ways. At a time when many industries used tenement house shops and workers' homes for some phase of production, the new bacteriology's emphasis on dust and fomite infection intensified concerns about the contagion-bearing properties of manufactured goods. For consumer and labor activists alike, the dangers posed by tainted clothing were a concrete example of how the "chain of disease" and the "socialism of the microbe" worked to tie together the fates of affluent and poor.[2]

These imagined connections between the home and the workplace became the locus for a variety of cross-class and cross-race alliances during the Progressive period. Above and beyond the personal struggles for protection chronicled in the preceding chapters, the cause of disease prevention took on added importance as a seemingly neutral ground where warring groups such as capital and labor, native-born and immigrant, and white and black could work toward common goals. Amid fears of class and race conflict, crusades against contagious disease, particularly tuberculosis, offered a big tent for a wide range of community concerns. In the political discourse of the Progressive period, the gospel of germs furnished a common language for debating the relationship between public health and social justice.

This chapter will look at the history of two Progressive-era reform efforts—the Joint Board of Sanitary Control in the New York City garment industry, and the Negro Anti-Tuberculosis Association of Atlanta—to trace how the chain of disease concept figured in the political landscape of the early twentieth century. Both groups emerged from the kind of wrenching conflicts that suggested the nation was on the brink of social disintegration; the Joint Board was organized after a series of violent strikes, the Negro Association after a horrifying race riot. Both originated in specific fears of the infective properties of clothing, in one case spread by garment workers, in the other, by black laundresses. Under the guise of eliminating the public health menace posed by diseased clothing, both groups invoked the socialism of the microbe in pushing for sweeping programs of social and economic reform.

Although nativism and racism provided much of their initial energy, these efforts at reform began a new and sustained dialogue between groups separated by class and race lines. It remained a lopsided debate in which the middle-class, Anglo-American reformers maintained the greater power, and its achievements were limited by deeply ingrained patterns of economic injustice and racial prejudice. Nonetheless, the histories of the Joint Board and the Negro Association remind us that the gospel of germs did not remain the exclusive property of any group and that those singled out as dangerous to the public health seized upon it to forge a two-edged sword of their own.

The Menace of the Sweatshop

As its founders often pointed out, the Joint Board of Sanitary Control represented a first in American labor history: a serious and largely successful collaboration of owners and workers in an effort to reduce job-related disease and injury. As such, its origins and early operation provide a revealing insight into the common ground that disease prevention, particularly tuberculosis control, offered to the warring parties of management and labor during the Progressive period. That such common ground could emerge in 1910 was due in large part to the popular hygiene agitations of the preceding two decades, which had heightened fears of germ-laden products. Between 1890 and 1910, consumer leagues, antituberculosis societies, and labor unions each worked in their own way to make the threat of clothing-borne disease one of the "worries and torments" afflicting the American consumer, particularly the housewife. That the Joint Board was initially founded to supervise sanitary conditions in the ladies' garment industry, and not the men's, no doubt reflected reformers' special efforts to awaken women consumers to the menace of disease.

The well-founded perception of the garment industry as a "stronghold" of the white plague, as it was termed by the novelist Ernest Poole in 1903, reflected its unusual pattern of economic development. Unlike the steel or food processing industries, which had consolidated under the competitive pressures of the late 1800s, the garment industry had grown more diffuse and diverse as it expanded. Although the larger factories, known as "inside shops," employed new technology and management techniques to make clothing more efficiently, they continued to use small subcontractors and primitive sweatshops for certain phases of production, such as finishing or trimming the garment. Given the widespread reliance on subcontracting, even garments sold by big manufacturers were likely at some point in their production to have been worked on in tenement-based shops that doubled as living spaces for workers and their families.[3]

The flood of immigrants into turn-of-the-century New York City ensured that tenement-shop bosses could find plenty of workers willing to put in long hours for low pay and that they could replace these workers easily should they become sick or protest their working conditions. In response, the more skilled garment workers organized unions to protect their interests; the International Ladies Garment Workers' Union (ILGWU), founded in 1900, was among the strongest. But the larger manufacturers resisted union demands for higher wages and shorter work hours, claiming that they could not compete with the sweatshop producers. Nor were New York State's overworked factory inspectors much help in regulating working conditions in the garment industry; there were simply too few of them to visit the thousands of tenement shops involved in clothing manufacture.[4]

Given late-nineteenth-century assumptions about how contagious diseases spread, the unsanitary conditions typical of these sweatshops constituted a serious public health problem. The Consumers' League of New York City, which was formed in 1891 to improve working conditions for women salesclerks, first realized the extent of the danger during the 1898 garment workers' strike. Maud Nathan reported, "We found that nearly all the fashionable ladies' tailors, while they provided elaborate showrooms and fitting rooms for their customers, gave many of their garments to be made up in the tenement homes of the workers." Consumers' League members were sure that if middle-class women knew about the deplorable sanitary conditions under which their family's clothing was made, they would pay more for products made under fairer and safer conditions.[5]

A tract issued in 1900, *The Menace to the Home from Sweatshop and Tenement-Made Clothing*, illustrates how consumer activists employed the chain of disease concept. Noting that in 1899 alone New York's immigrant workers made over $160 million worth of clothing, the tract emphasized how the advent of mass production and distribution had allowed infected goods made in one unsanitary locale to contaminate homes all over the country. Few consumers understood, the tract warned, that "the woman who shops in Concord, New Hampshire, is as likely to buy a tenement-made garment as though she shopped at a department store in New York City." Moreover, women consumers had no way of detecting the presence of infection in their clothing. Certainly cost was no indicator, for as the league discovered, 90 percent of the most expensive custom tailoring was finished in tenement shops. The menace of infected clothing presented a classic example of the "whited sepulcher" theme; the dainty baby's nightgown or the stylish cloak gave no visible evidence of its impurity. Invoking the new bacteriology's findings about house dust, fomites, and casual contact, the league's writings continually asked the woman consumer to picture the germ-ridden bodies and household furnishings that had touched the very clothing she now placed upon her children's bodies.[6]

In the wake of 1898 strike, the Consumers' League proposed to protect consumers against such horrors by organizing a "White Label campaign" patterned on the union label movement. The league would allow manufacturers who enforced good sanitary working conditions in their shops to place the label in their garments, and it would educate women to buy only labeled goods, even if they cost a little more. In this fashion, the "good" manufacturers would be rewarded for their compliance. The league was encouraged in the plan by the fact that some garment firms had already begun to see the value of the hygienic argument in promoting their goods. For example, Nathan noted that in the wake of the 1898 strike, one prominent manufacturer had voluntarily started to put in its garments labels that read, "Sanitary, non-sweatshop make." Likewise, a maker of baby clothes advertised them as "safe to wear" because they were made in "light, clean workrooms" and warned that "many garments offered for sale at low prices, and sometimes even the better grades, are manufactured in tenements or other unwholesome places, under conditions that make it hazardous to wear them, especially for infants and small children."[7]

With these positive signs in mind, the league first broached a trial of its white label plan to the manufacturers of ladies' and childrens' underwear. This branch of the industry had the most to worry about regarding the contagion fears because their garments were worn directly next to the skin. In 1898, an agent hired by the league found that seventeen of the twenty-two underwear manufacturers questioned said they would adopt the label if the league could assure a demand for it. A year later, the New York league helped to form the National Consumers' League in order to coordinate nationwide support for the white label movement.[8]

But the white label idea faced considerable resistance from the unions. Samuel Gompers, president of the American Federation of Labor (AFL), stated the unions' objections clearly in a 1903 convention speech. He feared that the "well meaning, philanthropic ladies" who led the movement would be willing to issue the label solely on the basis of improved sanitary conditions, "without regard to any consideration as to wages, hours, and other conditions of employment, and in some instances in rivalry to the union label of the organization of the craft." From his perspective, the white label concept represented first and foremost the interests of middle-class consumers—interests that its advocates were willing to advance at the workers' expense.[9]

The problem was not that labor unionists were uninterested in disease prevention or oblivious to consumer anxieties about germ diseases. On the contrary, they had long argued that low wages and long hours were destroying the health of working men and women, and they had tried whenever possible to play the disease card to advance their own position. Gompers himself had used the contagion argument during his early career as head of the New York City cigar-makers' union. Echoing doctors' warnings about the dangers of venereal disease, he had warned male smokers that nonunion tenement-house workers used their spit to roll cigars and thus impregnated them with the germs of deadly disease. Indeed, the term "white label" had originated among union cigar makers in San Francisco to differentiate their product from that produced by presumably disease-ridden Chinese workers. Nor did the issue of disease prevention interest only conservative craft unionists such as Gompers. More radical labor leaders also appealed to public health concerns to advance their causes. For example, in *The Jungle,* Socialist Upton Sinclair highlighted the unsavory conditions in the meat packing industry, and organizers of the radical International Workers of the World called attention to poor sanitary conditions in hotels and restaurants as part of their efforts to build the "one big union" of American workers.[10]

Moreover, in marked contrast to their reservations about the white label movement, union leaders responded positively to the overtures of the anti-tuberculosis movement. Starting in the early 1900s, anti-TB societies in large cities such as New York, Chicago, and Philadelphia began to court organized labor, with considerable success. Not only were union leaders concerned about labor's vulnerability to the white plague, but they also realized the potential for using the tuberculosis problem to support union demands for better working conditions. Yet although they accepted advice and assistance from the mainstream anti-TB movement, the unions put a noticeably different spin on the precepts of "tuberculosis religion," one that helps to explain their reservations about the white label idea.[11]

Even at its most conservative, organized labor's gospel of germs differed in important ways from the doctrines promoted by the mainstream public health movement, as evident in the anti-TB tract prepared and distributed by the AFL in 1906. To be sure, the tract conveyed the usual messages about the disease being caused by "living germs too small to be seen" and spread by spitting and intemperate living. But the code of individual protection was recast as a form of worker solidarity. The brochure emphasized that individual workers owed each other this carefulness, in order to do no harm to their workmates. Labor's version of tuberculosis religion also presented joining the union as the first and most important step of disease prevention. The motto "JOIN A LABOR UNION AND HELP STAMP OUT CONSUMPTION" was emblazoned in big type on the front cover of the pamphlet. The tract went on to explain, with key words in capital letters for emphasis, "In the same degree that the TRADE UNION MOVEMENT becomes powerful will it establish such IMPROVED CONDITIONS that will check and eliminate the ravages of CONSUMPTION." The agenda for fighting tuberculosis began with "a SHORTER WORKDAY" and "higher MINIMUM PAY" and continued with a long list of demands, including public parks and playgrounds, half a day off on Saturday for recreation, cheaper streetcar fares, and suburban residences for workers. As the list made clear, the AFL saw the link between disease and economic justice in even more expansive terms than did the most zealous middle-class anti-TB worker.[12]

Not surprisingly, union explanations for the tuberculosis problem tended to assign the greatest responsibility to the bosses. In asking all of its locals to pledge money for anti-TB work, a circular prepared in 1910 by the Central Labor Union of Philadelphia proclaimed, "The principal causes of our members being infected with this destructive disease are generally because of the hard conditions under which we are obliged to labor and our slim chances for good living." The circular left no doubt that it was "the owners of our workshops, factories, and department stores, our builders and our manufacturers" who really had the power to change the conditions that bred the white plague.[13]

Although in theory tuberculosis prevention fit the larger goals of the labor movement, in practice union leaders found the issue difficult to press with the rank and file. The Central Labor Union of Philadelphia, for example, found that the locals were far more willing to contribute money toward sick benefits and sanatorium care than to pay a "tuberculosis tax" for preventive initiatives in the workplace. As a workplace issue, disease control simply seemed less urgent than did higher wages and shorter hours. At a time when many locals had to strike in order to get the most basic concessions, anti-spitting crusades seemed more a luxury than a necessity. In a speech to the AFL in 1907, Paul Kennaday of the New York Committee on Prevention of Tuberculosis acknowledged the difficulties: "We recognize that these unions must at times have matters to consider of far more importance to them than tuberculosis."[14]

For these same reasons, the white label movement initially had little appeal to union leaders, who feared that middle-class consumer activists would overlook bread and butter issues such as wages and hours in order to obtain sanitary improvements. Yet as union leaders and social reformers agreed, the larger issue of disease prevention offered a potentially attractive area for collaboration. As Kennaday put it, "You gentlemen of organized labor and we of an organized movement for the prevention of tuberculosis have a common purpose, and that is that we all wish to improve conditions." The challenge that remained, he concluded, was "how we can best get together."[15]

The Joint Board of Sanitary Control

Labor leaders' increasing willingness to appeal to concerns about public health coincided with a period of unusual vigor and militancy in the American labor movement. Far from being a symptom of political weakness or lack of militancy, labor's appropriation of the "socialism of the microbe" rhetoric came at the high-water mark of its pre–World War I strength. The convergence of rising militancy and public health consciousness is nowhere better illustrated than in the New York City garment industry and the origins and early operation of the Joint Board of Sanitary Control. The formation of the Joint Board of Sanitary Control in 1910 demonstrates how consumer anxieties about contagion and worker concerns about sweatshop labor finally came together in a more workable, albeit uneasy, alliance. This alliance emerged in the wake of a long series of strikes in the New York City garment industry—strikes that reached a violent peak in the fall of 1909 with the legendary "rising of the 20,000," the largest strike by women workers in American history. Although the women's strike, which was led by the International Ladies Garment Workers' Union, proved unsuccessful, it laid the foundation for a more successful effort by the male workers of the cloakmakers' local the next year.[16]

The cloakmakers originally struck to gain recognition of their union, but as the conflict dragged on, they also called attention to the deplorable sanitary conditions common in all phases of the garment industry. During the lengthy negotiations to end the strike, Julius Cohen, a lawyer for the garment manufacturers' association, proposed that manufacturers and union locals donate money to establish a joint board of sanitary control composed of two members each from their ranks and three members representing the public. The joint board would then appoint a staff to inspect the garment factories and force management and workers alike to adhere to a mutually agreed upon sanitary code.[17]

What made supporting sanitary reform so appealing to both labor and management was the growing clout of the third party to their negotiations: the "public" whose health needed protecting from the dangers of sweatshop goods. The large manufacturers could more easily afford to upgrade their sanitary facilities than could their smaller competitors; they would thereby profit from the consumer preference for clean, contagion-free goods. The ILGWU could claim credit for securing improved working conditions, as well as better wages and hours, and thereby diminish the appeal of the nonunion shop. Thus in 1910, when the warring parties finally agreed upon their celebrated "Protocol of Peace," also known as the "White Protocol," the establishment of a joint board was one of its key provisions. Even though they soon disagreed over other aspects of the protocol, both management and labor continued their support for the joint board idea and gradually extended its authority over the entire ladies' garment industry.[18]

The Joint Board began its work with a budget of $7,000, which had been contributed in equal shares by the garment manufacturers' association and the ILGWU. The board's first act was to commission an investigation of sanitary conditions in the garment industry, which was completed in 1911 by George Price, who subsequently became the board's first director. Based on the commission's findings, the board members then agreed upon a "Code of Sanitary Standards" to be required of all garment manufacturing establishments. A staff of salaried inspectors was employed to visit each firm twice yearly and issue sanitary certificates bearing grades, the best being an *A,* the worst being a *C.* Shops that failed to meet the minimum standard were denied any certificate at all until they corrected their errors.[19]

The Joint Board's sanitary code covered a wide range of industrial hygiene issues such as proper lighting, safety devices on machinery, and fire escapes (a particular concern after the Triangle Shirtwaist fire in 1911). But the control of contagious diseases, and primarily tuberculosis, was among its chief objectives. The board's standards covered the basics of tuberculosis prevention, including proper ventilation and a spitting ban. In carefully prescribing the care of toilets, it also carried the overlay of older sanitarian concerns with sewer gas.[20]

The board's inspectors modeled their visits on those of state factory inspectors, but there were important differences. Unlike state factory inspectors, the board's inspectors were only advisors to management and unions and, in the words of one annual report, "had to depend, not upon threats, fines or penalties, but upon counsel, persuasion and education" to correct the "dense ignorance of sanitary and health matters" that existed in the industry. Compared to regular factory inspectors, the Joint Board's inspectors focused much more exclusively on sanitary practices. As one inspector observed, the minute she headed for the toilets, everyone on the shop floor knew she was from the Joint Board.[21]

By virtue of education and professional expertise, the Joint Board's early members were well suited for the work. The "public" was represented by three members drawn from the cream of New York City's Progressive community: the chair, William Jay Schieffelin, was head of a prominent wholesale drug company and active in the anti–Tammany Hall Citizens' Union. Lillian Wald, another public representative on the board, was a prominent reformer and founder of the Henry Street Nurses' Settlement. The board's secretary, Henry Moskowitz, was a protégé of Wald's who had earned a Ph.D. in philosophy from a German university and had considerable experience in tenement house reform. One of the first union representatives, and later director of the board, was George Price, who had an M.D. from the New York University School of Medicine.[22]

At the same time, the personal backgrounds of Joint Board members and staff made them more receptive than many of their Anglo-

American peers to the garment workers' point of view. Wald came from a German-Jewish family, and Moskowitz and Price were Eastern European Jews who had immigrated to the United States as youths. Price had particularly deep roots in the immigrant community, which enhanced his credibility with the workers. After stints as a newspaper reporter and a tenement house inspector, he received his medical degree in 1895 and set up a private practice on the Lower East Side. The first sanitary inspectors had equally strong ties to the garment workers. Pauline Newman, who served as an inspector and assisted Price for many years, was a Russian-born Jew who spoke Yiddish fluently. Like many in the ILGWU leadership, she was a fervent Socialist. Her close friend Rose Schneiderman, the influential Socialist labor organizer, also served as an inspector for the Joint Board.[23]

The philosophy of the Joint Board combined an admiration for scientific progress with a genuine commitment to radical reform. In essence, what George Price and his colleagues did was promote a labor-oriented version of the "socialism of the microbe." As Price put it in an article for the *Survey* in 1911, "A common ground had been created for capital and labor to stand upon in harmony without being accused of treachery to their respective causes." To Price and other labor reformers, the Joint Board's work beautifully illustrated an industrial philosophy rooted in the reciprocal rights and duties of capital and labor. Writing in the Board's *Bulletin* for 1912, Price explained, "In our campaign for industrial civilization, the slogan of the Joint Board of Sanitary Control has been, 'A Safe and Sanitary shop is the worker's birthright.'" But, as he noted, "There are no rights, says Mazzini, without corresponding duties," and the Joint Board principle "obligates both sides to realize the sanitary standards formulated by the Board in every shop."[24]

Although the board's approach tended to assume that workers were more sinned against than sinners when it came to sanitary infractions, it continually emphasized how important workers' cooperation was for bettering working conditions. Price praised the ILGWU locals for their support: "For the first time organized labor conceded that the demand for 'safe and sanitary shops' is equal in importance, if not superior, to the common demands for increase in wages and decrease in hours of labor." Only if workers were willing to go out on a "sanitary strike" over hygiene issues, as they had shown themselves willing to do, would their safety be assured.[25]

The *Workers' Health Bulletin,* published by the Joint Board in 1915 for distribution among garment workers, reinforced the idea of reciprocal rights and duties concerning public health issues. Like the AFL's tuberculosis tract, the Joint Board's catechism cast disease prevention in terms of union principles. Workers had to protect each other from disease; the section on the rules of tuberculosis protection was headed, "HOW TO AVOID GIVING TUBERCULOSIS TO OTHERS." Workers also had the right to demand from the owners a clean and sanitary shop, but in return employers had a right to ask workers to keep the shop clean. In dividing up the practical day-to-day work of what it termed "sanitary self-control," the Joint Board worked out a calculus of responsibility for the "safe and sanitary shop." Owners were held responsible for failing to ventilate toilets or to screen windows against flies, for permitting the use of the common cup or roller towels, and for not washing down shop walls and floors. Owners and workers were held jointly responsible for stopped-up toilets and floors littered with cast-off food crumbs or cloth.[26]

Although the Joint Board continually praised labor's effort "to control its own sanitary destiny," as Price phrased it, the reports from the shop floor suggest that his ideal of labor cooperation was not so easily accomplished. In theory, workers had everything to gain from becoming more conscientious about protecting themselves and their mates from tuberculosis and other contagious diseases. Yet in practice, workers did not necessarily share the board's assumptions about the urgency of disease prevention. Rose Schneiderman explained the problem succinctly in 1912: "In my shop talks I often hear, 'Never mind sanitation; what we want is bread,' and it takes a good many arguments to convince them that the body is nourished by fresh air, cleanliness, and sun-light as well as food."[27]

Just as immigrant housewives made a distinction between inside and outside, garment workers tended to view the shop as the boss's domain, a degraded space not worth keeping clean. "In the struggle for existence," Rose Schneiderman observed, "the workers, and especially our Jewish workers, have come to look upon the workshops as a place which must of necessity be dirty, without air, and unsafe." Pauline Newman also recalled workers' disdain toward her pleas about keeping the shop clean. During the workers' lunch hour, when she climbed up on the cutter's table to discuss the board's work, the men would say to her, "Why bother, it's just the shop." In response, she reminded them that they spent as much time in the shop as they did in their own homes. "You're spending here so much time. To put things into the toilet that the flush won't take it down, isn't going to do you any good. You'll have the smell because you're using it. You're not going home that early." As in tenement houses, the toilets became a sanitary battleground over responsibility. They would stop up, the shop owners would fix them, and the workers would stop them up again. The situation suggested either a real lack of comprehension about what would go down a flush toilet, akin to that Kittredge observed among tenement housewives, or a form of sanitary sabotage designed to annoy the boss and possibly the board inspectors as well.[28]

The most overt resistance to the sanitary argument was evident in union members' continued suspicion of the white label plan. The Joint Board began campaigns for its own sanitary label soon after its founding. An article in its 1912 *Bulletin* proclaimed that the Protocol label "would be virtually a Union label, for the existence of the Joint Board of Sanitary Control is the result of Union recognition"; but at the same time it would be "broader than a Union label, as it is issued by a Board in which the employers and the public are represented."[29]

In a special convention held in Yonkers in 1913, ILGWU delegates met to consider the board's white label plan. Although the more militant members opposed the idea, which they felt involved too much cooperation with the capitalists, a motion favoring the Protocol label passed by a vote of seventy-two to thirty-four. But none of the locals carried through on the plan, presumably because member sentiments in favor of it continued to be lukewarm. Garment workers still preferred the *union* label, which stood for fair wages and hours and an end to competition from nonunion shops, over the sanitary label, which represented a consumer interest that they still did not see as harmonious with their own.[30]

Not until more than ten years had passed did the Joint Board get its sanitary label, and even then, it was over the ILGWU's protests. In 1924, when the garment industry appeared headed for another prolonged strike, Governor Al Smith appointed an advisory commission to negotiate a settlement. In its list of demands, ILGWU leaders pointedly asked for a union, not a sanitary label, but the commission ignored them and endorsed the idea of a white label to be overseen by the Joint Board. Henry Moskowitz, who was appointed head of the board's new "Prosanis" label division, defended the commission's choice. "Faced with the problem of controlling the cut-throat competition of non-Union shops with the responsible employers," he explained, they had chosen the best solution. "Devices other than the Label have been futile," because manufacturers still send work out to be done in "unclean and unsafe shops." Only by enlisting the consumer's sanitary scruples could this practice be put to a stop, he argued.[31]

In contrast to the ILGWU's lukewarm response, a coalition of women's groups hailed the Prosanis label as an effort to protect "not only the workers, but those purchasing the garments, from the dangers of disease-breeding garments made in unclean and unsafe shops." The label campaign was launched not with a labor rally but with a "Prosanis fashion show" for the city's fashion designers and buyers. When the first label was issued in April 1925, it was ceremonially sown on a dress by the national president of the Junior League and Governor Smith's daughter. By 1926, over 10 million labels had been issued for use in 2,176 garment manufacturing shops.[32]

George Price and Henry Moskowitz continued to portray the union label idea as "narrow" in contrast to the "broad" appeal of the sanitary label. But given the list of union demands in the years before World War I for benefits such as better parks, recreation, and public transportation, the narrowness of labor's reform objectives is not so obvious in retrospect. What *was* narrow was the base of support for the union label. Not only were middle-class consumers and voters ambivalent about the organized labor movement, but the movement itself only represented around 4 million workers as of 1920. Moreover, the unions were controlled by men with little understanding of women's concerns as either workers or consumers. Recalling her years of working for the union label, Pauline Newman noted that she got better results from speaking to "society women and women's clubs" than to union meetings dominated by men. Linking sanitary reform with middle-class women's consumer interests was simply the most pragmatic way of drawing attention to disease in the workplace.[33]

Whether the Prosanis label and others like it really made a difference in middle-class women's consumer habits is hard to measure.

Did they buy clothes with the Prosanis label in order to avoid the risk of infectious disease, or for other reasons, such as style or price? Henry Moskowitz observed in 1926 that "skeptics have declared that women respond more sensitively to style than to a label having a social appeal," and they may well have been right. The consumer idea had more power in theory than in practice; the reality of women's resolve to buy sanitary clothes may have been far less than the reformers asserted.[34]

Yet for rhetorical purposes, the specter of an army of worried mothers determined to avoid germ-ridden sweatshop goods certainly had its uses in the labor politics of the early twentieth century. Popular understandings of the dust and fomite theories of germ transmission allowed consumer advocates and union leaders alike to insist that the "public" wanted germ-free clothes, and that in order to provide these clothes there had to be an end to exploitative tenement-house labor. For all its imperfections, that argument served the needs of both manufacturers and union leaders in the garment industry.

But what of the workers themselves? It seems fair to say that the Joint Board's activities did indeed improve their overall working conditions and access to health care. From 1910 to 1930, rates of tuberculosis among workers in the New York City garment industry declined steadily, a trend for which the board happily took credit. In addition to haranguing workers about their hygiene habits on the shop floor, the Joint Board also began to provide them with good, inexpensive health care. In 1914, George Price started to offer free physical examinations for garment workers; from that work eventually grew the Union Health Center, the first union-sponsored health care facility in the United States. Funded by the so-called tuberculosis tax of one dollar per member, which was paid by participating locals, the center provided annual physical exams as well as dental and medical care. It also expanded the program of health education that the Joint Board had begun in the 1910s. By the late 1920s, the Union Health Center had become a major provider of health care for the entire Lower East Side.[35]

The irony here, as with the whole antituberculosis movement, is that the growing consciousness of infection developed in the name of labor solidarity led to an increasing ostracization of sick workers. The rule soon became that once workers were diagnosed with tuberculosis, they had to leave the shop floor; although consumptives were eligible for treatment in union clinics and sanitariums, they were nonetheless eliminated from the workforce. With seemingly no sense of this irony, Pauline Newman proudly described how "years of painstaking effort to bring the message of good health to ILGWU members" had made garment workers more disease-conscious: "Where in years past few if any workers paid attention to a coworker who coughed or had a skin condition, today they refuse to work with him unless the Union Health Center certifies that the illness is not contagious." As was so often the case with the gospel of germs, a new sense of solidarity among the well members of society developed only by isolating and shunning the contagious.[36]

The Negro Anti-Tuberculosis Association of Atlanta

Just as the principle of "sanitary self-control" brought together leaders of management and labor in the garment industry, the antituberculosis movement opened up common ground for white and black reformers, particularly in the South. There the "chain of disease" concept exposed the thorny issue of what was euphemistically termed the "negro problem." Death rates from tuberculosis were three to four times higher for blacks than whites. These statistics could not be ignored given the fact that "germs know no color line," according to a popular slogan of the time. As Charles P. Wertenbaker, a white doctor in the U.S. Public Health Service, explained in 1909, "The negro is the disease reservoir of the South from which our supply of diseases is being constantly augmented."[37]

The antituberculosis crusade in Atlanta coincided with an exceptionally bleak period for race relations in what was then the most segregated city in Georgia. African Americans, who made up 40 percent of the city's total population, lived in a few overcrowded, unhealthy neighborhoods. Starting in the early 1900s, the city's white politicians conducted a vicious campaign to reduce black voting rights. In the midst of this push for disenfranchisement, Atlanta endured one of the worst race riots in southern history. For several days in September 1906, white mobs roved the streets, beating African Americans and destroying their homes and property; the city was eventually put under martial law. In the wake of the riot, legal disenfranchisement of black voters became virtually complete.[38]

The Atlanta Anti-Tuberculosis and Visiting Nurse Association was founded by a group of prominent white Atlantans in 1907, only a year after the riots. An outgrowth of the charity organization movement in the city, the society was deeply concerned with the high rates of the disease among African Americans, chiefly because of the threat that they represented to the white population. The clinic opened by the society that same year treated white and black patients on different days. To meet the great need for treatment, a separate "colored" clinic was opened in 1909 and put under the oversight of a special "Negro Race Committee." When local African-American leaders expressed an interest in running the clinic themselves, the Negro Race Committee suggested the formation of a "Negro auxiliary" to raise money for the existing clinic. On paper such an auxiliary was founded, but it accomplished little.[39]

Meanwhile, white anxieties about tuberculosis became focused on the disease threat supposedly represented by the city's black laundresses. In Atlanta, as throughout the South, many white families sent their laundry out to be washed and ironed by black women in their own homes. In 1909, the Anti-Tuberculosis Association's visiting nurses compiled a grim collection of case histories, many illustrated with photographs, illustrating how black consumptives were living and dying next to piles of white customers' laundry. For example, one case described a thirty-three-year-old man in an advanced stage of consumption whose wife did laundry in their one-room home. "There the ironing goes on—the clothes are piled on the bed etc.—and then carried into white homes," the commentary noted. It concluded, "This man . . . is a menace to *us*." In another case, the nurses described a laundress caring for her twenty-two-year-old consumptive daughter, who "*waited upon her & washed her bedding* & the *cloths used* on the *suppurating glands* of her neck in the *same tub* with her *regular washings* for her *white patrons*."[40]

The Atlanta Women's Club, no doubt horrified by such revelations, pressed the city council to pass an ordinance requiring medical inspections of all laundresses, followed by forcible isolation of all those found to be consumptive. Ironically, given that its own visiting nurses had done so much to start the clubwomen's crusade, the Atlanta Anti-Tuberculosis Association did not support the proposal; they pointed out that until the state provided a sanitorium for colored consumptives, it "would work a great hardship upon the families." The city ordinance did not pass then, nor again in 1912, when it was proposed during the mayoral campaign.[41]

Given the somber mood of race relations in the preceding decade, it would have been hard to predict in 1910 that Atlanta was about to give birth to one of the most successful black antituberculosis movements in the United States. Yet out of the ashes of the 1906 race riot and the 1909 laundress controversy, black and white community leaders managed to convert growing anxieties about tuberculosis into a remarkable effort at biracial collaboration. The Negro Anti-Tuberculosis Association of Atlanta, as it came formally to be called, was largely the inspiration of Lugenia Burns Hope, a noted leader of the African-American community.

Hope came to Atlanta in 1898 when her husband John was hired as a professor at the Atlanta Baptist College, which was later renamed Morehouse College. During her youth in Chicago, she had been active in a variety of reform groups. Watching the Morehouse professors frantically searching for firearms to keep white mobs from burning the college during the 1906 riot, Hope became convinced that race solidarity was the only safety against white injustice. To that end, in 1908 she started a group she called the Neighborhood Union, which was dedicated to the "moral, social, and educational improvement" of the race.[42]

Through her work with the Neighborhood Union, Hope quickly became aware that tuberculosis was a terrible blight on Atlanta's black community. She may also have known that in other parts of the country, including Washington, D.C., and Savannah, Georgia, other African-American leaders had started societies to fight the inappropriately named "white plague" in their neighborhoods. Just as union leaders had developed a distinctive analysis of the TB problem, these black-led antituberculosis societies interpreted the "negro problem" in terms quite different from those employed by the mainstream anti-TB movement.[43]

To race leaders, the tuberculosis epidemic among African Americans was stark evidence of the twin burdens of poverty and racism. In 1896, at an Atlanta University conference on black mortality in the cities, Dr. Henry Rutherford Butler of Atlanta pointedly noted that black Atlantans were barred from public parks, forced to live in the filthiest parts of the city, and too poor to afford a doctor. "Is it any wonder that we die faster than our white brother when he gets the first and best attention, while we are neglected on all sides?" he concluded.[44]

In a speech made a year before the great laundress debate, Lugenia Burns Hope made a similar argument, calling for sanitary investigations that would place the "responsibility of disease and crime where it justly belongs, on the landlords." She argued, "They are responsible for the conditions of the houses they rent, and when they insist on family after family moving into filthy[,] broken down[,] poorly ventilated houses[,] they are spreading disease to hundreds of families, and it is no wonder that mortality is so high, that hundreds of tuberculosis patients die every year—while these men receive their 'blood money' and care nothing for the people or community." She concluded, "The Negro has been censured for years as a breeder of all that is vile—a menace to any community—but the truth is we are neglected."[45]

In 1912, Hope decided to pay a visit to the Atlanta Anti-Tuberculosis and Visiting Nurse Association's offices and explore the prospects for collaborating on an anti-TB campaign. During that visit, she met Rosa Lowe, the secretary for the association, who listened to her proposition with interest. The Georgia-born daughter of a Methodist minister, Lowe came from a background not that different from Hope's. After graduating from Scarritt Bible and Training College, an institution known for its socially liberal views, she had supported herself as a nurse. Lowe had come to Atlanta in 1903 to head the Wesley Settlement House, which served workers employed by the Fulton Bag and Cotton Mill. She left that job in 1907 to work full-time for the antituberculosis society.[46]

It took some months for Lowe to follow up on Hope's proposal, but when she did, her approach differed quite noticeably from the Atlanta Association's previous dealings with local black leaders. Lowe wrote Hope a friendly letter, proposing that they work "hand in hand"—a significant choice of language, given the prevailing anxiety about hand-borne contagion in that era—to develop a joint anti-TB program, and the two women agreed to call a general meeting to discuss how the collaboration should proceed.[47]

In a statement that Lowe wrote in 1914 outlining the new initiative, she took a strikingly different tone from previous discussions of the "negro problem." She began, "All health workers recognize the importance of enlisting the interest of every group and individual in the community in the prevention of disease," and because blacks made up at least one-third the population in most southern cities, "any program of work which ignores them is an incomplete one." Since emancipation, she noted, the southern freedmen had worked hard to better themselves, and in Atlanta, there were at that time "many well organized institutions for help of the individuals [sic] as well as the community." Lowe enumerated the many black-led groups, including "ministers, doctors, insurance agencies, teachers, kindergarten associations, and nurses" who were willing to assist in the work. "The most difficult obstacle in the work among the Colored people," she observed, "is the establishment of confidence, for Negroes as a whole have lost faith in the White Race and vice versa, and it is only individuals here and there who have retained this trust." Lowe blamed racial discrimination for the breakdown in trust, for "neither in politics, education or religion has the Negro received a fair deal."[48]

Much as George Price had identified public health as neutral ground for capital and labor, Lowe wrote, "It remains to hygienists to establish a friendly relation between these two Races living in one community, upon both of whom the responsibility rests for a healthful city." To this end, she outlined a sweeping program of reform, which included improving recreational facilities, sanitation, street lights, housing, schools, hospitals, and medical care. "All these things enter into a program of health and where the field is not occupied by others the fighters of tuberculosis should enter, if not to do the actual work, to enlist others in order that this proportion of our citizens be protected also."[49]

As Lowe's statement made clear, she shared Hope's conviction that the "negro problem" resulted from neglect rather than innate racial defect. The two women saw the antituberculosis issue as a way to press for public health services that black Atlantans could justly claim, even in a period when their political and social rights were being sharply circumscribed. If white Atlantans were so concerned about the "disease reservoir" that existed in the homes of their black neighbors, Hope and Lowe argued, let them help rather than hinder black residents' efforts to live according to the gospel of germs. In terms similar to George Price's notion of industrial rights and duties, they outlined a kind of public health citizenship based on reciprocal duties and obligations among the individual household, the neighborhood, and the city. Black Atlantans stood ready to demonstrate their willingness to make the household and neighborhood changes necessary to diminish the disease threat; in return, they expected the same municipal services already enjoyed by white neighborhoods.

With this shared vision in mind, the old white-dominated Negro Race Committee was replaced by the black-run Negro Anti-Tuberculosis Association, which had Rosa Lowe as its secretary. Over the next decade, a productive pattern of cooperation developed between the Negro Association, also referred to as the "Colored Branch," and the white Atlanta Association. Using the same zone system set up by the Neighborhood Union, which had divided the city's black neighborhoods into sixteen sections, local committees were created to organize the antituberculosis work.[50]

In setting up these committees, Hope drew heavily upon her middle-class peers. By virtue of being the "First Lady" of Morehouse, she was well connected to Atlanta's black educational elite, including the faculty and alumni of Morehouse, Spellman College, and Atlanta University. The Negro Association also courted influential churches, appointing leading African-American ministers to its board of directors and consulting closely with local congregations in its work. Black life insurance companies, who like their northern counterparts had a strong profit motive to lengthen policyholders' lives, also became a vital source of financial support.[51]

In other words, the Negro Association was dominated by what W. E. B. Du Bois called the "talented tenth"—the educated middle-class elite of the black community. Like many such associations in the Progressive period, the Negro Association's approach emphasized self-improvement and strict standards of personal morality. The evangelical fervor and middle-class virtues of "tuberculosis religion" appealed to them no less than to their white counterparts in the anti-TB movement. By combining hygiene reform with moral uplift, the agents of the Negro Association gave owners of Atlanta dance halls and saloons good reason to fear their visitations. At the same time, their moralism was infused by a commitment to racial solidarity and social justice that was often lacking from white TB crusaders' interactions with the poor of their own race. The group identified far more closely with the recipients of its benevolence than did the Joint Board of Sanitary Control's members with the garment workers.[52]

The Negro Association's determination to resist discrimination took it immediately in very different directions from the agenda pursued by the white-dominated anti-TB societies. In the first place, Lugenia Burns Hope put a new twist on the conceptualization of tuberculosis as a "house disease." As used by the white visiting nurses in 1909, the "home survey" served only to stigmatize consumptives as disease threats to white families. In contrast, Hope turned the home survey into an opportunity to educate the occupants about disease prevention. If every African-American housewife in Atlanta could be informed about the threat of contagion from spitting, dust, impure milk, flies, and the like, she reasoned, they might make a tremendous difference in preventing illness in their families.[53]

Simultaneously, Hope used the home survey to try to force landlords and city health officials to fulfill their legal and moral obligations to black households. Her zone committees compiled lists of housing and community nuisances and took photographs of "unsanitary places" in an effort "to get the Board of Health interested," as she wrote. The information was passed on to the more politically powerful white anti-TB association, which then pressured the health board to have the defects corrected. Following the home-to-world line of influence, the practice of household inspections was eventually broadened to include other places where diseases might be spread, such as grocery stores, ice cream stands, and lodge halls.[54]

Starting in 1917, Hope's conception of the neighborhood sanitary watch was incorporated into the spring cleanup crusade sponsored by the National Negro Business League. In the early 1910s, local chambers of commerce had started sponsoring cleanup weeks as a form of civic improvement, and African-American business and community leaders quickly expanded the idea to include their neighborhoods. In Atlanta, cleanup week evolved into a coordinated plan of home surveys, special garbage collections, and educational programs in churches and lodges. The very first year that the National Negro Business League sponsored a prize for the best citywide program, Atlanta won the award. The outreach achieved during these campaigns was indeed impressive: the Negro Association reported that in 1917, its agents visited 3,786 homes with an estimated 13,000 occupants; in 1919, they surveyed 5,406 homes, reaching 23,771 occupants. Assuming that their statistics were accurate, the anti-tuberculosis crusade reached nearly one-fifth the city's African-American population during those two years alone.[55]

Unfortunately, the Negro Association's success in getting black citizens to collect trash far outstripped the city of Atlanta's willingness to haul it away. Rosa Lowe had anticipated this problem before the 1917 campaign began, and had visited the health officer of Atlanta to get "the promise of his cooperation." But as of 1919, Lugenia Hope was still complaining that "this Association has worked long and faithfully on the City Sanitary Department to do its duty, but little has been done for the six years we have been working together."[56]

Besides promoting sanitary privies and clean yards, the Negro Association took up causes that involved a much broader conception of enhancing public health. In their hands, the cause of tuberculosis prevention stretched to provide a remarkably capacious rationale for racial justice and self-improvement. For example, in 1916 Hope and Lowe began making annual inspection tours of the city's segregated school system and meeting with representatives of Atlanta Board of Education to demand improvements in the "colored" schools. The Negro Association justified these demands by arguing not only that overcrowded, unsanitary schools provided fertile breeding grounds for disease, but also that African-American children needed a decent education to master the rudiments of modern health citizenship. Besides badgering the school board, the Negro Association saw to it that the Modern Health Crusade was introduced in the city's segregated schoolrooms; as a result, the number of children treated in the TB clinics steadily increased.[57]

Other issues raised by the Negro Association had even less to do with conventional notions of public health. A case in point was its involvement in the public library controversy. In 1904, the Carnegie Foundation offered to donate funds to build a "negro library," but the local library committee placed so many restrictions on where the new facility could be located that the project was virtually blocked. In the mid-1910s, the Negro Association started a campaign to pressure the library committee into action. Its president, Henry H. Pace, collected letters from other cities, including Louisville and Savannah, that "highly recommend the establishment of libraries for negroes as an aid in building up good citizenship," in his words. At the suggestion of one of the Atlanta Association's members, Julian V. Boehm, Pace wrote a letter to the *Atlanta Constitution,* which Boehm forwarded to the paper's editor, Clark Howell, who not only published it but also wrote an editorial in support of the library. Embarrassed by the negative publicity, a representative from the library board immediately agreed to meet with Pace and Boehm, and the library eventually opened in 1921.[58]

The Negro Association raised similar complaints about the dearth of playgrounds and parks for the city's African-American population. The antituberculosis movement celebrated the gospel of sunshine and fresh air, yet Atlanta's growing system of public parks existed for the use of whites only. Raising this issue with the city's park manager, Dr. Loring B. Palmer of the Negro Association was told that the manager "would be willing to recommend to [the city] Council that a certain property be bought for this purpose in case the Anti-Tuberculosis Association would back the movement by its moral support." The white association agreed to do so, and it enlisted the help of the Rotary Club to implement the plan.[59]

It is apparent from both the library and park campaigns that members of the white tuberculosis society played important roles as go-betweens with the city's white politicians and civic leaders. The white tuberculosis organization seemed willing to support what were to be segregated facilities—that is, a "colored" public library and a "colored" park. But as the Negro Association worked to broaden the meanings of tuberculosis prevention, their dialogue with their white counterparts moved into far less comfortable territory. The tensions between the two groups were evident at a special joint meeting called on July 27, 1917, to discuss the mass migration of southern blacks to work in northern factories during World War I. In the South, leaders of both races feared that the "Great Migration" would diminish the region's economic and social progress. At the same time, African Americans were all too aware of why so many of their brethren wanted to leave.[60]

At the joint meeting, members of the Negro Association spoke freely about the many problems that were driving black Georgians northward. The Reverend Peter James Bryant noted that "immigration started down near Albany where the colored people had been knocked down and killed, the churches burned etc." Other members mentioned the low wages, Jim Crow restrictions on public transportation and recreation, dilapidated schools, police harassment, and municipal taxation without political representation. The minister Lorenzo H. King warned, "The Negro was losing faith in the religion of the white man as was demonstrated in the attitude of the white man to the Negro." Making a careful distinction between social intimacy and citizenship rights, he explained that "the Negroes do not want to be taken into the homes and bosoms of the whites,—that such a thought is as disagreeable to them as it is to the whites." Rather what they wanted was to be given the same opportunities as whites had for political and economic improvement.[61]

The response from the whites present was restrained yet sympathetic. The minutes recorded that a speaker identified only as Judge Tindall "arose to remind the colored people present that the white people are up against a hard proposition because they have so many ignorant white people among them who stand in the way of any radical changes." Tindall praised the self-improvement ethic in the African-American community and "held out to them the strong hope that within a short period great changes would be made in the conditions now existing." Kendall Weisiger added that the meeting "had opened his eyes and made him look on it in a different light entirely" and that "every white person present would be a better friend to the Negroes for having heard their statements and would exert their influence to bring others to see the situation in the same light." The meeting concluded with an agreement "that the two races get closer together in the matter of cooperation" and that their leaders "bring about a better feeling and better conditions, keeping closer together on the vital subjects of the day."[62]

But although backing from the white anti-TB society certainly helped the Negro Association, its achievements rested first and foremost on the support of the African-American community. Indeed, one reason that the Negro Association could pursue such a broad agenda was its financial independence from the white anti-TB society. Here the black insurance companies played a particularly important role. Under the leadership of Henry H. Pace, who headed the Standard Life Insurance Company, the "colored insurance committee" raised the money to hire the first salaried African American on the association's staff, a position explicitly created for Lugenia Burns Hope in 1919, and to purchase the educational literature and other materials passed out during cleanup week and the Christmas seal campaigns.[63]

After World War I, the Negro Association sought to extend its financial base by organizing tuberculosis leagues in black churches. The first was founded in 1921 in one of the city's biggest congregations, Big Bethel Baptist. Church members paid one dollar to join; half of this dollar went to the church to be used as a relief fund, the other half to finance the Negro Association's educational activities. In their 1922 annual report, the Negro Association's board expressed its hope that the new source of funding would put it "in a position to push certain sanitary improvements in the housing conditions of the Negro population," which was "a fundamental requirement upon which any prevention of the spread of this dire disease must be firmly based." When in a planning meeting someone questioned whether the Negro Association could raise $1,000, both the minister of Big Bethel and Lugenia Burns Hope "expressed a belief that this could be done, especially in view of the fact that the Colored People feel more kindly toward the Anti-Tuberculosis Association than any other organization in the city, and realize that it has been a great help to them in their activities."[64]

The Negro Association continued to work vigorously in the 1920s and became one of the most active such societies in the country. Its social service "institutes" to train antituberculosis workers became the nucleus of the first black social work program in the United States, which later grew into the Atlanta School of Social Work. Noting that Atlanta's African-American population had one of the lowest death rates from TB in the nation, Kendall Weisiger, president of the Atlanta Association, observed in 1926 that "these figures are not accidental" but rather reflected the work of the Negro Association, the "only one of its kind in the world."[65]

Like the Joint Board of Sanitary Control, the Negro Anti-Tuberculosis Association of Atlanta achieved an unusual degree of success. A strong local leadership, working in partnership with sympathetic whites, built a vigorous grassroots movement around the issue of tuberculosis prevention. Yet although the Atlanta group was particularly successful, it was by no means unique in the strategies and themes that it employed. Between 1900 and 1930, similar "colored" tuberculosis societies were formed in both the North and the South. Working with the "National Negro Health Week" program, which operated out of the Tuskegee Institute (now Tuskegee University) in Alabama, they developed an ambitious program of popular health education and reform.[66]

The African American anti-TB movement used white anxieties about infectious diseases to some constructive purpose. The concept of the "disease reservoir" proved to be a classic example of the "two-edged sword" provided by the new belief in the communicability of consumption. So long as whites believed that "germs had no color line," they could ill afford to let their laundresses, cooks, and servants fall prey to disease. Given even that slight opening to assert their needs, African-American reformers parlayed the imperatives of the gospel of germs into a much broader program of social uplift.

Still, the success of the Atlanta antituberculosis crusade brought its own painful ironies. The spirit of cooperation between white and black reformers in Atlanta exposed the realities of racism in the larger anti-tuberculosis movement. Despite the NTA's rhetoric of inclusiveness, the Atlanta society continually bumped up against the national movement's own internal color line. In 1915, Rosa Lowe wrote to the NTA's educational director Philip Jacobs about an upcoming regional conference "to ascertain whether or not the negroes and whites were to be allowed to assemble in the same hall for the meetings, and whether the negro delegates could have the advantage of the subjects discussed by the whites by being present at the sessions where the general subjects were discussed." Again in 1924, when the NTA held its national convention in Atlanta, the Negro Association requested that meetings for black delegates not be held separately, on the grounds "that the discussion of the work for colored groups would be very beneficial to the white people, who were promoting different phases of development."[67]

In more subtle ways, the great white plague reinforced the association between racial stereotypes and disease risk. While Northern anti-t.b. societies were beginning to show some cultural sensitivity to the "new immigrants," putting images of Venice on their posters and printing their "Don't cards" in many languages, the children in Atlanta's segregated schools were still learning their hygiene lessons from the Modern Health Crusade literature, with its images of little blonde crusaders dressed in white. In a letter to Lugenia Burns Hope written in 1920, Leet Myers, the director of "Negro work" at the National Child Welfare Association, acknowledged, "It hardly seems fair that the Negro child should have all his teaching and his idealism presented in terms of white people." Child health organizations gradually began to develop educational materials using pictures of African Americans, but the vast majority of black children in the 1920s and 1930s still got their health catechisms in a form that suggested whiteness and cleanliness were one and the same. The only way to make the faces of good health look like their own was to color them by hand, as was often done in the children's drawings submitted as part of National Negro Health Week.[68]

In order to get the most basic municipal services for black communities, reformers were forced to appeal to, and thus reinforce, white perceptions of blacks as dangerous disease carriers. As the immigrant-reformer Mary Antin put the matter bluntly in a 1922 letter to Lugenia Hope, "That there should be such need is pitiful." It was incredible, she lamented, that in a land of supposed freedom and democracy, "there should have to be a militant organization to beg for lights for the people!"[69]

What successes black Atlantans enjoyed in this regard were overshadowed by the continued injustices and indignities of life under the Jim Crow laws. The fact that tuberculosis rates fell in Atlanta did nothing to reverse the pattern of segregation entrenched in every aspect of its daily life. When hygienic improvements were offered African Americans, they came in the bitter guise of segregation: the "colored" library, the "colored" park. Likewise, when they finally arrived in the South, public drinking fountains and sanitary toilets came in segregated pairs, becoming among the most visible and detestable symbols of a segregated society. Not until the 1960s did the civil rights movement begin to dismantle the practice of hygienic segregation, along with the many other forms of discrimination entrenched in the South.

Still, for all its glaring weaknesses, the two-edged sword of disease consciousness provided both union leaders and race leaders some small leverage for change. However meager the results may seem now, the "socialism of the microbe" argument had considerable power and appeal as a consensus-building tactic in the early twentieth century. At a time when housing and working conditions for poor Americans, including southeastern European immigrants and African Americans, were desperately poor, the specter of infectious germs moving across class and race lines became a useful ploy for calling attention to their plight. The issue of disease prevention offered a vision of public health citizenship that disadvantaged groups might more easily claim than political influence or economic parity. As the Georgia Federation of Colored Women's Clubs asserted in 1921, when demanding an equal share of taxpayers' funds for "sanitary or health improvements," "Colored people are citizens, too, having the same needs as the rest of the citizens and the same rights for consideration."[70]

Precisely because germs seemed to observe no color or class line, groups tarred with the brush of contagion could assert a compelling claim to provide for their public health needs. The universalistic aspects of the germ—that it preyed with the same malice upon all bodies, male or female, white or black, rich or poor—lent it a potentially subversive air: here was an agent for change capable of showing up the growing inequality and injustices rampant in early twentieth-century America. In the hands of individuals such as George Price, Pauline Newman, Lugenia Burns Hope, and Rosa Lowe, the gospel of germs gave rise to a wider conception of public health morality, that is, the basic obligations that members of modern industrial society owe one another to prevent the spread of contagion.

For groups denied the most basic essentials of sanitation, such as bacteria-free water or public sewers, the notion of public health citizenship provided a potent means to highlight the larger society's failure to provide for their basic human needs. To the extent that adherence to the "gospel of germs" became a sine qua non of modern citizenship, the failure to provide its essential elements—pure air, clean water, and nourishing food—became a social problem that could be ignored only at great risk to the body politic. The inconsistency of fearing the poor as disease reservoirs while tolerating the material conditions that doomed them to tuberculosis provided the underprivileged a stick with which to beat at the complacency of middle-class Americans. If the stick proved too frail to batter down the economic and racial injustices of the day, at least it left a mark.

· IV ·

The Gospel in Retreat

10 · The Waning of Enthusiasm

In 1914, the bacteriologist Charles-Edward Amory Winslow wrote an article for *Popular Science Monthly* titled "Man and the Microbe" that summed up the current status of the gospel of germs. What the experimental research of the last two decades had shown, he wrote, was that "it is people, primarily, and not things that we must guard against." In the vast majority of cases, Winslow asserted, communicable diseases resulted not from exposure to dust or fomites, but from immediate contact with other people. This contact was not always direct; germs could circulate through the media of water or food, insect carriers, and shared objects such as drinking glasses recently touched by the sick. But "back of all such material agents of transmission . . . lies the human being, and the nearer to this source we get,—the more direct and rapid the transfer,—the greater is the danger," Winslow wrote. Those few cases of disease not readily explainable by contact with the sick were most likely the work of the "healthy carrier," whose discovery Winslow termed one of the "great contributions to sanitary science in the last ten years."[1] To account for the mysterious outbreaks of typhoid or diphtheria that cropped up from time to time, modern science now suggested searching for a "Typhoid Mary" rather than for a defective sewer trap or a contagion-bearing cloak.[2]

Winslow's diminution of the dangers from sewer gas, dust, and fomites in favor of an emphasis on contact infection and the healthy carrier represented a movement that came to be known as the "new public health." The first signs of this shift in ideology appeared in the early 1900s, when several leading public health authorities expressed skepticism about certain aspects of the preventive gospel. Charles Chapin wrote a series of articles criticizing the persistence of the old "filth theory of disease" and its accompanying "fetish of disinfection" in modern public health practice. At around the same time, Winslow himself reported on a series of meticulous experiments proving that air drawn from the foulest of rooms and the dankest of sewers contained no live bacilli capable of causing disease. In Cuba, an American commission led by army surgeon Walter Reed demonstrated that the bedding and clothing of yellow fever patients could not convey the disease to a well person, no matter how laden these fabrics were with their discharges. He concluded that the mosquito alone bore the malaria plasmodium to its victims.[3]

At first, these voices of dissent made little impression. Philanthropist Albert G. Milbank complained about an experimental study of ventilation that his own foundation had funded: "It was rather absurd to expend so much money in proving that bad air is not harmful" because "even if they proved it, we would not believe it." But the up-and-coming generation of experimentalists was not so easily discouraged. Slowly but surely, they argued for revising the gospel of germs in light of new lessons from the laboratory, particularly regarding the frequency of infection by dust and fomites.[4]

A social worker named Bailey Burritt caught the new mood of scientific skepticism in a 1916 letter he wrote to the venerable housing reformer Lawrence Veiller. In the letter, Burritt recommended revisions in the hygiene tract distributed by the New York Charity Organization Society. "The statement that germs 'thrive and flourish' in a dark room is scarcely substantiated by facts," he began. Moreover, the section "Things That Cause Sickness" listed dirty sinks, piles of rubbish, garbage, filthy floors and the like, whose "connection with sickness is very vague to say the least." It would be better, he concluded, to eliminate these items and put others in their place, such as spitting, flies, dust (he was still a believer in the dust theory), neglect of hand washing, and the dangers of the common cup and towel. It was important to correct the tract, Burritt urged Veiller, because these sorts of statements "perpetuate a sanitary fallacy and lead your readers back to the fomites theory of disease and away from the modern conception of contact infection and the necessity of personal hygiene."[5]

As Burritt's letter makes evident, advocates of the new public health meant not to discredit the entire gospel of germs, but rather to excise the vestiges of sanitarian belief that still clung to it. Theirs was a mission of focusing and refining: in place of the sewer gas hypothesis, they stressed the fecal contamination of water. Instead of rejecting the dust theory of tuberculosis infection outright, they demoted it to an occasional cause, one far less significant than the hazards of droplet infection. The belief in the long-term infectiveness of fomites was recast as the more immediate transfer of germs through the recently touched doorknob or drinking glass.[6]

Although these refinements may seem subtle, they nonetheless justified an important redirection of the public health movement's efforts. As outlined in Hibbert Winslow Hill's influential 1913 statement *The New Public Health,* the preventive implications of the most recent experimental work were clear: disease control should focus more on isolating the sick and less on ridding the environment of potentially germ-laden dust and dirt. Exacting cleanliness might be sought for moral and aesthetic reasons, Hill conceded, but public health funds and individual energies would be better spent on identifying and isolating the contagious and on inspecting public water and food supplies.[7]

The emphasis on isolation and case finding reflected a growing sense of confidence in laboratory-derived methods of diagnosis, treatment, and prevention. Between the late 1890s and the early 1910s, researchers developed the Widal test for typhoid fever, the Wassermann test for syphilis, and the Schick test for immunity to diphtheria. In time, the X ray, which was invented in the late 1890s, became an excellent tool for detecting tubercular lesions at a much earlier stage of the disease than could the old sputum test. As diagnostic techniques improved and health departments gained more legal powers, the possibilities for early identification and isolation of the sick and of healthy carriers greatly expanded. The development of effective immunizations and treatments for specific diseases complemented the extension of diagnostic powers. The success of the diphtheria antitoxin, the discovery of Salvarsan as a treatment for syphilis, and the development of a typhoid vaccination all reinforced the conviction that laboratory discoveries would continue to strengthen and refine the efforts of organized public health.[8]

In embracing these trends, proponents of the new public health also sought to distance themselves from the evangelical tone and social causes embraced by the Progressive-era disease crusades. For the public health movement to progress along more scientific lines, they believed it needed to be placed more firmly under the direction of individuals like themselves who were fully trained in experimental methods and thus able to sort the preventive wheat from the chaff. Advocates of the new public health still saw a great need for popular health education, especially of women and children. As Hill wrote, "The infectious diseases in general radiate from and are kept going by women" and thus "to teach women, girls, prospective mothers, that they may practice in their household and in turn teach their children to war on invisible germ-foes is one of the functions of public health bacteriology." But cultivating this "sanitary con-science," as Winslow termed it, was best done under the careful supervision of public health experts and not left to hygienic amateurs who were more interested in social causes than in experimental truths.[9]

To improve their professional credibility, the leaders of the public health movement felt that it was essential to adhere to a more strin-gent, laboratory-based standard of knowledge. The limitations of the previous generation of reformers, especially their slipshod scientific thinking and their soft-hearted humanitarianism, seemed painfully evident. As Iago Gladston, a public health physician in New York City, put it, "Enthusiasms and excess zeal often lead into grave errors." These same sentiments were expressed in two popular works of the 1920s, Sinclair Lewis's portrait of the young research scientist in *Arrowsmith,* and Paul de Kruif's classic history of the trials and tribu-lations of the germ theory, *The Microbe Hunters.*[10]

Arrowsmith, published in 1924, was written in the same debunking spirit as were Lewis's earlier works, *Mainstreet* and *Babbitt.* With de Kruif's assistance, Lewis made the novel into a morality play about modern medicine, one peopled with classic types such as the money-grubbing surgeon, the hardworking rural practitioner, and the ideal-istic scientist. Among the most foolish characters in the novel was Dr. Almus Pickerbaugh, the head of a public health department where young Arrowsmith works briefly. Described as bearing a marked resemblance to Theodore Roosevelt, Pickerbaugh uses the health department as a bully pulpit for his moronic brand of health boos-terism, which consists chiefly of ridiculous jingles—such as "Boil the milk bottles or by gum, You better buy your ticket to Kingdom Come"—delivered to housewives and the chamber of commerce.

Instead of doing scientific work in the health department's laboratory, as he had hoped, Arrowsmith is forced to investigate citizens' complaints about sewer gas and to harass public spitters. Pickerbaugh, who knows less science than does a visiting nurse (the most cutting comparison Lewis can imagine), blocks efforts to implement really useful measures, such as pasteurizing the city's milk supply or pulling down its tuberculosis-ridden tenements. Representing himself as a "sidekick of Darwin and Pasteur," he finally wins election to Congress, leaving Arrowsmith to conclude sagely that the real spirit of science consists in "the faith of being very doubtful" and "the gospel of not bawling gospels."[11]

A year later, Paul de Kruif continued the assault on the old-style sanitarian faith in *The Microbe Hunters,* a jaunty account of the germ theory's trials and triumphs from Antoni van Leeuwenhoek's invention of the microscope to Paul Erhlich's discovery of Salvarsan. De Kruif's narrative venerated the new generation of researchers and their agnosticism concerning the faith of "the silly sanitarian." For example, he wrote that thanks to a "real" scientist like Walter Reed, the "bubble of the belief that clothing can transmit yellow fever was pricked by the first touch of human experimentation." Like Lewis, de Kruif portrayed the heroics of the modern "microbe hunters" to consist of skepticism and detachment, not engagement with the messy world and its troubles. "Science is cruel, microbe hunting can be heartless," he concluded, and the "relentless devil that was the experimenter" would continue to ask questions about beliefs that everyone else took for granted.[12]

The "new public health" represented not only a rejection of a feminized "sentimental sanitarianism" in favor of a more manly laboratory science, but also a growing desire to remove medicine from partisan politics. The gospel of germs had spread hand-in-hand with the Progressive movement; when its reform spirit faltered after World War I, so did the allure of broad-based social interventions. After the Red Scare of 1919, the "socialism of the microbe" ceased to have much rhetorical appeal. The violent strikes and race riots that followed World War I, as well as the failures of Prohibition, encouraged many physicians to retreat from the political and social entanglements of the Progressive period. In the conservative climate of the 1920s, it was more congenial to think in terms of public health surveillance than to advocate social transformation. Disease and health issues by no means disappeared from the political arena, but the range of reforms offered in the official name of public health became considerably more circumspect. Debates about the meaning of public health citizenship shifted to different battlegrounds, such as improving access to medical care and expanding maternal and child health services.[13]

Mass health education continued to be a high priority in the interwar decades, but its methods and goals changed in important ways. Leaders in both public health and organized medicine keenly appreciated the value of health propaganda, especially the power of film and radio, and sought to bring popular health education under closer medical supervision. For example, in the early 1920s, the American Medical Association created a separate Bureau of Health and Public Instruction, and in 1923 began to publish its own lay health magazine, *Hygeia,* the forerunner of *Today's Health.* Through these kinds of activities, the AMA sought to ensure that the American public received accurate information that also supported their faith in the medical profession. At the same time, the work of popular outreach remained largely the domain of female-dominated professions, including a new specialty known as "health education." Although a small minority of women continued to find work as bacteriologists, the majority found their chief opportunities as interpreters of scientific knowledge generated by a predominantly male elite. Thus the gendered division of labor within medicine and public health became even more rigid and clear-cut during the interwar period.[14]

Although the prevention of infectious diseases remained an important goal of mass health education, its agenda broadened to include many other health issues as well. In large part, this shift was dictated by dramatic changes in the disease environment itself. By the late 1920s, heart disease and cancer had replaced communicable diseases such as tuberculosis, influenza, and pneumonia as the leading causes of death. As fewer and predominantly poorer people died of tuberculosis, typhoid, or smallpox, it became far easier to conceive of controlling infectious diseases by finding and isolating the infected person—a strategy that lessened the need for broad-based crusades to combat their spread. Simultaneously, health educators extolled a new ideal of "positive health" that moved beyond the absence of contagious disease to envision higher levels of overall fitness and mental well-being. New understandings of nutrition, particularly the discovery of vitamins and their importance to health, focused greater attention on diet as a factor in resisting infection. For all of these reasons, popular health directives began to focus more on the virtues of regular exercise, good eating habits, and mental balance.[15]

The public health movement's ability to envision an ideal of positive health beyond the "mere" absence of infectious disease reflected the increasing institutionalization of germ protections. As municipal sanitary services and the isolation of contagious individuals became more effective, dodging germs became less and less necessary. Although the availability of municipal sanitary services continued to be uneven, especially during the Great Depression, they eventually penetrated even the most marginalized urban and rural neighborhoods. Likewise, the regulation and modernization of food processing gradually made the advantages of pasteurized milk and germ-free groceries affordable for most Americans. With each passing year, the worry of securing a germ-free existence slowly passed from the individual household to the public health department and the private corporation. Writing in 1932, the prominent bacteriologist Stanhope Bayne-Jones likened these institutional protections to the sterile stoppers that scientists used to protect their cultures from accidental contamination; so long as "man" kept them in place, "he will be safe and physically prosperous behind these 'cotton plugs' which keep microbes out of his system."[16]

Given all these changes, the anti-TB societies had to redefine drastically their priorities. In the 1920s and 1930s, they began to concentrate more on funding basic research on the disease, encouraging early detection through physical exams and X-ray screenings, and building clinic and hospital facilities. As death rates from tuberculosis continued their steady decline, the societies eventually broadened their mission to include the prevention of other lung diseases, particularly cancer. In 1954, the National Tuberculosis Association changed its name to the American Lung Association, and in the face of rising deaths from lung cancer subsequently launched a whole new crusade to convince Americans to give up the cigarette.[17]

The "battle with bacteria" continued to be a prominent theme in the education of American women during the interwar period, yet the teaching of household bacteriology changed in important ways. As public health science grew more sophisticated, the specialty of household or "practical" bacteriology, as it came to be called, became increasingly the province of those with advanced scientific degrees. Practical bacteriology was a particularly attractive field for the growing numbers of women earning graduate degrees in science. Texts on household science written after 1915 tended to be written by women with master's degrees in science, who often collaborated with more senior male bacteriologists.[18]

Although household bacteriology remained a fixture of the home economists' curriculum, they faced increasing competition in the 1920s from health educators who also sought to popularize public health information. As home economists worked to strengthen their position in the research university during the interwar period, textiles and nutrition offered more promising fields in which to do independent research and find well-paid positions. Under the new banner of "positive health," they found that their skills in nutrition were more in demand than was their expertise in household sanitation.[19]

The content of household bacteriology changed in important ways as well. When one compares S. Maria Elliott's 1907 volume on the subject with interwar textbooks, a striking change is immediately evident: the lengthy chapters on domestic plumbing and sewer traps had disappeared by the 1920s, reflecting the final demise of the sewer gas theory. Other aspects of the gospel of germs remained strong. Although public health experts downplayed dust infection in the transmission of TB, many preventive hygiene texts continued to stress the dust theory into the 1940s. The amount of space devoted to the management of contact infection and sanitary food preparation also increased substantially in the interwar hygiene texts.[20]

In this slightly revised form, the principles of "practical bacteriology" continued to be widely disseminated among American girls and women. "Keeping clean in these days means keeping free from troublesome germs as well as visible dirt," the Girl Scout handbook for 1922 affirmed, and through countless courses in home economics, first aid, and child care, the future homemakers of America were taught that defending against germs by careful dusting, food preparation, and personal hygiene were important duties of the modern wife and mother.[21] Yet over time, the institutionalization of germ protection gradually reduced the urgency of these precautions. Improvements in municipal services and food production lessened the need for constant vigilance while shopping or cooking. More and better hospitals and insurance plans lifted the burden of nursing relatives through life-threatening bouts of infectious diseases. Increasingly, the housewife's most important public health duties came to consist of immunizing her children and schooling them in the avoidance of contact infection.[22]

Keeping the Gospel Alive

Although public health crusades against infectious diseases lost intensity in the interwar period, anxieties about contagion by no means disappeared. The gospel of germs was kept relevant in large part due to the dawning of popular consciousness about a new germ scourge known as the virus. In the late 1890s, researchers first began to suspect the existence of microbes too small to be seen with the conventional microscope; due to their ability to pass through the superfine filters that trapped larger bacterial forms, they were often referred to as "filterable viruses." By the 1930s and 1940s, the growing recognition of the importance of viruses, as well as of other nonbacterial agents such as protozoa and fungi, led to the adoption of the term "medical microbiology," instead of the more narrow "bacteriology," to denote the scientific study of microscopic disease agents.[23]

Until the invention of the electron microscope in the late 1930s, no one could actually see a virus. Their causal relationship to specific diseases, however, had been confirmed by animal experimentation in the early 1900s. Scientists assumed, correctly as it turned out, that highly contagious diseases for which no bacterial agent had been found, such as smallpox and influenza, were caused by viruses. Despite their smaller size, viruses were just as deadly as their larger bacterial relatives. In the microbial survival of the fittest, yet another cruel and mysterious combatant had been identified.[24]

Even as confidence in the "new public health" was building, the Spanish influenza pandemic of 1918–1919 appeared to dispel any illusions that deadly contagious diseases were a relic of the past. During the pandemic, which killed over 21 million people, researchers suspected but could not prove that the causal microorganism was a virus; it is now believed to have been the work of a particularly virulent mutation of the common influenza virus. In the

United States, an estimated 675,000 Americans, many of them young and healthy adults, died amid scenes reminiscent of the bubonic plague: sick relatives and neighbors were abandoned to die alone, and small children were discovered huddled by the bodies of their dead parents. Large public gatherings where the disease could spread were canceled, and people donned face masks when they had to travel on public transportation. After the war, a curious collective amnesia set in about the great influenza epidemic, but for those who survived it, the experience reinforced anxieties about contagion and the need for personal vigilance to escape it.[25]

These anxieties were also kept alive by recurrent outbreaks of another frightening viral disease—infantile paralysis, also known as polio. Polio epidemics began to appear in Europe and the United States in the late 1800s. By 1909, researchers were sure that the agent was a virus. The 1916 epidemic brought the disease to national notoriety in the United States, and for the next forty years, the unpredictable appearance of polio, usually during the summer months, reinforced parental dedication to practicing the gospel of germs.[26]

Ironically, polio was a disease made more threatening by the gospel's very success. Epidemiologists now believe that prior to the twentieth century, most people were exposed to the virus as infants, when its effects were mild and conferred lasting immunity. (Mothers can also pass on natural antibodies to their children through breast milk.) As milk and water supplies improved, the affluent became less and less likely to acquire this natural immunity early in life. If they then encountered the virus in adolescence or young adulthood, as happened to Franklin Delano Roosevelt in 1921, they were far more likely to die or to be left permanently paralyzed.[27]

Given prevailing public health assumptions, the notion that higher standards of cleanliness put people at greater risk from polio was difficult to assimilate; instead, preventive advice emphasized the probable agency of fly infection and indirect forms of contact. At a time when serious childhood diseases were beginning to disappear among the affluent, the wild card of polio infection contributed to the continued salience of the germ menace. The preventive guidelines issued for polio in the 1930s and 1940s kept alive the pre–World War I gospel of germs and added to it new concerns about contagion from public swimming pools and the like.[28]

The polio outbreaks contributed to the gospel of germ's continuance as a centerpiece of interwar child health education, as did new studies of tuberculosis that pointed to the importance of childhood exposure as a factor in developing the disease. Having long realized that it was easier to mold good habits in children than to break bad habits in adults, public health educators now had even more reason to concentrate their efforts on a young audience. The success of the Modern Health Crusade in the 1910s gave rise to a vigorous child health education movement in the 1920s, which made preventive hygiene a standard feature in the public school curriculum. As school attendance through the eighth grade became mandatory, teachers had ample time to indoctrinate children in the gospel of germs. Thus the work begun by the prewar antituberculosis groups and infant welfare societies shifted increasingly to the public school classroom.[29]

Although in affluent white America the gospel of germs became increasingly the domain of children, African Americans still had a pressing need for more broad-based, community-wide methods of disease education. Due to poverty and racism, their battle against bacteria had scarcely begun by the 1920s, when the new public health came into dominance. The persistence of high rates of illness and death made control of infectious diseases a continued high priority for the black community between the two world wars. In the rural South, "moveable schools" conducted by African-American extension agents instructed families in basic survival skills such as building sanitary privies, screening windows against flies, and making sputum cups out of newspaper. In 1921, Dr. Robert R. Moton, head of the Tuskegee Institute, convinced the U.S. Public Health Service to support the National Negro Health Week program, and over the next decade, its good-health campaigns grew increasingly ambitious. In towns and cities throughout the United States, race leaders and local health departments collaborated on anti-fly campaigns, playgrounds, and well-baby clinics. In its scale of community involvement and range of health issues addressed, the "Negro health movement" continued to equal, even to surpass, the zeal of the Progressive-era antituberculosis crusades.[30]

Selling the Germ in the Interwar Period

Another potent force keeping "antisepticonsciousness" alive in the interwar period was advertising. Since the late nineteenth century, germ fears had fueled the promotion of many products, and manufacturers were not about to abandon such a profitable merchandizing strategy. Indeed, as American advertising became more psychologically sophisticated in the interwar period, the microbe took on even greater salience in the world of commerce. Yet as the menace of death from typhoid and tuberculosis steadily faded, advertising appeals to germ consciousness changed in subtle ways.[31]

The marketing of Listerine illustrates one strategy for domesticating the germ menace that proved particularly successful in the interwar period. In the early 1920s, Listerine ads began featuring stories about pretty young women and bright young men whose hopes for happiness were dashed by "halitosis," a pseudo-scientific term for bad breath supposedly caused by mouth bacteria. In these "sociodrama" advertisements, germs mattered not as a cause of real disease, but of social dis-ease, the stigma of "always a bridesmaid, never a bride." Using this approach, the Lambert Pharmacal Company boosted its yearly sales from $100,000 in 1920 to $4 million dollars in 1927.[32]

But although they widened the germ's menace to include social ostracism, advertisers by no means dropped the dangers of deadly disease. Along with the famous halitosis series, Lambert Pharmacal ran other, more public-health-oriented advertisements touting Listerine's ability to deal "instant death to germs of disease." These advertisements often mentioned specific germs using awesome-sounding scientific names such as "streptococcus hemolyticus," "staphylococcus aureus," and "bacillus typhosus." They also featured emblems of the laboratory such as sketches of microscopes and petri dishes, which they clearly expected American audiences to recognize and respect after several decades of popular education about the wonders of bacteriology. One 1934 advertisement even included a short lesson about filterable viruses, explaining that they were the suspected cause of the common cold. The fact that Listerine was strong enough to kill the "bacillus typhosus" made it all the more credible as a preventive against the less fearsome communicable diseases of everyday life, such as the "street car" cold and the flu.[33]

Listerine advertisements illustrated beautifully the new public health emphasis that people, not things, represented the greatest menace to health. The whited sepulcher theme was now played out in terms of the human body: one advertisement showed a picture of an attractive woman, framed by the question, "What lies beyond these lips . . . and yours?" and accompanied by close-up drawings of four different kinds of germs that lived in the human mouth. Listerine advertisements also focused on the dangers carried by the human hand. One informed the reader that there were "17 diseases carried by the hands . . . many of them dangerous"; another, aimed at mothers, explained, "If you could look at your hands under a microscope you would hesitate to prepare or serve baby's food, or give him a bath, without first rinsing the hands with undiluted Listerine."

The Du Pont Company used similar strategies in promoting the use of cellophane, a flexible cellulose film invented in 1908 by a Swiss chemist. After buying the manufacturing and marketing rights to cellophane in 1923, the parent company set up a subsidiary, Du Pont Cellophane, to develop the new product. Cellophane found its first big users in industries that had long been concerned with germ protection, namely the candy, baking, and processed meats industries. As Du Pont chemists succeeded in making the product more moisture proof and less expensive, its use expanded rapidly to include cigarettes, chewing gum, and textiles.[34]

Cellophane's success reflected Du Pont's aggressive, sophisticated marketing rather than its superiority to other forms of sanitary packaging. As one historian of the campaign observed, "The rise of Cellophane is a selling rather than a manufacturing achievement." Du Pont invested heavily and wisely in promoting the product to both manufacturers and consumers; advertising expenditures rose from $65,000 in 1926 to $385,000 in 1933. The focal point of this lavish advertising campaign was *germ*-protection, and more specifically, the "flies, fingers, and food" trio so central to the gospel of germs. To manufacturers, the Du Pont Cellophane sales force pitched the line that women consumers wanted goods untouched by the human hand. For example, the sales material developed for the baking industry explained "why women fell for the new bread" with quotes reportedly culled from women shoppers, who supposedly bought the loaf because, in the words of one, "I knew it hadn't been touched."[35]

Du Pont Cellophane also pitched its argument directly to the consumer through extensive advertising spreads in such popular magazines as the *Saturday Evening Post, Good Housekeeping,* and the *Ladies' Home Journal.* In addition, it promoted cellophane on the *Cellophane Radio Show,* which featured the etiquette writer Emily Post and aired in the morning right before housewives went to shop. In these consumer-oriented campaigns, Du Pont played upon the germ phobias associated with human touch. As one advertisement began ominously: "Strange Hands, Inquisitive hands. Dirty Hands. Touching, feeling, examining the things you buy in stores. Your sure protection against *hands-across-the-counter* is tough, clear, germ-proof Cellophane." Another ad featured an array of improbable-looking "bugs" (far less realistic looking than those in Listerine ads) and warned, "An unprotected piece of candy—or some other food—is the germs' chance to get acquainted with you."[36]

In promoting the idea of cellophane as a "sanitary sentinel," Du Pont focused on consumer products with a long history of association with germ danger such as bread, candy, meat, tobacco products, clothing, and baby layettes. Although it cast itself as an ally of the public health movement, the company also tended to play upon the older fears of dust, dirt, and fomites that the new public health was trying to unseat. In a letter sent to almost four hundred public health departments in 1933, Du Pont's director of sales explained that "by graphic illustrations, colorful subjects and sincere straight-from-the shoulder copy, we are attempting to register with the buying public the importance of selecting protected food . . . free from the danger of contamination from dirt, dust, and germs."[37]

Thanks in large part to Du Pont's campaign to make America "cellophane-conscious," consumers became increasingly conditioned to pay for packaging as well as product. In general, the quest for germ protection contributed to an ever-greater tolerance for the concept of disposability. An editorial in the *Independent* noted in 1911, "As we advance in civilization the use of personal and perishable utensils and apparel becomes more extended." Nowhere was this principle more clearly demonstrated than in the demand for hygienic packaging; to avoid other people's germs, consumers purchased ever more goods wrapped in otherwise useless materials, which were then consigned to the garbage.[38]

This trend was dramatically evident in the growing popularity of sanitary paper goods. As we saw earlier, prior to World War I, the idea of paying even a penny for a paper drinking cup had seemed too much to many consumers. Disposable "crepe handkerchiefs" had only a limited market among consumptives and other invalids. Commercially made sanitary napkins had a somewhat wider following, but were still too expensive for most women to afford. The only disposable paper product with a really broad market was toilet tissue; for this sanitary item, the dangers of fecal contamination appear to have offset concerns about cost. The first year that toilet tissue was listed as a separate area of manufacture in 1919, production already equaled almost 80,000 tons. Still, many poorer American families continued to make do with scrap paper, old mail order catalogues, and corncobs.[39]

As steady advances in manufacturing processes made paper products cheaper and more versatile in their uses, paper manufacturers worked hard to tilt in their favor the calculation of cost versus convenience. The sanitary angle was especially useful in justifying the onetime use of paper products. For example, the Cup and Container Institute, which was started by paper cup manufacturers, stressed the dangers of unsterile crockery and drinking glasses; it warned in a 1935 publication that the growing popularity of travel and eating out "add[s] enormously to the opportunity for a salivary exchange, with all the health hazards which that implies." The best safeguard against "this mutualization of saliva," the institute concluded, was to demand paper drinking cups and plates.[40]

It was women consumers who first responded to the sanitary appeal of Kotex and Kleenex. In the late nineteenth century, women had begun buying commercially made menstrual pads as well as constructing homemade versions from inexpensive materials such as cloth, cotton gauze, and cheesecloth. In 1914, researchers at the Kimberly-Clark Company, a Wisconsin paper manufacturer, developed a new, less expensive process for making a soft tissue, which they called "cellucotton." Introduced as a surgical dressing during World War I, nurses in American field hospitals soon realized that it made excellent menstrual pads. Kimberly-Clark saw the potential for a disposable "sanitary napkin," and in 1920 introduced the "Kotex" line. As prices declined and advertising campaigns stressed its hygienic virtues, Kotex and its imitators came into wide use among American women during the interwar period.[41]

In 1924, Kimberly-Clark introduced "Kleenex" for use as a "sanitary cold cream remover." With growing numbers of women wearing makeup, the company saw a market for a disposable tissue to replace what it christened the "unsanitary" cold cream towel. Then, in the late 1920s, a consumer survey done in Peoria, Illinois, revealed that women had come up with another, more extensive use for Kleenex:

they were blowing their noses with it. The company quickly developed an advertising campaign urging consumers to give up the "uncivilized" and unsanitary cloth handkerchief, and sales of the product more than doubled the next year.[42]

Pax Antibiotica

The germ menace's hold on the popular imagination only really started to weaken with the "antibiotic revolution" that began in the late 1930s. The long-hoped-for discovery of effective antibacterial drugs created a new sense of freedom from infectious diseases and encouraged the belief that they would soon cease to constitute a serious threat to American society. As soon as scientists had these "magic bullets" at their command, the consequences of transgressing the gospel of germs began to seem less and less serious.[43]

For almost fifty years, researchers had been frustrated by repeated failures to find effective chemotherapies for infectious diseases. Those "internal antiseptics" capable of killing pathogenic microbes inside the body had too many serious side effects to make them useful. Paul Ehrlich's much-heralded remedy for syphilis, the arsenical drug Salvarsan, was a case in point. Even in the less toxic form of Neosalvarsan, the drug was so punishing in its effect on the body that many patients refused to finish the course of treatment. As a consequence, the development of vaccines and antitoxins seemed a much more promising route of research during the early decades of the twentieth century.[44]

Still, a few researchers continued to search doggedly for chemical or biological substances that could kill microbes without destroying their human host. The first breakthrough came in 1935, when Gerhard Domagk, a German research chemist, discovered that a newly patented chemical dye called "Protonsil" cured streptococcal infections in mice. The active agent in the dye was later discovered to be a substance called sulfonamide. Sulfonamide and its derivatives, such as sulfadiazine and sulfapyridine, which became known collectively as the "sulfa drugs," proved effective against streptococcal infections, gonorrhea, and pneumonia.[45]

The discovery of Protonsil revived hope for finding more magic bullets, a hope that was soon fulfilled by the even more spectacular success of penicillin. Penicillin was the first true antibiotic, a term coined for naturally occurring, as opposed to chemically synthesized, germicides. The English researcher Alexander Fleming discovered the drug in 1928, when in a serendipitous mistake one of his gelatin plates became contaminated by a strange mold, which "lysed," or disintegrated, a colony of staphylococcus adjacent to it. It took many more years of work before another pair of researchers, Howard Florey and Ernst Chain, harnessed penicillin's therapeutic properties into an easily usable form. First introduced as an experimental drug during World War II, penicillin demonstrated an almost magical ability to save soldiers near death. Toward the end of the war, the "miracle drug" was made available for civilian use, and the press began hailing penicillin as one of the great discoveries of modern science.[46]

Even before the euphoria produced by penicillin's success had subsided, laboratories started announcing the discovery of more antibiotic drugs. In 1943, two soil biologists at Rutgers University, Albert Schatz and Selman Waksman, isolated streptomycin sulfate, the first drug found to combat the dreaded *Mycobacterium tuberculosis*. In the late 1940s, Paul Burkholder of Yale University found another soil microorganism, which produced a compound called chloramphenicol, marketed as Chloromycetin, that proved effective against the organisms responsible for typhoid and rickettsial infections such as typhus fever. At around the same time, Benjamin Duggar of Lederle Laboratories discovered chlortetracycline hydrochloride, the first of the powerful tetracycline family, which worked against the same organisms as well as those responsible for various forms of pneumonia.[47]

From the outset, the new antibiotics had their limitations. Some, including streptomycin and chloramphenicol, had serious side effects; others, like penicillin and the sulfa drugs, caused potentially dangerous allergic reactions in a small minority of patients. Researchers also soon noted the troubling tendency of bacteria to develop drug resistant strains. Further, none of the new germicidal drugs proved effective against viral diseases. Still, given how few weapons medicine possessed against infection prior to the late 1930s, physicians and lay people alike understandably hailed the first generation of antimicrobial drugs with enormous optimism.[48]

It is hard to overestimate the excitement and drama generated by the discovery of these new "miracle drugs." In 1943, *Reader's Digest* dubbed penicillin "yellow magic" (the unrefined drug was a yellowish color), and reported amazing stories of its ability to revive patients who had been "as good as dead" from horrible infections. In 1949, in an article on the expanding pharmaceutical industry, *Nation's Business* referred to penicillin as "the most glamorous drug ever invented" and noted that "in just seven years, penicillin has turned in an incredibly brilliant record of lifesaving."[49]

The inventors of these new drugs were hailed as national heroes, the modern-day descendants of Louis Pasteur and Robert Koch. Selman Waksman, for example, the codiscover of streptomycin, was deluged with letters from grateful parents thanking him for saving their children's lives. As one parent wrote of streptomycin, "To me this is a magical name because the doctors tell me that this is the drug that was responsible for saving my baby's life." Waksman himself expressed a sense of awe concerning his discovery's power: "How can I describe the impressions left upon me by the first sight of a child . . . who had been saved from certain death by the use of a drug in the discovery of which I had played but a humble part?"[50]

The optimism of the 1950s was further heightened with the development of not one but two effective polio vaccines. The often bitter competition between their inventors, Jonas Salk and Albert Sabin, hardly exemplified the ideal of scientific cooperation, and the clinical trials, especially of the Salk version, were dogged by problems. Yet the introduction of the polio vaccine was celebrated as another major victory of modern medical science. By the early 1960s, parents at long last were free of the worry about polio that since 1916 had come with every summer.[51]

With the array of drugs and vaccines available by 1965, the need to guard against contact infection understandably relaxed. Americans quickly came to believe that with a few soon-to-be-cured exceptions, modern medicine and public health had "conquered" epidemic disease. Young physicians in the 1960s were advised, "Don't bother going into infectious diseases," and to concentrate on cancer and heart disease instead. Reflecting the same sense of confidence in the *Pax antibiotica,* the U.S. surgeon general announced in 1969 that "it was time to close the book on infectious diseases" because they no longer represented a serious threat to America's health.[52]

Meanwhile, Americans continued to practice the habits associated with the gospel of germs and to use products such as mouthwash, household disinfectant, paper cups, and facial tissues that had once been recommended as barriers to infectious disease. But with each passing year of relative freedom from the scourge of TB and polio, the link between these habits of cleanliness and the prevention of life-threatening diseases gradually disappeared. By the mid-1960s, customers at roadside diners did not connect their paper cups to the old fear of the common drinking cup and the "loathsome" diseases that might be contracted from using it, nor did people blowing their noses into a Kleenex tissue realize that the "crepe napkin" had once been seen as an indispensable sanitary aid for the consumptive.

Still, the old germ consciousness lingered on, especially among older Americans. Those who grew up during the peak years of the early twentieth-century disease crusades felt an instinctive, emotional connection between cleanliness and safety from disease, which they often passed on to their children. Oral histories of second- and third-generation immigrant women done in the 1970s, at the height of public confidence about the so-called conquest of epidemic diseases, attest to the continued strength of earlier teachings about the "Three D's": dirt, disease, and death. Mixed in with remembrances of childhood illnesses and the death of siblings, the interviewees referred to their own compulsions to keep their homes clean. As one Italian-American woman noted, for most of her adult life, "I felt everything had to be dusted every day of the week."[53]

And so the rituals of germ avoidance remained in practice, a little-noticed part of everyday life, until the early 1980s, when the appearance of a deadly new infectious disease, acquired immune deficiency syndrome (AIDS), once again brought them to public consciousness. Within a few years, the AIDS epidemic, in concert with a number of other historical forces, would destroy the confidence bred by the Pax antibiotica and usher in another era of anxiety about the microbe. The gospel of germs was born again in the age of AIDS.

Epilogue: The Gospel of Germs
in the Age of AIDS

In 1984, a thirteen-year-old Indiana boy named Ryan White learned that the blood transfusions used to treat his severe hemophilia had given him AIDS. What happened next illustrates the worst consequences of the popular education campaigns dedicated to controlling the spread of germs. When word of his diagnosis got out in his hometown of Kokomo, Indiana, the boy and his family were shunned. Despite repeated assurances from his doctor and state public health officials to the contrary, his neighbors refused to believe that the virus was not spread by casual contact. Ryan was subjected to the full gamut of fears instilled by the gospel of germs: for example, people refused to shake hands with him or to use the bathroom after he did. Rumors circulated that he deliberately spat on vegetables at the grocery store to spread the disease. His illness was attributed to his mother's failure to keep her house clean. The owner of a diner where he ate made the waitress throw away the dishes he and his family had used. Even in his church, Ryan and his family were asked to sit in a special pew so that members could avoid his coughs. Eventually the only people who would spend time with him were the reporters sent to cover his story for the national media.

The campaign of harassment and abuse intensified when Ryan insisted on returning to high school, and after a bullet was fired through his living room window, the Whites moved to another town. There Ryan's life took a turn for the better: town officials began an intensive program of AIDS education in the public school that he was slated to attend, and his new neighbors dealt more calmly with their fears of infection. In the meantime, Ryan's determination to live a normal teenage life made him a national celebrity. Through his testimony before the 1988 AIDS commission and a television film about his life, Ryan White helped to educate a whole generation about the limitations of the gospel of germs in the age of AIDS.[1]

Today, after many years of intensive public health education, we can hope that Ryan White's experience in the early days of the AIDS epidemic would not happen again. But there is no denying that many Americans continue to fear this disease irrationally. People with AIDS have lost their jobs, been shunned by friends, and even had their homes burned down. The AIDS epidemic has exposed the worst aspects of our modern-day beliefs about the germ. When applied indiscrimately and fueled by homophobia and racism, there can be no crueler punishment of the sick than shunning and fearing them in the name of germ avoidance.

Ever since its recognition in the early 1980s, AIDS has revealed both the strengths and weaknesses of modern scientific medicine. Laboratory researchers have demonstrated remarkable scientific virtuosity in isolating the human immunodeficiency virus (HIV) and developing a diagnostic test for its presence in the body. The funding poured into research on HIV in the last decade has greatly advanced the understanding of viruses in general and retroviruses in particular. Recent progress in vaccine development and combined drug therapies holds out new hope for prevention and treatment of HIV infection. Yet despite the rapid advance of scientific knowledge, researchers have been unable to duplicate the miracle of penicillin with a sure-fire cure for AIDS, or for that matter, for any viral disease. Like the brilliant successes of bacteriology in the late 1800s, the advance of modern virology has only underlined the desirability of avoiding infection in the first place.[2]

Thus prevention has been the keynote in the public health campaign against AIDS, as it was against the great white plague a century ago. Within a few years of the epidemic's outbreak, researchers were able to pinpoint the way the virus is transmitted through exchanges of blood and semen. But the disease's early association with homosexual men and intravenous drug users compounded the traditional problems of public education regarding sexually transmitted diseases. Social conservatives objected to frank discussions of anal sex and needle sharing, whereas gay-rights advocates battled public health restrictions on bathhouses.[3]

In many ways, the AIDS epidemic has taxed the limits of the concept of public health citizenship forged during the Progressive era. Social reformers have often invoked the ideal of disease prevention as a civic value around which even the most diverse groups can rally. So long as a health issue appears to affect everyone indiscriminately, as tuberculosis did in the late nineteenth century, this tactic can be quite effective. But if a disease affects only some segments of society, especially those already stigmatized for other reasons, its prevention potentially arouses far more hostility and conflict. To the extent that AIDS was initially linked to homosexual men, intravenous drug users, and Haitian immigrants, efforts to prevent it have become entangled with virulent forms of prejudice.

In addition to the "us versus them" problem, educating Americans about AIDS has run up against the very beliefs in contact infection that previous generations of public health reformers had worked so hard to inculcate. Despite the fact that a bare handful of cases have been verified in which infection did not involve unprotected sex, needle pricks, or blood transfusions, many Americans, like Ryan White's neighbors in Kokomo, have found it difficult to accept that the AIDS virus is not spread by less intimate forms of contact.

As White's experience underlines, the lessons of the "gospel of germs" are everywhere apparent in the pattern of fears expressed toward people with AIDS. The exaggerated concern about contact with their saliva clearly shows the influence of the "spit and death" slogans repeated relentlessly during the antituberculosis campaigns. The anxiety about sharing glasses or dishes resonates with the "common cup" controversy. The worry that mosquitoes can carry the AIDS virus reflects the successful equation of insects and microbial "bugs."

In the early days of the AIDS epidemic, public health workers had essentially to undo the lessons associated with the gospel of germs. Health educators refer to this process as "negative" education because it involves explaining which behaviors do *not* spread the disease, such as sharing glasses or shaking hands. The list of mistaken beliefs spotlighted in AIDS advertisements are a mirror image of early twentieth-century health catechisms: under the category of "None of these will give you AIDS" appear a toilet, a set of dishes, a handshake, and a doorknob—in other words, the classic sources of contagion highlighted by the gospel of germs.[4]

In a more general sense, our deepening acquaintance with the AIDS virus has revived the sense of awe and fear that many Americans first felt when they learned of the germ's existence a hundred years ago. HIV demonstrates in chilling detail how fragile are the boundaries between our bodies and the invisible universe of microbes with which we cohabit. By destroying the immune system, this postmodern virus makes its host prey to opportunistic infections from ordinary microorganisms that exist in and on the human body. The ravages that can be wrought by such garden-variety organisms as candida or herpes simplex are a vivid reminder of the delicate balance of power between host and parasite. In short, AIDS has reawakened our consciousness of the "microbial survival of the fittest."[5]

Ironically, because of their impaired immune systems, people with AIDS are at much greater risk from breaches in sanitary protection than are the healthy folk who shun them. Their very lives can depend upon avoiding the unguarded sneeze carrying the *Mycobacterium tuberculosis,* the tap water polluted with cryptosporidia, or the salmonella contained in an improperly cooked meal. More than any disease of the late twentieth century, AIDS has reawakened the fear of the one false step, the thoughtless behavior that leads to death—whether it be the unprotected sex or needle sharing that allows HIV access to the body, or the random encounter with some microbe that ultimately overloads the immune system's ability to function. From start to finish, the course of AIDS illustrates the horrifying consequences of transgressing the rules of germ protection.[6]

The lesson so often repeated by late-nineteenth-century hygiene reformers, that "little things are no trifles," has been reinforced by a host of other New Age microorganisms. The environmental and economic upheavals of the last twenty years have created unprecedented opportunities for viruses to jump species, especially in areas bordering on the rich biological terrain of the subtropical rain forest. The global transportation system, which makes it possible to travel across several continents in the time that it takes a deadly microbe to incubate, has seemingly made the whole world vulnerable to diseases originating in the most remote places. Popular accounts of these so-called emerging viruses, particularly the hemorrhagic fevers, have familiarized Americans with a new vocabulary of terms such as "hot zone," "amplification," and "crash and bleed out."[7]

Accounts of the techniques observed in "Level 4 biocontainment" laboratories, where the most lethal viruses such as Ebola are handled, reinforce the "one false step" motif. Their extreme contagiousness is reflected in the elaborate disinfection, special space suits, and ventilation systems used in these so-called hot zones. The technology of the new virology makes the exacting precautions pioneered by Louis Pasteur and Robert Koch seem like child's play in comparison. The hazards posed by the careless puncture or cut have not only captured the imagination of writers and filmmakers, but have also led to real life dramas in which researchers have been infected through seemingly minor breaches of laboratory protocol.[8]

The clinical and epidemiological course of highly infectious viruses such as Marburg and Ebola further underlines the risks of casual contact with the ill. Shedding the virus via copious, bloody discharges, a single person, or "index case," has been known to start an epidemic in which hundreds of people died in only a few days. Like the bubonic plague and smallpox in previous eras, the emerging viruses illustrate the chilling truth that contact with the bodily byproducts of the ill can be lethal. Perhaps more surprisingly, theories about how the emergent viruses spread have revived aspects of the "house disease" concept. In outbreaks of Bolivian hemorrhagic fever, for example, researchers suspect that wild mice living in the villagers' homes shed the virus in their urine, which mixed with the dirt on the floor; dry sweeping may have stirred up the virus into the air and infected the house's inhabitants, a mode of transmission remarkably reminiscent of T. Mitchell Prudden's dust theory of infection.[9]

Similarly, outbreaks of exotic diseases have refocused attention on the perennial risks of cohabitation with animals and insects. The appearance of the Hantavirus on Indian reservations has been attributed to the ubiquitous presence of rodents, which may provide a reservoir for the disease. Lyme disease, dengue hemorrhagic fever, and equine encephalitis have reminded the public of the role ticks and mosquitoes can play in spreading serious, often deadly ailments.[10]

At the same time that these more exotic aliments have made headlines, previously "tamed" species of microorganisms have reemerged in recent years as a public health threat. Tuberculosis is the most publicized example of this phenomenon. After decades of declining mortality rates, TB began an upswing in large cities such as New York during the late 1980s, as cuts in social services left more people hungry and homeless and AIDS rendered more immune systems vulnerable. Public health authorities estimated in 1992 that 40 percent of all AIDS patients had active cases of tuberculosis. To make matters worse, during the mini-epidemic doctors began to see new drug-resistant strains of TB as patients failed to complete the long course of antibiotic treatment. Margaret Hamburg, then health commissioner of New York City, warned in 1992, "If drug resistance is not curbed we could quickly find ourselves back in a world before modern medications were available."[11]

In addition to tuberculosis, physicians are now encountering drug-resistant strains of streptococcus and staphylococcus, and newspaper and magazine articles have proclaimed, "Common Bacteria Said to Be Turning Untreatable." Beyond the problem of drug-resistance, microorganisms, like their human hosts, are also constantly evolving, so that old, milder strains may become more virulent all on their own. For example, a common form of the *Escherichia coli* bacteria found in the human gut has turned dangerous by "borrowing" a gene from the organism responsible for Shiga dysentery. Epidemics involving this new strain of *E. coli* have recently led to deaths in both the United States and Japan.[12]

The steady deterioration of protective public health and social services in the 1980s and 1990s has amplified the risks presented by AIDS and other superbugs. Since the nineteenth century, the first and most efficacious line of defense against infectious diseases has been prevention, not cure. But since 1980, sharp reductions in federal funding to states and cities have resulted in declining expenditures on basic public health services. The consequences of these breaches in the "cotton plugs" that shield us against infectious diseases have not taken long to appear.[13]

For example, numerous cities across the United States have experienced serious problems with their water supplies in the last few years—problems that have undermined public confidence in the safety of their drinking water. The worst outbreak to date occurred in Milwaukee, Wisconsin, in 1993, when an estimated 200,000 people were sickened by drinking water contaminated with cryptosporidia, a species of protozoa that causes diarrhea. During this and other urban water emergencies, health departments have warned inhabitants to boil tap water before drinking it, to use boiled water for washing infants, and to allow dishes to dry thoroughly before using them—all precautions central to the old gospel of germs. Health authorities attribute the growing frequency of these incidents to aging facilities and cutbacks in routine inspection, which suggests that they will only become more common. Meanwhile, many households have returned to practices that would have been quite familiar to our Victorian ancestors—namely, buying bottled water and installing home water filtration systems.[14]

The safety of the food supply has also become a major concern due to antiquated methods of federal meat inspection and reduced surveillance of food handling establishments. In 1993, hamburgers tainted with the new virulent strain of *E. coli* sickened four hundred people who ate in a Jack in the Box restaurant in Washington State; several children died as a result, and almost two hundred people were hospitalized. This incident, along with numerous reports of salmonella-infected chickens and eggs, has focused new attention on the dangers of improperly processed and cooked food. Here again, the failure of institutional protections against food contamination has fostered a new emphasis on vigilance in the household. In the past few years, the U.S. Department of Agriculture has conducted a widespread publicity campaign to acquaint consumers with protocols for the safe storage and cooking of meat, the need to keep cutting boards and kitchen counters salmonella free, and other lessons reminiscent of the old "household bacteriology."[15]

The reappearance of tuberculosis and other supposedly "conquered" diseases also underlines a point well known to Progressive-era reformers—that inadequate nutrition, housing, and health care weaken natural resistance against disease. Since 1980, the rate of poverty has increased dramatically in the United States. Recent studies of Western industrial nations show that the United States has both the greatest degree of income inequality and the most children living in poverty. As conservatives in Congress push to reduce welfare and other services to the poor, such inequities will surely worsen, creating conditions ideal for the spread of old and new diseases alike.[16]

Concerns about the larger environmental context of human health add yet another layer of worry to current anxieties regarding infectious diseases. Today Americans are hearing more and more about the potential health hazards of the deteriorating ozone layer. Prolonged exposure to ultraviolet rays damages the human immune system, thereby increasing the risk of both cancer and infectious diseases. In July 1996, the United Nations issued a report warning that global warming "could have a wide range of impacts on human health, most of which would be adverse." To cite but one example, climatic changes due to global warming could lead to the spread of malaria into many regions where it does not currently exist.[17]

Literally closer to home, the findings of environmental health sciences have led to a curious rebirth of the house disease concept. Researchers have uncovered a host of dangers lurking in the American household, including asbestos, lead paint, pesticide-tainted water, radon gas, and electromagnetic waves. Although it is framed more in terms of a chemical than a microbial threat, the modern notion of the "sick building" has striking parallels to the late-nineteenth-century conception of the "house disease." Moreover, the preventive strategies recommended for combatting household disease risks are eerily reminiscent of those advocated a century ago in the name of the germ. Whereas prospective buyers and renters were once warned to find out the "consumptive history" of a house, consumers today are counseled to test for radon and other environmental dangers. To guarantee their family's safety, homeowners consult the modern-day equivalent of the sanitary engineer to install "environmental control systems." In tones reminiscent of turn-of-the-century scientific housecleaning, guides to reducing allergens in the home dictate an exacting regimen of cleaning in which the prime suspects of the old germ warnings—carpets and upholstered furniture—still figure prominently, but now as the haunts of the dreaded dust mite. Like the Germicide and the Sy-Clo toilet in years past, sales of air filters, dehumidifiers, vaporizers, home test kits for radon and lead, and the like are booming in the wake of popular anxieties about health protection. In the midst of all this anxiety, advertisers report that the "germ sell" has become newly popular again.[18]

In retrospect, much of the anxiety now being expressed about the "revenge of the superbugs" will probably turn out to have been unwarranted. Clearly, the same kind of commercial interests that kept alive the fear of sewer gas for so many years are hard at work trying to use the current germ panic to sell books, movie tickets, water filters, and household disinfectants. The rediscovery of the germ is probably also part of a larger coming-to-terms with the globalization of late-twentieth-century culture, a response similar to the Progressive generation's reactions to the "incorporation of the germ." Some commentators even link the apprehensions about germs to the end of the Cold War and the collapse of the Soviet Union, which left Americans needing a new public enemy number one.[19]

Yet although the germ menace has certainly been exaggerated, there is no denying that it reflects some very real and very serious problems. A combination of threats stemming from how humans interact with their environment—the "natural" evolution of microbial life, the decay of the public health infrastructure, and the overall degradation of the environment—have brought us back full circle to a sense of vulnerability that our great-grandparents knew well. For baby boomers, who were raised at the high point of medical optimism about infectious diseases, the AIDS epidemic and its sequelae have been particularly humbling experiences. The unprecedented era of security from disease they enjoyed from having been born after the *Pax antibiotica* seems as irretrievably lost as is the Great Society or the spirit of Woodstock. We can only begin to imagine the long-term psychological consequences of the AIDS epidemic for the children of the 1980s and 1990s, who have been raised in its tragic shadow.[20]

In this climate, the Darwinian rhetoric of the microbial "survival of the fittest," so popular at the turn of the century, is once again coming into wide use. But noticeably lacking in contemporary accounts is the optimism that many late-nineteenth-century Americans expressed in the wake of the germ theory's acceptance. The secure faith evident in a 1924 *Scientific American* editorial—that "the natural outcome of the struggle between mankind and microbe has always favored mankind"—has been replaced by a much gloomier mood. Now the revenge of the superbugs appears as an all-too-appropriate punishment for the hubris of the twentieth century—that is, for the blind faith that science, technology, and capitalism can solve all problems, including the threat of infectious disease.[21]

Laurie Garrett's *The Coming Plague* illustrates well this sense of impending apocalypse. She describes a microbial universe hauntingly like our own, "a frantic, angry place, a colorless, high-speed pushing and shoving match that makes the lunch-hour sidewalk traffic of Tokyo seem positively poky." Using images strikingly similar to those used by the first germ theorists, she writes, "In the microbial world warfare is a constant. The survival of most organisms necessitates the demise of others." Garrett concludes with this grim message: "While the human race battles itself, fighting over ever more crowded turf and scarcer resources, the advantage moves to the microbes' court. They are our predators and they will be victorious if we, *Homo sapiens,* do not learn how to live in a rational global village that affords the microbes few opportunity. It's either that or we brace ourselves for the coming plague."[22]

The implication of Garrett's work, as well as of many other popular accounts of the new germ warfare, is that the "bad" microorganisms are gaining ground too quickly. Despite the extraordinary gains in molecular biology, microbiology, and immunology, "the bad guys, the pathogens, particularly the newly recognized ones, seem to the general public to have become nastier faster than scientists have become smarter," as the medical and environmental historian Alfred Crosby notes. While we continue to hope for some miracle drug to cure AIDS, the larger breakdown of barriers against disease, from the water supply to the ozone layer, seem less amenable to the quick fixes of science and technology.[23]

For all these reasons, the crusades of a century ago to spread the gospel of germs do not seem nearly as silly as the Arrowsmith generation once portrayed them. Many Americans, wondering if electrical power lines or radon gas represent "real" disease risks, can sympathize with the fears of sewer gas and pathogenic wallpaper that tormented our great-grandparents. As we rush to install water filters and radon detectors, we can identify with the immigrant housewives' urge to clean house as a way to protect their families against a menacing outside world.

I hope that the current debate will lead to greater appreciation of past public health achievements. The frequency with which health departments today resort to public education about private protection in order to prevent dangerous encounters with HIV, *E. coli,* and cryptosporidia reinforces one of the central themes of this book: the private side of public health is not just an important social phenomenon; it has also protected people from disease. The contempt that Sinclair Lewis and his generation felt for the Dr. Pickerbaughs of the early disease crusades should not obscure the very important role that education can play in disseminating useful knowledge about avoiding infectious diseases.

The history of the gospel of germs also suggests the need to rethink the current assault on federal and state governments. In the past, progress in U.S. public health has come about chiefly through collaboration between the private and public sectors, not through an exclusive reliance on one or the other. Consider the Progressive period: at no point in American history, before or since, have private reform initiatives been so numerous or so effective. The vibrancy of America's "civil society" at the turn of the century will perhaps never again be equaled. But it was precisely this same generation of reformers who realized that the private side of public health could not sufficiently restrain the sanitary abuses prevalent in an unregulated marketplace. Private reform paved the way for the state to become an active guardian of sanitary safety, one that performed regulatory services that were far beyond the capabilities of any one group. In the rush to dismantle "big government" in the late twentieth century, we would do well to remember that a *partnership* of public and private initiatives built the public health institutions that until very recently served most Americans well.

Finally, in the current climate of political distrust and divisiveness, the Progressive-era concept of the "chain of disease" also inspires admiration. To be sure, the gospel of germs did its share to divide Americans along lines of class, gender, race, and ethnicity. The great disease crusades of a century ago undoubtedly increased stigmatization of the sick and discrimination against foreign-born and non-white Americans. For Typhoid Mary, who spent over thirty years incarcerated on North Brother Island, or the black sharecroppers denied treatment during the infamous Tuskegee syphilis experiment, the new science of bacteriology hardly brought a greater spirit of tolerance.[24]

Yet there was another side to the great disease crusades of the Progressive period that merits our respect and even emulation. The fact that the microbe preyed upon all bodies, regardless of their racial and social characteristics, opened up some avenues of moral and social discourse that did not simply encourage suspicion and hate. The belief that "germs know no class or color line" had positive as well as negative consequences. The notions of public health citizenship developed by grassroots labor organizers and African-American community leaders represented their creative use of anxieties about disease to address fundamental injustices and prejudices.

For all their self-righteousness and ethnocentrism, the middle-class Anglo-American reformers dedicated to spreading the gospel of germs felt a sense of responsibility for the health of other people that is often conspicuously lacking in modern-day discussions of infectious disease. The "imagined community" created by their beliefs in the chain of disease and the socialism of the microbe led them to try to rectify the worst abuses of industrial progress. As a result, their vision ranged far beyond the privileged reaches of their own homes into the workplaces and households of the marginal, outcast members of their society. Consider how much easier it would be today for celebrity crusaders to awaken middle-class Americans to the economic exploitation of garment workers both in the United States and abroad if public health theory still supported the belief in fomite infection. Would consumers be willing to purchase clothing made in sweatshops where women and children are held in virtual slavery if they thought they were "buying smallpox," in Florence Kelley's colorful phrase?[25]

As soon as we acknowledge our own foibles, the great disease crusades of the turn of the century serve to educate and inspire rather than simply anger or amuse us. As we begin a new century, we face many serious challenges in the control of infectious diseases, both in the United States and around the world. We are struggling to become more aware of the health consequences of our increasingly irreversible effects on the global environment, to balance the strengths and weaknesses of modern medicine, and to reverse race and class prejudices that sap the foundations of public health. I believe that the historical failures and successes of the gospel of germs have some valuable insights to offer as we fight those good fights.

Notes

I have been highly selective in my references to the vast secondary literature concerning the history of the germ theory and its influence on American culture. Beyond citing those works that have most directly and deeply influenced my own thinking, I have sought primarily to direct the reader to a few of the most recent and comprehensive treatments of the many subjects I cover.

Abbreviations

AHS   Atlanta Lung Association Papers, Atlanta History Center, Atlanta, Georgia

ALA   Archives, American Lung Association, New York, New York

BHM   *Bulletin of the History of Medicine*

BJBSC   *Bulletin of the Joint Board of Sanitary Control*

BNTA   *Bulletin of the National Tuberculosis Association*

CHS   Chicago Historical Society, Chicago, Illinois

CPP   Historical Collections, College of Physicians of Philadelphia, Philadelphia, Pennsylvania

CRCFW   *Cornell Reading-Course for Farmers' Wives*

CSS   Community Service Society Papers, Rare Book and Manuscript Library, Columbia University, New York, New York

CU   Records, Office of the Dean, New York State College of Home Economics, Group 749, Division of Rare and Manuscript Collections, Cornell University Library, Ithaca, New York

CUE   Extension Records, New York State College of Home Economics, Group 919, Division of Rare and Manuscript Collections, Cornell University Library, Ithaca, New York

DC   R. G. Dun and Company Collection, Baker Library, Harvard University Graduate School of Business Administration, Cambridge, Massachusetts

GH	*Good Housekeeping*
GMM	George Meany Memorial Archives, Silver Spring, Maryland
HC	Trade Catalog Collection, Hagley Museum and Library, Wilmington, Delaware
HD	Records of the Du Pont Cellophane Company, Series 2, Part 2, Archives of the E. I. du Pont de Nemours and Company, Hagley Museum and Library, Wilmington, Delaware
HL	Theodore Roosevelt Collection, Houghton Library, Harvard University, Cambridge, Massachusetts
HM	*The Hotel Monthly*
ILR	Collection 60, Union Health Center, International Ladies Garment Workers' Union, Records, Kheel Center for Labor-Management Documentation and Archives, Cornell University, Ithaca, New York
JBSC	Joint Board of Sanitary Control
JHE	*Journal of Home Economics*
JHM	*Journal of the History of Medicine and Allied Sciences*
JOL	*Journal of the Outdoor Life*
JWT	J. Walter Thompson Company Archives, Special Collections Library, Duke University, Durham, North Carolina
KC	Oral Histories, Corinne Krause Collection, Library and Archives, Historical Society of Western Pennsylvania, Pittsburgh, Pennsylvania
LHJ	*Ladies' Home Journal*
NA	Record Group 42, Series 87, Correspondence and Other Records, Office of Public Buildings and Public Parks of the National Capital, National Archives, Washington, D.C.
NLM	Historical Division, National Library of Medicine, Bethesda, Maryland
NTA	National Tuberculosis Association
NU	Neighborhood Union Collection, Series 14-B, Special Collections and Archives, Atlanta University, Atlanta, Georgia
NYH	*New York Herald*
NYT	*New York Times*
PC	Department of History, Presbyterian Church (U.S.A.), Philadelphia, Pennsylvania
PO	United States Patent Office, Alexandria, Virginia
PSM	*Popular Science Monthly*
PSPT	Pennsylvania Society for the Prevention of Tuberculosis
SCH	School of Home Economics, Records, 1900–1972, Record Group

22.2, the College Archives, Simmons College, Boston, Massachusetts

SCP   School of Public Health Nursing, School of Nursing Records, 1902–1970, Record Group 22.1, the College Archives, Simmons College, Boston, Massachusetts

SE   *Sanitary Engineer*

SEP   *Saturday Evening Post*

TU   Tuskegee University Archives, Tuskegee, Alabama

UA   Central Labor Union Minutes, Records of Philadelphia Council, AFL-CIO, microfilm reels 1 and 2, Urban Archives, Temple University, Philadelphia, Pennsylvania

UPA   University of Pennsylvania Archives, Philadelphia, Pennsylvania

VP   Special Collections, University of Pennsylvania Library, Philadelphia, Pennsylvania

WC   Warshaw Collection, Smithsonian Institution Archives Center, National Museum of American History, Washington, D.C.

Introduction

1. "Ann Landers," *Philadelphia Inquirer,* Sept. 13, 1989, p. 2C.

2. For a compelling account of the fears experienced by people who act as "buddies" to AIDS patients, see Emily Martin, *Flexible Bodies: Tracking Immunity in American Culture from the Days of Polio to the Age of AIDS* (Boston: Beacon, 1994), pp. 135–139.

3. Germ-related beliefs and behaviors are good examples of what the sociologist Pierre Bourdieu refers to as "habitus," that is, modes of thinking and acting that are acquired in childhood and rarely questioned in later life. See Pierre Bourdieu, *Outline of a Theory of Practice* (New York: Cambridge University Press, 1977), pp. 80–81.

4. For a classic account of traditional notions of contagion, see Daniel Defoe, *A Journal of the Plague Year,* ed. Louis Landa (1722; reprint, New York: Oxford University Press, 1969). On conceptions of atmospheric infection, see Alain Corbin, *The Foul and the Fragrant: Odor and the French Social Imagination* (Cambridge, Mass.: Harvard University Press, 1986).

5. On how dirty most Americans were before the Civil War, see Suellen Hoy, *Chasing Dirt: The American Pursuit of Cleanliness* (New York: Oxford University Press, 1995), pp. 3–16. Personal diaries and letters suggest how rarely ordinary Americans worried about catching diseases through casual contact with other people, even those who were obviously sick. See, for example, Laurel Thatcher Ulrich, *A Midwife's Tale: The Life of Martha Ballard, Based on Her Diary, 1785–1812* (New York: Alfred A. Knopf, 1990), pp. 40–46.

6. On gentility and the impulse toward certain kinds of cleanliness, see Richard L. Bushman, *The Refinement of America: Persons, Houses, Cities* (New York: Alfred A. Knopf, 1992); and John F. Kasson, *Rudeness and Civility: Manners in Nineteenth-Century Urban America* (New York: Hill and Wang, 1990).

7. On the changing views of consumption, see Sheila M. Rothman, *Living in the Shadow of Death: Tuberculosis and the Social Experience of Illness in American History* (New York: Basic Books, 1994).

8. Edward Trudeau, *An Autobiography* (New York: Doubleday Page, 1916), pp. 29–31.

9. Sherrill Redmon, "The Poisoned Wedding," unpub. paper in the author's possession.

10. On the death of the Lincolns' son, see Jean Baker, *Mary Todd Lincoln: A Biography* (New York: Norton, 1987), pp. 208–209. On the English royal family's encounters with typhoid, see Balthazar W. Foster, *The Prince's Illness: Its Lessons* (London: J. & A. Churchill, 1872). See Chapter 1 for an account of Martha Roosevelt's death from typhoid.

11. In this book, I confine my focus to forms of prevention that required substantial changes in individual and household behavior. I thus leave aside other important preventive strategies that emerged in this time period, including immunization programs and screening programs based on physical exams, diagnostic tests, X rays, and the like. In her valuable study of the press coverage of diphtheria, typhoid, and syphilis between 1870 and 1930, Terra Ziporyn provides a good overview of these contemporaneous developments. See Terra Ziporyn, *Disease in the Popular American Press: The Case of Diphtheria, Typhoid Fever, and Syphilis, 1870–1920* (New York: Greenwood, 1988).

12. Frank Buffington Vrooman, "Public Health and National Defence," *The Arena* 69 (Aug. 1895): 425–438, quotation from p. 425.

13. T. Mitchell Prudden, *Dust and Its Dangers* (New York: G. P. Putnam's Sons, 1890), pp. 93–94. On popular religious culture, see David D. Hall, *Worlds of Wonder, Days of Judgment* (Cambridge, Mass.: Harvard University Press, 1990). The Puritan minister Cotton Mather used the phrase "invisible world" to refer to the supernatural workings of Satan and his agents during the celebrated 1692 witchcraft outbreak in Salem Village. See Cotton Mather, *The Wonders of the Invisible World* (Boston, Mass.: Benjamin Harris, 1693).

14. William Gilman Thompson, "The Present Aspect of Medical Education," *PSM* 27 (1885): 589–595, quotation from p. 590; H. G. Wells, "The War of the Worlds," *The Works of H. G. Wells*, vol. 3 (New York: Charles Scribner's Sons, 1924), pp. 207–492; the description of the bacteria's triumph is on pp. 436–437. Comparisons between "primitive" and "modern" views were commonplace in science writing at that time. See John Burnham, *How Supersti-*

tion Won and Science Lost: Popularizing Science and Health in the United States (New Brunswick, N.J.: Rutgers University Press, 1987).

15. My conception of the different versions of the germ theory has been influenced by the work of Gerald Geison on Pasteur and Christopher Lawrence and Richard Dixey on Lister. See Gerald L. Geison, *The Private Science of Louis Pasteur* (Princeton, N.J.: Princeton University Press, 1995); and Christopher Lawrence and Richard Dixey, "Practising on Principle: Joseph Lister and the Germ Theories of Disease," in *Medical Theory, Surgical Practice: Studies in the History of Surgery,* ed. Christopher Lawrence (New York: Routledge, 1992), pp. 153–215.

16. Ellen Richards quoted in Laura Shapiro, *Perfection Salad: Women and Cooking at the Turn of the Century* (New York: Farrar, Straus and Giroux, 1986), p. 181.

17. See Abraham Benenson, ed., *Control of Communicable Diseases in Man: An Official Report of the American Public Health Association,* 15th ed. (Washington, D.C.: APHA, 1990).

18. These various literatures are cited at more appropriate places in the text. Among the few sources I found that tried to address the more collective conception of the germ as it developed in the United States was Andrew McClary, "Germs Are Everywhere: The Germ Threat as Seen in Magazine Articles, 1890–1920," *Journal of American Culture* 3 (1980): 33–46. I also found inspiration in Bruno Latour, *The Pasteurization of France,* trans. Alan Sheridan and John Law (Cambridge, Mass.: Harvard University Press, 1988). Even though she focuses on only three diseases, Ziporyn's *Disease in the Popular American Press* is also a very useful overview of the evolution of popular attitudes during this period.

19. Burnham, *How Superstition Won;* Roger Cooter, *The Cultural Meaning of Popular Science: Phrenology and the Organization of Consent in Nineteenth-Century Britain* (New York: Cambridge University Press, 1984); Latour, *Pasteurization of France;* and Martin S. Pernick, *The Black Stork: Eugenics and the Death of "Defective" Babies in American Medicine and Motion Pictures since 1915* (New York: Oxford University Press, 1996). See also the excellent review essay by Roger Cooter and Stephen Pumfrey, "Separate Spheres and Public Places: Reflections on the History of Science Popularization and Science in Popular Culture," *History of Science* 32 (1994): 237–267. They explore the concept of ethnoscience on p. 243.

20. See Burnham, *How Superstition Won,* esp. Chap. 1.

21. Sinclair Lewis, *Arrowsmith* (1924; reprint, New York: Harcourt Brace Jovanovich, 1952); Paul de Kruif, *Microbe Hunters* (1926; reprint, New York: Harcourt, Brace and World, 1953).

22. Thomas McKeown, *The Modern Rise of Population* (New York: Academic Press, 1976).

23. Simon Szreter, "The Importance of Social Intervention in Britain's Mortality Decline, c. 1850–1914: A Reinterpretation of the Role of Public Health," *Social History of Medicine* 1 (1988): 1–37; Anne Hardy, *The Epidemic Streets: Infectious Disease and the Rise of Preventive Medicine, 1856–1900* (Oxford, Eng.: Clarendon Press, Oxford University Press, 1993). For the United States, see Gretchen A. Condran, Henry Williams, and Rose A. Cheney, "The Decline in Mortality in Philadelphia from 1870 to 1930: The Role of Municipal Services," *Pennsylvania Magazine of History and Biography* 108 (1984): 153–177; and Samuel H. Preston and Michael R. Haines, *Fatal Years: Child Mortality in Late-Nineteenth-Century America* (Princeton, N.J.: Princeton University Press, 1991). See also the excellent symposium on *Fatal Years* published in *BHM* 68 (1994): 86–128.

24. Douglas C. Ewbank and Samuel H. Preston, "Personal Health Behaviour and the Decline in Infant and Child Mortality: The United States, 1900–1930," in John Caldwell et al., eds., *What We Know about Health Transition: The Cultural, Social and Behavioural Determinants of Health* (Canberra: Australian National University Press, 1990), pp. 116–149, quotation from p. 128.

25. Mary Douglas, *Purity and Danger: An Analysis of the Concepts of Pollution and Taboo* (1966; reprint, Boston: ARK Paperbacks, 1984), p. 2. Elias, *The Civilizing Process,* p. 159, makes much the same argument.

26. The surgeon general's remarks are quoted in Barry Bloom and Christopher J. L. Murray, "Tuberculosis: Commentary on a Reemergent Killer," *Science,* Aug. 21, 1992, p. 1055.

27. This point is nicely made by Charles E. Rosenberg. See his "Framing Disease: Illness, Society, and History," in *Explaining Epidemics and Other Studies in the History of Medicine,* ed. Charles E. Rosenberg (New York: Cambridge University Press, 1992), pp. 305–318.

28. The "narrowing thesis" is clearly stated in Paul Starr, *The Social Transformation of American Medicine: The Rise of a Sovereign Profession and the Making of a Vast Industry* (New York: Basic Books, 1982), p. 189. A similar argument has been made by Robert Gottlieb in his *Forcing the Spring: The Transformation of the American Environmental Movement* (Washington, D.C.: Island Press, 1993). On the germ theory's association with the forces of exclusion and repression, see Alan M. Kraut, *Silent Travelers: Germs, Genes, and the "Immigrant Menace"* (New York: Basic Books, 1994); Rothman, *Living in the Shadow,* esp. pp. 179–193; and Judith Walzer Leavitt, *Typhoid Mary: Captive of the Public's Health* (Boston: Beacon, 1996).

29. My interest lies primarily in the ways the germ theory was invoked in discussions of infectious disease and its prevention. JoAnne Brown's work in progress will, I believe, provide a broader view of how the germ theory was employed as a metaphor in many dimensions of American political and

cultural life. See, for example, JoAnne Brown, *The Definition of a Profession: The Authority of Metaphor in the History of Intelligence Testing, 1890–1930* (Princeton, N.J.: Princeton University Press, 1992), pp. 78–81. Much more research needs to be done on the *interrelationships* among the many popular public health crusades of the early twentieth century. Martin Pernick's work raises important questions about the overlapping agendas of infectious disease control and the popular eugenics movement. See Pernick, *The Black Stork,* esp. pp. 50–53, 58–59.

30. For England, see for example, Hardy, *Epidemic Streets,* and for France, David S. Barnes, *The Making of a Social Disease: Tuberculosis in Nineteenth-Century France* (Berkeley: University of California Press, 1995). On the colonial experience, see Warwick Anderson, "Immunities of Empire: Race, Disease, and the New Tropical Medicine, 1900–1920," *BHM* 70 (1996): 94–118; Mark Harrison, *Public Health in British India: Anglo-Indian Preventive Medicine, 1859–1914* (New York: Cambridge University Press, 1994); Randall M. Packard, *White Plague, Black Labor: Tuberculosis and the Political Economy of Health and Disease in South Africa* (Berkeley: University of California Press, 1989); and Mary P. Sutphen, "Not What but Where: Bubonic Plague and the Reception of Germ Theories in Hong Kong and Calcutta, 1894–1897," *JHM* 52 (1997): 81–113.

31. Two more systematically comparative works that support these generalizations are Georgina D. Feldberg, *Disease and Class: Tuberculosis and the Shaping of Modern North American Society* (New Brunswick, N.J.: Rutgers University Press, 1995); and Alisa Klaus, *Every Child a Lion: The Origins of Maternal and Infant Health Policy in the United States and France, 1890–1920* (Ithaca, N.Y.: Cornell University Press, 1993). On advertising and hygiene, see the brief but provocative comparative overview in Adrian Forty, *Objects of Desire* (New York: Pantheon, 1986), esp. Chap. 7. On the "selling health" theme, see Elizabeth Toon, "Selling Health: Consumer Education, Public Health, and Public Relations in the Interwar Period" (paper delivered at the Berkshire Conference on Women's History, Chapel Hill, N.C., June 1996). For two insightful discussions of the complex meanings of disease rhetoric, see Pernick, *The Black Stork,* and Guenter Risse, "'A Long Pull, a Strong Pull, and All Together': San Francisco and Bubonic Plague, 1907–1908," *BHM* 66 (1992): 260–286.

32. "Report of the Committee on Disinfectants," *Transactions of the American Medical Association* 17 (1866): 129–155, quotation from p. 129.

1. Apostles of the Germ

1. Corinne Roosevelt Robinson to H. F. Pringle, Sept. 18, 1930, Pringle Notes, HL.

2. Elliott Roosevelt quoted in David McCullough, *Mornings on Horseback* (New York: Simon and Schuster, 1981), p. 285.

3. On typhoid, the healthy carrier concept, and the life of Mary Mallon, see the excellent work by Judith Walzer Leavitt, *Typhoid Mary: Captive of the Public's Health* (Boston: Beacon, 1996).

4. Roger S. Tracy, *Handbook of Sanitary Information for Householders* (New York: D. Appleton, 1884), p. 14.

5. Frances Theodora Smith Parsons, *Perchance Some Day* (Privately printed, 1951), pp. 26, 44–45, HL; McCullough, *Mornings on Horseback*, pp. 66–67, 127, 135–136, 244–245.

6. The Roosevelt family papers (HL) contain no record of the explanations that family doctors offered for Martha Roosevelt's illness. By 1880, it was common for physicians to order inspections of household plumbing following a typhoid outbreak. See, for example, William Keating, "An Epidemic of Typhoid Fever from Defective Drainage," *Transactions of the College of Physicians of Philadelphia,* ser. 3, vol. 4 (1879): 85–125.

7. These statistics are taken from Gretchen Condran, "Changing Patterns of Epidemic Disease in New York City," in *Hives of Sickness: Public Health and Epidemics in New York City,* ed. David Rosner (New Brunswick, N.J.: Rutgers University Press, 1995), pp. 27–41. See also Samuel H. Preston and Michael R. Haines, *Fatal Years: Child Mortality in Late Nineteenth-Century America* (Princeton, N.J.: Princeton University Press, 1991). For a good overview of nineteenth-century public health issues, see John Duffy, *The Sanitarians: A History of American Public Health* (Chicago: University of Illinois Press, 1990).

8. William H. Mays, "On the Supposed Identity of the Poisons of Diphtheria, Scarlatina, Typhoid Fever, and Puerperal Fever," *San Francisco Western Lancet* 9 (1880–1881): 110–115, quotation from p. 110.

9. Useful overviews of the germ theory in relation to the history of disease etiology can be found in Lester S. King, *Transformations in American Medicine: From Benjamin Rush to William Osler* (Baltimore, Md.: Johns Hopkins University Press, 1991); Robert P. Hudson, *Disease and Its Control: The Shaping of Modern Thought* (New York: Praeger, 1983); Margaret Pelling, "Contagion/Germ Theory/Specificity," in *Companion Encyclopedia of the History of Medicine,* eds. W. F. Bynum and Roy Porter (New York: Routledge, 1993), vol. 1, pp. 309–334; and Oswei Temkin, "An Historical Analysis of the Concept of Infection," in *The Double Face of Janus* (Baltimore, Md.: Johns Hopkins University Press, 1977), pp. 456–471.

10. For a more extensive discussion of these medical debates, see Nancy Tomes, "American Attitudes toward the Germ Theory of Disease: The Richmond Thesis Revisited," *JHM* 52 (1997): 17–50. My understanding of the American debate is heavily indebted to several works on mid-nineteenth-century disease theory: John K. Crellin, "The Dawn of the Germ Theory:

Particles, Infection and Biology," in *Medicine, and Science in the 1860s,* ed. F. N. L. Poynter (London: Wellcome Institute of the History of Medicine, 1968), pp. 57–76; John M. Eyler, *Victorian Social Medicine: The Ideas and Methods of William Farr* (Baltimore, Md.: Johns Hopkins University Press, 1979); Margaret Pelling, *Cholera, Fever, and English Medicine, 1825–1865* (New York: Oxford University Press, 1978); and Pelling, "Contagion/Germ Theory/Specificity."

11. Lionel S. Beale, quoted in W. D. Foster, *A History of Medical Bacteriology and Immunology* (London: Heinemann, 1970), p. 16.

12. For the etymology of "germ," see the *Oxford English Dictionary,* 2d ed., vol. 6 (Oxford: Oxford University Press, 1989), pp. 467–468.

13. See Tomes, "American Attitudes," for a more extended discussion of this medical civil war in the United States. Although I disagree with much of her interpretation, an interesting account of the debate can be found in Phyllis Allen Richmond, "American Attitudes toward the Germ Theory of Disease (1860–1880)," *JHM* 9 (1954): 58–84. K. Codell Carter has written a number of valuable articles on the germ theory and its effects on medical thinking during this period, of which I found must useful "Ignaz Semmelweis, Carl Mayrhofer, and the Rise of Germ Theory," *Medical History* 29 (1985); 33–53; and "Koch's Postulates in Relation to the Work of Jacob Henle and Edwin Klebs," *Medical History* 29 (1985): 353–374.

14. Karl Liebermeister, "Introduction," in *Cyclopedia of the Practice of Medicine,* vol. 1: *Acute Infectious Diseases,* ed. Hugo von Ziemssen, trans. Thomas Satterthwaite (New York: William Wood, 1874), pp. 1–33, quotation from p. 6. For useful overviews of the modern germ theory's antecedents, see William Bulloch, *The History of Bacteriology* (London: Oxford University Press, 1960), Chaps. 1 and 2; Pelling, "Contagion/Germ Theory/Specificity"; Richard Harrison Shryock, "Germ Theories in Medicine Prior to 1870," *Clio Medica* 7 (1972): 81–109; and Catherine Wilson, *The Invisible World: Early Modern Philosophy and the Invention of the Microscope* (Princeton, N.J.: Princeton University Press, 1995), Chap. 5.

15. Liebermeister, "Introduction," pp. 6, 7. On the early history of the microscope and microscopy, see Wilson, *Invisible World.*

16. Liebermeister, "Introduction," p. 7.

17. On the history of popular microscopy, see Stella Butler, R. H. Nuttall, and Olivia Brown, *The Social History of the Microscope* (Cambridge, Eng.: Whipple Museum of the History of Science, n.d.), esp. p. 10; and John Harley Warner, "'Exploring the Inner Labyrinths of Creation': Popular Microscopy in Nineteenth-Century America," *JHM* 37 (1982): 7–33. On the growth of medical microscopy, see James H. Cassedy, "The Microscope in American Medical Science, 1840–1860," *Isis* 67 (1976): 76–97; and Deborah Jean Warner, "Medical Microscopy in Antebellum America," *BHM* 69 (1995): 367–386.

18. On Justus von Liebig's influence on zymotic theories of disease, see Eyler, *Victorian Social Medicine,* pp. 100, 103. My summary of Pasteur's work relies heavily on Gerald L. Geison, "Louis Pasteur," in *The Dictionary of Scientific Biography,* vol. 3, ed. Charles C. Gillispie (New York: Charles Scribner's Sons, 1974), pp. 350–416; and Gerald L. Geison, *The Private Science of Louis Pasteur* (Princeton, N.J.: Princeton University Press, 1995).

19. René Dubos, *Louis Pasteur: Free Lance of Science* (1960; reprint, New York: Da Capo, 1986), p. 233.

20. For an account of Pasteur's part in the debate on spontaneous generation, see Geison, *Private Science,* pp. 110–142. See also John Farley, *The Spontaneous Generation Controversy from Descartes to Oparin* (Baltimore, Md.: Johns Hopkins University Press, 1977).

21. See Geison, *Private Science,* Chap. 5.

22. Ibid., pp. 32–33, 90–91. On Pasteur's early writings on the germ theory, see also Dubos, *Louis Pasteur,* pp. 233–266.

23. The best general survey of this early research is Bulloch, *History of Bacteriology.*

24. John C. Dalton, "The Origin and Propagation of Disease," in *Annual Report of the Board of Regents of the Smithsonian Institution . . . for the Year 1873* (Washington, D.C.: Government Printing Office, 1874), pp. 226–245, quotation on p. 299. On parasitology and germ theory, see John Farley, "Parasites and the Germ Theory of Disease," in *Framing Disease: Studies in Cultural History,* ed. Charles E. Rosenberg and Janet Golden (New Brunswick, N.J.: Rutgers University Press, 1992), pp. 33–49. Farley argues here that the scientific study of parasites had little influence on early formulations of the germ theory. But in my reading of American sources, I found the model of parasitism to be employed much more frequently than he suggests.

25. Roy MacLeod, "John Tyndall," in *Dictionary of Scientific Biography,* vol. 13, pp. 521–524.

26. Like Pasteur, Lister originally thought more in terms of a germ theory of putrefaction rather than a germ theory of disease. See Christopher Lawrence and Richard Dixey, "Practising on Principle: Joseph Lister and the Germ Theories of Disease," in *Medical Theory, Surgical Practice: Studies in the History of Surgery,* ed. Christopher Lawrence (New York: Routledge, 1992), pp. 153–215.

27. The first article with the phrase "germ theory" in the title was Jabez Hogg, "The Organic Germ Theory of Disease," *Medical Times and Gazette* (1870); 659, 685. In general, there are very few items on the topic dated before the late 1860s. See *Index Catalogue of the Library of the Surgeon-General's Office, United States Army* ser. 1, vol. 5 (Washington, D.C.: Government Printing Office, 1880–1895), pp. 385–388. Initially, the phrase "germ theory of disease" was also used to refer to the work of the British physician and

microscopist Lionel Beale, who argued that disease germs were "bioplasts"—immature human cells that multiplied too quickly and displaced the healthy cells—a theory not unlike the modern view of cancer. The Pasteur/Lister theory was often called the "parasitic" germ theory to distinguish it from Beale's hypothesis.

28. Thomas J. MacLagan, "Correspondence," *Lancet,* Feb. 19, 1876, pp. 295–296; quotation from p. 295. MacLagan was commenting on the ongoing debate between John Tyndall and H. Charlton Bastian. On the many uses of the word "germ," see Pelling, "Contagion/Germ Theory/Specificity," esp. p. 314.

29. For a more extended discussion of this point, see Tomes, "American Attitudes."

30. See, for example, H. Charlton Bastian, "An Address on the Germ Theory of Disease," *Lancet,* Apr. 10, 1875, pp. 501–509.

31. Edward P. Hurd, "On the Germ Theory of Disease," *Boston Medical and Surgical Journal* 91, no. 5 (July 1874): 97–110, quotation from p. 100.

32. Ibid., 101–102, 106.

33. Frank J. Davis, "Atmospheric Germs and Their Relation to Disease," *Chicago Medical Examiner* 12 (1871): 191–199, quotation on p. 198.

34. For a discussion of Salisbury's work, see Richmond, "American Attitudes," pp. 62–64.

35. Liebermeister, "Introduction," p. 8.

36. Hurd, "On the Germ Theory," pp. 105, 109.

37. For an overview of Koch's work on anthrax, see Thomas Brock, *Robert Koch: A Life in Medicine and Bacteriology* (Madison, Wisc.: Science Tech, 1888), pp. 27–37.

38. Ibid.

39. On the tradition of public science, see Jan Golinski, *Science as Public Culture: Chemistry and Enlightenment in Britain, 1760–1820* (New York: Cambridge University Press, 1992); and Larry Stewart, *The Rise of Public Science: Rhetoric, Technology, and Natural Philosophy in Newtonian Britain, 1660–1750* (New York: Cambridge University Press, 1992). Both sides of the germ theory debate employed such metaphors and everyday allusions to make their case, as I discuss at greater length in Tomes, "American Attitudes." My focus here is on the strategies of representation employed by the theory's early advocates because they most directly shaped the public's conceptions of the germ. For an interesting discussion of the varied metaphors used to comprehend the nature of infection, see Pelling, "Contagion/Germ Theory/Specificity," pp. 313–315.

40. The homely quality of these investigations is nicely conveyed in John Tyndall, "Spontaneous Generation," *PSM* 12 (1878): 476–488, 591–604. He recounts the trip to the Turkish bath on p. 594.

41. Dr. [William] Roberts on Spontaneous Generation, *PSM* 9 (1876): 638–639. The references to turnip slices and mutton chops can be found in Tyndall, "Spontaneous Generation," pp. 484, 597.

42. For a good example of this line of defense, see P. Schutzenberger, "Air-Germs and Spontaneous Generation," *PSM* 9 (1878): 91–102.

43. "Professor [John] Tyndall's Recent Researches," *PSM* 8 (1876): 686–699, quotation on p. 696.

44. John Tyndall, "Fermentation and Its Bearings on the Phenomena of Disease," *PSM* 9 (1876): 129–154, quotations from pp. 141, 148.

45. The Ehrenberg quotation is in Fredinand Papillon, "Ferments, Fermentation, and Life," *PSM* 5 (1874): 542–556, quotation on p. 551. Other quotations are from Tyndall, "Spontaneous Generation," p. 486; and "Tyndall and Roberts on Spontaneous Generation," *PSM* 10 (1877): 758.

46. Quoted in a book review of Antoine Magnin, *The Bacteria,* in "Literary Notices," *PSM* 19 (1881): 706.

47. Joseph Richardson, *The Germ Theory of Disease, and Its Present Bearing upon Public and Personal Hygiene* (Philadelphia: Philadelphia Social Science Association, 1878), p. 9. Richardson's statistics here are best read as attempts to impress the reader rather than literal statements of fact, since it is hard to imagine how only 20,000 spores could be lined up to make an inch, whereas 50 million could fit on a period.

48. Quoted in "Disadvantages and Advantages of Bacteria," *PSM* 21 (1882): 709. On the tendency to categorize animals into good and bad species, see Keith Thomas, *Man and the Natural World: A History of Modern Sensibility* (New York: Pantheon, 1983); and Thomas R. Dunlap, *Saving America's Wildlife: Ecology and the American Mind, 1850–1990* (Princeton, N.J.: Princeton University Press, 1988).

49. Papillon, "Ferments, Fermentation, and Life," p. 551.

50. On the "seed and soil" metaphor, see Vivian Nutton, "The Seeds of Disease: An Explanation of Contagion and Infection from the Greeks to the Renaissance," *Medical History* 27 (1983): 1–34.

51. Dalton, "Origin and Propagation of Disease," p. 243.

52. Mays, "On the Supposed Identity," p. 111; Stephen Smith, "Practical Tests of the Antiseptic System," *Transactions of the Medical Society of New York for 1878* (Albany, N.Y.: Medical Society of New York, 1878), pp. 106–131, quotation on p. 130; Richardson, *The Germ Theory,* p. 4.

53. For a brief but useful discussion of the connections between Darwinian theory and germ theory, see W. F. Bynum, "Darwin and the Doctors: Evolution, Diathesis, and Germs in Nineteenth-Century Britain," *Gesnerus* 40 (1983): 43–53, esp. pp. 49–52. On Darwinism and American medicine generally, see John S. Haller, *American Medicine in Transition, 1840–1910* (Urbana: University of Illinois Press, 1981), Chap. 8. Germ theorists by no means

employed evolutionary theory in identical ways. For example, Kenneth W. Millican, in his *Evolution of Morbid Germs: A Contribution to Transcendental Pathology* (London: H. K. Lewis, 1883), used Darwin's theory of evolution to argue that disease germs of one fever could evolve into the germs of another.

54. Henry Gradle, *Bacteria and the Germ Theory of Disease* (Chicago: W. T. Keener, 1883), p. 2.

55. Ibid., p. 56.

56. See, for example, Papillon, "Ferments, Fermentation, and Life."

57. William Preston Hill, "An Essay on the Origin of the Germ Theory" (M.D. thesis, University of Pennsylvania, 1885, [unpaginated]), VP.

58. Geison, *Private Science,* Chaps. 6, 7, 8, 9; Brock, *Robert Koch,* Chap. 18; and John K. Crellin, "Internal Antisepsis or the Dawn of Chemotherapy?" *JHM* 36 (1981): 9–18.

59. My interpretation here is similar to Latour's argument about the French hygienists. See Bruno Latour, *The Pasteurization of France,* trans. Alan Sheridan and John Law (Cambridge, Mass.: Harvard University Press, 1988), esp. pp. 25–26, 34. Latour himself seems to think that the French case differs in essential respects from the American and British; see p. 26, n. 17. Although it is certainly true that the Americans and British did not focus so exclusively on the figure of Pasteur, Latour's general point—that supporters of the germ theory translated the hygienists' precepts into their own terms and adopted their sanitary agenda—holds true for the United States generally. As does Latour, I believe the synthesis of sanitary science and germ theory initially strengthened both.

60. Lloyd Stevenson, "Science Down the Drain: On the Hostility of Certain Sanitarians to Animal Experimentation, Bacteriology, and Immunology," *BHM* 29 (1955): 1–26; and Charles E. Rosenberg, "Florence Nightingale on Contagion: The Hospital as Moral Universe," in *Explaining Epidemics and Other Studies in the History of Medicine,* ed. Charles E. Rosenberg (New York: Cambridge University Press, 1992), pp. 90–108. On Elizabeth Blackwell, see also Regina Morantz-Sanchez, *Sympathy and Science* (New York: Oxford University Press, 1985), pp. 186–191, and "Feminist Theory and Historical Practice: Rereading Elizabeth Blackwell," *History and Theory* 31 (1992): 51–69.

61. F. A. P. Barnard, "The Germ Theory of Disease and Its Relations to Hygiene," in American Public Health Association, *Public Health Reports and Papers . . . 1873,* pp. 70–87; quotation on p. 87. Much the same point is made in Howard Kramer, "The Germ Theory and the Early Public Health Program in the United States," *BHM* 22 (1948): 233–247.

62. René Vallery-Radot, *The Life of Pasteur,* trans. Mrs. R. L. Devonshire (New York: Doubleday, Page, 1920), p. 213. His daughters' deaths from typhoid are discussed on pp. 86, 130–131. See also Geison, *Private Science,* p. 48.

2. Whited Sepulchers

1. Charles Wingate, "The Unsanitary Homes of the Rich," *North American Review* 137 (1883): 172–184; quotations from pp. 173, 174. As a sanitary engineer, Wingate had a professional stake in convincing affluent Americans to fear defective household plumbing, a point I discuss at more length in Chapter 3.

2. The reference to "whitewashed tombs" can be found in Matt. 23:27 Revised Standard Version.

3. On parallel developments in late-nineteenth-century England, see Annmarie Adams, *Architecture in the Family Way: Doctors, Houses, and Women, 1870–1900* (Montreal: McGill-Queen's University Press, 1996).

4. For a sense of traditional plague precautions, see Philip Ziegler, *The Black Death* (New York: Harper and Row, 1969); and Carlo Cipolla, *Fighting the Plague in Seventeenth-Century Italy* (Madison: University of Wisconsin Press, 1981).

5. These traditional beliefs are well illustrated in Daniel Defoe, *A Journal of the Plague Year* ed. Louis Landa (1722; reprint, New York: Oxford University Press, 1969).

6. On the reluctance of hospitals to treat contagious illnesses, see Charles Rosenberg, *The Care of Strangers: The Rise of America's Hospital System* (New York: Basic Books, 1987), pp. 22–23, 30–31. A dramatic example of community censure is the harsh treatment of Martha Carrier and her children, who contracted smallpox in Andover, Massachusetts, in 1690 and were blamed for spreading the disease to others. Martha Carrier was later accused of witchcraft and executed during the Salem trials. See Carol Karlsen, *The Devil in the Shape of a Woman: Witchcraft in Colonial New England* (New York: Vintage, 1987), pp. 99–100.

7. Rush is quoted in Whitfield J. Bell, Jr., *The College of Physicians of Philadelphia: A Bicentennial History* (Canton, Mass.: Science History Publications, 1987), p. 28. On the new public health science, see James C. Riley, *The Eighteenth-Century Campaign to Avoid Disease* (New York: St. Martin's, 1987); and William Coleman, *Death Is a Social Disease: Public Health and Political Economy in Early Industrial France* (Madison: University of Wisconsin Press, 1982). My generalizations about health manuals are based on my reading of such standards as Bernhard C. Faust, *Catechism of Health for the Use of Schools, and for Domestic Instruction* (1794; reprint, New York: Arno, 1972), and Caleb Ticknor, *The Philosophy of Living, or The Way to Enjoy Life and Its Comforts* (New York: Harper and Bros., 1836). See also on this genre, Charles E. Rosenberg, "Catechisms of Health: The Body in the Prebellum Classroom," *BHM* 69 (1995): 175–197. Note that the human body was often likened to a house, a tradition that made the sanitarian teachings about domestic hygiene all the easier to understand.

8. *Report of the Council of Hygiene and Public Health of the Citizens Association of New York upon the Sanitary Condition of the City* (1866; reprint, New York: Arno, 1970), p. xcvi. On the importance of the Civil War, see Suellen Hoy, *Chasing Dirt: The American Pursuit of Cleanliness* (New York: Oxford University Press, 1995), Chap. 2. On hospital reform, see Rosenberg, *Care of Strangers,* Chap. 5. On urban sanitary reform, see Stanley K. Schultz, *Constructing Urban Culture: American Cities and City Planning, 1800–1920* (Philadelphia: Temple University Press, 1989).

9. For two accounts of the problem, one contempory, one historical, see William Paul Gerhard, *The Drainage of a House* (Boston: Rand Avery, 1888), p. 4; and Joel Tarr, James McCurley, and Terry F. Yosie, "The Development and Impact of Urban Wastewater Technology," in *Pollution and Reform in American Cities, 1870–1930,* ed. Martin L. Melosi (Austin: University of Texas Press, 1980), pp. 59–82. On the revolution in domestic plumbing, see May N. Stone, "The Plumbing Paradox: American Attitudes toward Late-Nineteenth-Century Domestic Sanitary Arrangements," *Winterthur Portfolio* 14 (1979): 283–309; Maureen Ogle, "Domestic Reform and American Household Plumbing, 1840–1870," *Winterthur Portfolio* 28 (1993): 33–58; and Maureen Ogle, *All the Modern Conveniences: American Household Plumbing, 1840—1890* (Baltimore, Md.: Johns Hopkins University Press, 1996).

10. Wingate, "Unsanitary Homes," p. 173. On the Windsors' trials with typhoid, see Balthazar W. Foster, *The Prince's Illness: Its Lesson. A Lecture on the Prevention of Disease* (London: J. A. Churchill, 1872), p. 16.

11. Robert Clark Kedzie, "Preventive Medicine," *Sanitarian* 3, no. 26 (May 1875): 86.

12. Joseph Edwards, *How We Ought to Live* (Philadelphia: H. C. Watts, 1882), pp. 151, 407.

13. On changing patterns of publishing and knowledge diffusion, see Richard D. Brown, *Knowledge Is Power: The Diffusion of Information in Early America, 1700–1865* (New York: Oxford University Press, 1989); and Carl F. Kaestle, et al., *Literacy in the United States: Readers and Reading since 1880* (New Haven, Conn.: Yale University Press, 1991). On medical advice literature, see Anita Clair Fellman and Michael Fellman, *Making Sense of Self: Medical Advice Literature in Late-Nineteenth-Century America* (Philadelphia: University of Pennsylvania Press, 1982). On the explosion of literature on domestic architecture and hygiene, see Adams, *Architecture in the Family Way;* Ogle, *All the Modern Conveniences;* and Ogle, "Domestic Reform," esp. p. 40.

14. Henry I. Bowditch, *Public Hygiene in America* (Boston: Little, Brown, 1877), p. 38.

15. William Eassie, *Sanitary Arrangements for Dwellings, Intended for the Use of Officers of Health, Architects, Builders, and Householders* (London: Smith Elder, 1874); Henry Hartshorne, *Our Homes* (Philadelphia: Presley Blakiston,

1880); and Frederick Castle, ed., *Wood's Household Practice of Medicine, Hygiene, and Surgery*, 2 vols. (New York: William Wood, 1880). The quotation from the Castle volume is taken from the title page. Volume 1 has several chapters on house construction and domestic hygiene. Only a few books and articles dating from before 1875 appeared under the heading of "habitations" in the first series of the U.S. surgeon general's catalogue; the bulk of the items listed date from the late 1870s and 1880s. See *Index Catalogue of the Library of the Surgeon General's Office, United States Army*, 1st ser. (Washington, D.C.: U.S. Government Printing Office, 1880–1895).

16. George Waring's articles were reprinted in book form as *The Sanitary Drainage of Houses and Towns* (New York: Hurd and Houghton, 1876). On his career, see James H. Cassedy, "The Flamboyant Colonel Waring: An Anticontagionist Holds the American Stage in the Age of Pasteur and Koch," *BHM* 36 (1962): 163–176. The alternative or sectarian medical press also joined in promoting the tenets of home hygiene. The earliest description of the germ theory I have found was in a sectarian volume [John Harvey Kellogg], *The Household Manual* (Battle Creek, Mich.: Office of the Health Reformer, 1875).

17. Broadsides and circulars had long been used to educate the public in times of epidemic diseases. Some typical American examples include: Sanitary Committee of the Board of Health of Philadelphia, *Sanitary and Preventive Measures: Disinfectants, How to Use Them, or What May Be Done by the Public to Guard against Yellow Fever* (Philadelphia: Markley and Son, 1878), CPP; and Massachusetts State Board of Health, *Suggestions for Preventing the Spread of Scarlet Fever* [n.d.; received at the surgeon's general's office in 1888], NLM. For examples of literature aimed at a more humble readership, see Citizens' Sanitary Society of Brooklyn, *Sanitary Tracts* (New York: E. P. Coby, n.d.); and Mary Armstrong, *Preventable Diseases*, Hampton Tracts for the People, Sanitary Series, no. 3 (Hampton, Va.: The Hampton Institute, 1878).

18. Kellogg, *Household Manual*, p. 16. This is the earliest mention of germs that I found in my sample of manuals. Note that a very brief account of how "microscopic plants" cause zymotic diseases appears in Catharine Beecher and Harriet Beecher Stowe, *The American Woman's Home* (1869; reprint, New York: Arno, 1971), pp. 421–422, but they do not use the word "germ."

19. *NYT*, Feb. 11, 1874, p. 2.

20. Armstrong, *Preventable Diseases*, p. 6.

21. Emma Hewitt, *Queen of the Home* (n.p.: Miller Magee, 1888), p. 225; and George Wilson, *Health and Healthy Homes: A Guide to Domestic Hygiene* (Philadelphia: Presley Blakiston, 1880), p. 117. The precocity of lay, as compared to medical, interest in the germ theory was first noted in Phyllis Allen Richmond, "American Attitudes toward the Germ Theory of Disease (1860–1880)," *JHM* 9 (1954): 58–84.

22. Armstrong, *Preventable Diseases*, p. 5. On the threat of "vitiated" air, see Gavin Townsend, "Airborne Toxins and the American House, 1865–1895," *Winterthur Portfolio* 24, no. 1 (spring 1989): 29–42.

23. Henry Hartshorne, *Our Homes* (Philadelphia: Presley Blakiston, 1880), p. 9. On Victorian concepts of decay and disease, see Christopher Hamlin, "Providence and Putrefaction: Victorian Sanitarians and the Natural Theology of Disease," *Victorian Studies* 28 (1985): 381–411.

24. This same "add-on" approach is evident in the thinking of hospital reformers. See Rosenberg, *Care of Strangers,* pp. 137–141.

25. In manuals written before 1880, the chief concern was unhealthy "ground air" arising from saturated soil—a focus that reflected the popularity of the German physician Max von Pettenkofer's theory that tainted "ground water" produced cholera. In later writings, the ground water argument took on a new association with microbes. See, for example, the unsigned editorial "Soil and Health," *SE* 5, no. 5 (Dec. 29, 1881): 100. The specifications for ventilation were heavily influenced by the literature of hospital construction, as described in Rosenberg, *Care of Strangers,* esp. pp. 139–141.

26. T. J. MacLagan, "How Typhoid Fever Is Conveyed," *PSM* 16 (1880): 460–467, quotation from pp. 462–463. For a typical description of a do-it-yourself window ventilator, see Roger S. Tracy, *Hand-book of Sanitary Information for Householders* (New York: D. Appleton, 1884), p. 17.

27. Hartshorne, *Our Homes,* p. 101.

28. Alfred Carroll, "The Enemy in the Air," *Sanitarian* 6, no. 63 (June 1878): 253–255, quotation on p. 255. The latter article was reprinted from the *Messenger,* a Staten Island, New York, missionary paper.

29. J. Pridgin Teale, *Dangers to Health: A Pictorial Guide to Domestic Sanitary Defects* (London: Churchill, 1879), p. 9. For a description of the peppermint test, see Tracy, *Hand-book,* pp. 63–64.

30. Hartshorne, *Our Homes,* p. 100.

31. Ellen H. Richards and Marion Talbot, *Home Sanitation: A Manual for Housekeepers* (Boston: Ticknor, 1887), p. 9. The careers of Richards and Talbot are discussed in more detail in Chapter 6.

32. MacLagan, "How Typhoid Fever Is Conveyed," p. 465. For a representative set of instructions on how to conduct a home hospital, see Edwards, *How We Ought to Live,* pp. 395–401. On the arrangements of the childbirth chamber, see Adams, *Architecture in the Family Way,* Chap. 4.

33. Massachusetts Board of Health, *Suggestions,* p. 1; Joseph F. Perry, *Health in Our Homes* (Boston: Thayer, 1887), p. 403. The Massachusetts circular emphasized that the germ theory was only a "hypothesis" but stressed that whatever the agent of scarlet fever, it could be spread by fomites. In Williams's story, a boy catches scarlet fever and the doctor insists that his

beloved stuffed bunny be burned because "it's a mass of scarlet fever germs." A fairy saves the bunny by turning him into a real rabbit. See Margery Williams, *The Velveteen Rabbit* (1922; reprint, New York: Knopf, 1985).

34. Hewitt, *Queen of the Home*, p. 112.

35. New Hampshire State Board of Health, *Disinfectants and Their Use* (Concord, N.H.: Parsons B. Cogswell, 1885), p. 5. See Hartshorne, *Our Homes*, pp. 130–136, for a standard discussion of disinfection.

36. On gentility, family, and class, see Richard L. Bushman, *The Refinement of America: Persons, Houses, Cities* (New York: Alfred A. Knopf, 1992); John F. Kasson, *Rudeness and Civility: Manners in Nineteenth-Century Urban America* (New York: Hill and Wang, 1990); Ogle, "Domestic Reform"; and Mary Ryan, *Cradle of the Middle Class: The Family in Oneida County, New York, 1790–1865* (New York: Cambridge University Press, 1981).

37. Clara Bloomfield-Moore, *Sensible Etiquette of The Best Society*, 10th ed. (Philadelphia: Porter and Coates, 1878), p. 257. Novels of the era underline the link between class and housing. In *The Age of Innocence*, set in New York society in the 1870s, Edith Wharton installed newlyweds Newland Archer and May Welland in a townhouse of "ghastly greenish-yellow stone" whose chief virtue was its "perfect" plumbing. The housing theme also figured prominently in William Dean Howells, *The Rise of Silas Lapham* (Boston: Ticknor, 1885).

38. On the importance of the home to Victorian conceptions of family and social life, see Clifford Edward Clark, Jr., *The American Family Home, 1800–1960* (Chapel Hill: University of North Carolina Press, 1986), esp. pp. 1–71; Louise L. Stevenson, *The Victorian Homefront: American Thought and Culture, 1860–1880* (New York: Twayne, 1991); and Stuart M. Blumin, *The Emergence of the Middle Class: Social Experience in the American City, 1760–1900* (New York: Cambridge University Press, 1989), Chap. 5. On the expansion of modern conveniences, see Stone, "Plumbing Paradox"; Ogle, *All the Modern Conveniences;* and Ogle, "Domestic Reform."

39. On the larger context of Victorian anxieties about gentility, see Kasson, *Rudeness and Civility;* and Karen Halttunen, *Confidence Men and Painted Ladies: A Study of Middle-Class Culture in America, 1830–1870* (New Haven, Conn.: Yale University Press, 1982).

40. [Robert Tomes,] *The Bazar Book of Health* (New York: Harper and Bros., 1873), p. 17; "Decomposition," *Sanitarian* 2, no. 7 (Oct. 1874): 316.

41. For an insightful discussion of the connection between woman, body, and house, see Adams, *Architecture in the Family Way*. See also Hoy, *Chasing Dirt*. On dirt and sexuality, see Phyllis Palmer, *Domesticity and Dirt: Housewives and Domestic Servants in the United States, 1920–1945* (Philadelphia: Temple University Press, 1989), esp. pp. 138–141.

42. Harriette M. Plunkett, *Women, Plumbers, and Doctors* (New York: D.

Appleton, 1885), p. 43, 10; Perry, *Health in Our Homes,* (Boston: Thayer, 1887) p. 65.

43. For a similar argument, see Adams, *Architecture in the Family Way,* esp. Chap. 3.

44. George E. Waring, *How to Drain a House* (New York: Henry Holt, 1885), pp. v–vi; Plunkett, *Women, Plumbers, and Doctors,* p. 10.

45. See Nancy Schrom Dye and Daniel Blake Smith, "Mother Love and Infant Death, 1750–1920," *Journal of American History* 73, no. 2 (Sept. 1986): 329–353. On the high rates of child mortality that persisted into the early 1900s across all classes, see Samuel H. Preston and Michael R. Haines, *Fatal Years: Child Mortality in Late Nineteenth-Century America* (Princeton, N.J.: Princeton University Press, 1991). On Victorian mourning customs, see Martha Pike and Janice Armstrong, *A Time to Mourn* (Stony Brook, N.Y.: The Museums at Stony Brook, 1980). For a personal account of how infant mortality affected one mother and her surviving children, see Kathryn Kish Sklar, *Florence Kelley and the Nation's Work: The Rise of Women's Political Culture, 1830–1900* (New Haven, Conn.: Yale University Press, 1995), pp. 27–31.

46. On the late-nineteenth-century "servant problem," see Faye E. Dudden, *Serving Women: Household Service in Nineteenth-Century America* (Middletown, Conn.: Wesleyan University Press, 1983).

47. Plunkett, *Women, Doctors, and Plumbers,* p. 10. On feminism and hygiene reform, see Adams, *Architecture in the Family Way,* Chap. 5.

48. Hewitt, *Queen of the Home,* p. 225.

3. Entrepreneurs of the Germ

1. On Garfield's wounding and subsequent illness, see Charles E. Rosenberg, *The Trial of the Assassin Guiteau* (Chicago: University of Chicago Press, 1968), pp. 1–12. On Waring's arrival in Washington, see George Waring (hereafter GW) to Almon F. Rockwell (hereafter AFR), Aug. 13 and 15, 1881, box 10, item 292, NA.

2. On the medical aspects of the case, see Stanley A. Fish, "The Death of President Garfield," *BHM* 24 (1950): 378–392.

3. On the broader evolution of sanitary entrepreneurship in this period, see Maureen Ogle, *All the Modern Conveniences: American Household Plumbing, 1840–1890* (Baltimore, Md.: Johns Hopkins University Press, 1996). For parallel developments in England, see Annmarie Adams, *Architecture in the Family Way: Doctors, Houses, and Women, 1870–1900* (Montreal: McGill-Queen's University Press, 1996).

4. On its symbolic importance, see Frank Friedel and William Pencak, eds., *The White House: The First Two Hundred Years* (Boston: Northeastern University Press, 1994). For its architectural history, see William Seale, *The*

President's House: A History, 2 vols. (Washington, D.C.: White House Association, 1986). The executive mansion did not officially become known as the White House until 1901, during Theodore Roosevelt's administration.

5. Hayward and Hutchinson to T. L. Casey, July 16, 1879, box 7, folder 271, NA; and M. M. Magruder to T. L. Casey, July 17, 1879, box 7, folder 264, NA. The chief engineer of the U.S. Army selected an engineer from the Quartermaster Corps to serve as officer in charge of the White House. For a summary of his duties, see the National Archives Inventory for Record Group 42, NA, compiled by Mary Jane Dowd.

6. The *Baltimore American* article was reprinted in *SE* 4, no. 13 (June 1, 1881): 304. On Rockwell's renovations, see Hayward and Hutchinson to AFR, Mar. 16, 1881, box 10, folder 86, NA. On Mrs. Garfield's illness, see Seale, *President's House,* vol. 1, pp. 519–520.

7. W. B. Allison to AFR, Apr. 8, 1881; Eppa Hunton to James A. Garfield, June 7, 1881; AFR to Eppa Hunton, June 20, 1881, all box 10, folder 213, NA.

8. Rosenberg, *Trial,* pp. 2–4, 8–9. Wound infection was so common prior to the careful practice of antisepsis that physicians assumed that it was a natural part of healing. The eminent surgeons called in to treat Garfield do seem to have followed antiseptic precautions, but in the hours after the shooting, the President's wound was probed by other, less scrupulous doctors, who may have established the infection while trying to determine the position of the bullet. See Fish, "Death of President Garfield."

9. *NYH,* Aug. 1, 1881, Garfield Scrapbook, vol. 2, p. 2, NLM; and *NYH,* Aug. 5, 1881, Garfield Scrapbook, vol. 2, p. 8, NLM. The first mention of the "malaria" thesis came in the *NYH,* July 29, 1881, Garfield Scrapbook, vol. 1, p. 68, NLM. This three-volume set collected newspaper clippings, most of them from the *NYH,* concerning Garfield's shooting and subsequent illness. After several White House staff members came down with malaria-like fevers in July, Garfield's physicians began to give him prophylactic doses of quinine. Note that in the late 1800s the term "malarial" denoted any fever that came and went intermittently, including but not confined to the more specific disease we now know as malaria. Because the word "malaria" means "bad air," the term was often used to refer to diseases supposedly caused by exposure to sewer gas.

10. A. Mead to AFR, Aug. 9, 1881, box 10, folder 348, NA; Ogilvie and Bennem to AFR, Aug. 20, 1881, box 10, folder 299, NA; and E. E. Rice to AFR, Aug. 13, 1881, box 10, folder 213, NA.

11. MacVeigh's suggestion concerning Waring is mentioned in AFR to Brigadier General H. G. Wright, Jan. 4, 1882, box 10, folder 436, NA. The President's attendants, including Colonel Rockwell, clearly were concerned about the sickroom atmosphere even before the *Herald* started its sewer gas crusade. Besides installing the special cooling device to allay the summer

heat and refresh the air, Rockwell had the President removed temporarily so that the draperies and carpets could be removed and the chamber given "a general cleaning and thorough fumigation," measures that suggested a concern about infection. See *NYT,* July 29, 1881.

12. George E. Waring, "Suggestions for the Sanitary Drainage of Washington City," *Smithsonian Miscellaneous Collections* 26 (1880): 1–23, quotation on p. 12. On Waring's career, see James H. Cassedy, "The Flamboyant Colonel Waring: An Anticontagionist Holds the American Stage in the Age of Pasteur and Koch," *BHM* 36 (1962); and Martin Melosi, *Pragmatic Environmentalist: Sanitary Engineer, George E. Waring, Jr.* (Washington, D.C.: Public Works Historical Society, 1977).

13. GW to AFR, Aug. 23, 1881, box 10, folder 303, NA. The text of the preliminary report is repeated virtually in full in the final report. See GW to AFR, Nov. 26, 1881, box 10, folder 411, NA. Waring also sent a copy to his friend John Shaw Billings; it is available at the NLM.

14. Wayne MacVeigh to AFR, Sept. 2, 1881, box 10, folder 316, NA. He wrote of Waring's report, "My daily mail satisfies me that, if it is of a character to do good, an intelligent summary of it is badly needed." The report was published only after the President's death, according to the *NYT,* Oct. 15, 1881.

15. For the autopsy results, see Fish, "Death of President Garfield," p. 388. Garfield himself wanted to go to New Jersey, and "the outside pressure upon the surgeons in favor of such a step was so urgent and came from such quarters," according to the *NYT,* Aug. 26, 1881, that they agreed. The *NYH,* Sept. 27, 1881, Garfield Scrapbook, vol. 3, p. 58, NLM, quoted one of Garfield's surgeons, Dr. Frank Hamilton of New York City, as discounting the malaria hypothesis. Although Hamilton attributed Garfield's decline to other factors, in other contexts he was a champion of the sewer gas threat. See Frank H. Hamilton, "Sewer-Gas," *PSM* 22 (1882): 1–20.

16. *SE* 4, no. 22 (Oct. 15, 1881): 524–525. See also *NYT,* Oct. 15, 1881.

17. John Bogart to AFR, Oct. 6, 1881, box 10, folder 353, NA; GW to AFR, Oct. 19, 1881, box 10, folder 381, NA. On Arthur's anxieties, see Seale, *President's House,* vol. 1, pp. 536–538.

18. Waring, "Report . . . on the Improvement of the Sanitary Condition of the Executive Mansion," enclosed with GW to AFR, Nov. 26, 1881, box 10, folder 411, NA. George E. Waring, *How to Drain a House* (New York: Henry Holt, 1885), pp. 138, states that the five toilets installed at the White House were his first use of the new model, which he describes at length on pp. 133–138. References to the White House appear in advertisements for Cudell's Patent Sewer Gas Trap, *SE* 5, no. 13 (Feb. 23, 1882): 272; and the Dececo Water Closet, *SE* 6, no. 7 (Sept. 21, 1882): 336. On the resolution of the conflict between Rockwell and Waring, see AFR to GW, Nov. 3, 1881, box

10, folder 381; GW to AFR, Nov. 10, 1881, box 10, folder 394; and William Paul Gerhard to AFR, Oct. 24, 1881, box 10, folder 358, all in NA. In his final report, Waring professed himself satisfied with work done by the local plumbing firm, Hayward and Hutchinson, under his assistant's supervision.

19. GW, "Report," box 10, folder 411, NA; Seale, *President's House,* vol. 1, p. 537.

20. The text of the Senate bill is recorded in the *Congressional Record,* June 20, 1882, pp. 5126–5128. See also Seale, *President's House,* vol. 1, p. 537.

21. On the pre–germ theory expansion of commercial sanitary services, see Maureen Ogle, *All the Modern Conveniences: American Household Plumbing, 1840–1890* (Baltimore, Md.: Johns Hopkins University Press, 1996); Maureen Ogle, "Domestic Reform and American Household Plumbing, 1840–1870," *Winterthur Portfolio* 28 (1993): 33–58, esp. pp. 43–53; and Nancy Tomes, "The Private Side of Public Health: Sanitary Science, Domestic Hygiene, and the Germ Theory," *BHM* 64 (1990): 509–539, esp. pp. 531–535.

22. On the range of disinfectants available by 1870, see "Report of Committee on Disinfectants," *Transactions of the American Medical Association* 17 (1866): 129–155. On the use of the germ theory to promote disinfectants, see James Harvey Young, *The Toadstool Millionaires: A Social History of Patent Medicines in America before Federal Regulation* (Princeton, N.J.: Princeton University Press, 1961), Chap. 10.

23. *Disinfection and Disinfectants: Preliminary Report Made by the Committee on Disinfectants of the APHA,* (1885), p. 3, pamphlet collection, CPP. In this study, Sternberg and his associates made a point only to test substances commonly available for household use. The final report was published as [American Public Health Association], *Disinfection and Disinfectants: Their Application and Use in the Prevention and Treatment of Disease, and in Public and Private Sanitation* (Concord, N.H.: Republican Press Association, 1888).

24. American Public Health Association, *Disinfection and Disinfectants,* pp. 12, 13. On p. 17, Sternberg complains about Wither's misuse of an earlier report on his study.

25. Automatic Fountain and Disinfecting Company, *The Botsford Automatic Fountain* [1893], Disinfectants, box 1, WC. On the importance of odor, see Alain Corbin, *The Foul and the Fragrant: Odor and the French Social Imagination* (Cambridge, Mass.: Harvard University Press, 1986). On the patent medicine industry, see Young, *Toadstool Millionaires.*

26. Tayman's Disinfectant and Fumigating Co., *Tayman's Disinfectors and Fumigators* (Philadelphia: privately printed, 1885), p. 4, HC.

27. Class 4-222, Patent 274,332, Mar. 20, 1883, PO; Class 4-221, Patent 334,158, Jan. 12, 1886, PO. Although inventors emphasized the need for germicidal action, their ordering of the pernicious elements in sewer gas usually placed germs last, suggesting that they were the least of its dangers.

28. Although researchers began testing drinking water for the presence of bacteria in the 1870s, authoritative studies really did not begin until after the mid-1880s. See Barbara Guttman Rosenkrantz, *Public Health and the State* (Cambridge, Mass.: Harvard University Press, 1972), esp. pp. 97–107. The Massachusetts State Board of Health opened the pioneering Lawrence Experiment Station in 1887. For developments in England that influenced American practice, see Christopher Hamlin, *A Science of Impurity: Water Analysis in Nineteenth-Century Britain* (Berkeley: University of California Press, 1990).

29. Sub-Merged Filter Company, Ltd., *A Perfect House Filter* (Philadelphia: n.p., [1885]), p. 4, HC; and The Hyatt Pure Water Co., *The Hyatt System of Water Purification* (New York: n.p., n.d., ca. 1890), p. 5, HC. In the Hyatt brochure, about fifty private homes in Philadelphia are listed on pp. 50–52. The Hyatt Company was based in New York but had branches in other cities. The NLM has similar brochures for the Gate City Stone Water Filter Company of New York and the Boston Water Purifier. The technology of water filtration was pioneered by brewers and other manufacturers who required a pure water supply and then adapted for home use.

30. See report on Edward J. Mallett, Jr., New York City, vol. 337, p. 2242, DC. I have located only Mallet's second patent, issued in 1882, which covered an improvement in the Germicide's "air disinfector." Class 4-222, Patent 253,400, Feb. 7, 1882, PO. That patent mentions a prior patent on the device he received in February 1880, Patent 9068. Other details are from credit reports filed concerning the Germicide Company of New York, New York City, vol. 373, p. 1201H, DC; and the Penn Germicide Company, Pennsylvania, vol. 162, p. 13, DC. Mallet may well have had tuberculosis, for the agent reported that "owing to lung trouble he spends a g[oo]d deal of his time in the West," a common place of residence for late-nineteenth-century consumptives chasing a cure.

31. *The Germicide, Endorsed by Science and Experience* [1882], p. 2, HC. The price list is attached to a form letter inside a later edition of the brochure for the Pennsylvania Germicide Co., *Information Regarding the Germicide* [1884], CPP.

32. *Germicide*, pp. 3, 4, 5. Emphasis in original. Cohn cited 1837 as the year when "the connection between putrefaction and germ life was first proved and promulgated," which was probably a reference to Cagniard de la Tour's work on yeast. In an equally vague reference, most likely to Louis Pasteur's work, Cohn continued, "Twenty five years later, it began to develop in its wider medical sense."

33. Ibid., pp. 6–7. Testimonials for the Germicide appear on pp. 13–16. The Chicago Board of Health commissioned an experimental study of the device's effectiveness in conjunction with the Illinois Microscopical Society,

and it wrote an open letter to the citizens of Chicago recommending the Germicide's use in every household. This was powerful confirmation of how seriously these kind of toilet attachments were taken in the early 1880s. See *Report of the Health Department of the City of Chicago on the Germicidal Action of Zincic Chloride* (Chicago: Jones Stationary and Printing, 1881), HC.

34. The Pennsylvania Germicide Company, *Information Concerning the Germicide and Its Protective Influence* 1884, pp. 14–36, CPP, lists names and gives testimonials. Richardson's name appears on p. 16, the *NYH*'s on p. 30. This copy contained a form letter, directed specifically at physicians, that C. L. Cohn wrote to accompany the brochure. Hartshorne gave a speech in 1886 praising the device, which was reprinted in The Germicide Company of New York, *Sewer-Gas and the Remedy* (n.d.), pp. 21–27, Disinfectants, box 1, WC.

35. Penn Germicide Company, Pennsylvania, vol. 162, p. 13, DC. Entries are dated from May 1884 to June 1888. R. G. Dun and Company stopped filing the reports in ledgers at about this time, so I could not trace further the company's history.

36. New York Scientific Sanitary System, New York City, vol. 337, p. 2282, DC.

37. See Henry C. Meyer, *The Story of the Sanitary Engineer* (New York: Jaques, 1928). The *SE* took a very broad view of its mission to promote sanitary science; it covered such topics as food adulteration and tenement house reform.

38. *SE* 5, no. 1 (Dec. 1, 1881): 1. On reports of bacteriological findings, see, for example, *SE* 4, no. 15 (July 1, 1881): 349 (on typhoid and Eberth's bacillus); and *SE* 5, no. 21 (Apr. 20, 1882): 433 (Koch's research on disinfectants).

39. *SE* 6, no. 24 (Nov. 9, 1882): 485. The combat between doctors and sanitary engineers was most evident in debates over what constituted a safe sewer trap. See, for example, *SE* 5, no. 17 (Mar. 23, 1882): 337.

40. For an overview of the plumbing trade, see Martin Segal, *The Rise of the United Association: National Unionism in the Pipe Trades, 1884–1924* (Cambridge, Mass.: Wertheim Committee, Harvard University, 1970), pp. 1–16. My characterizations of master plumbers are also based on the proceedings of the National Association of Master Plumbers (hereafter NAMP), which are available at the Library of Congress.

41. NAMP, *Proceedings of the First Annual Convention*, 1883, p. 23. The NAMP was formed by representatives of twenty local associations of master plumbers in the East and Midwest.

42. NAMP, *Proceedings of the Eighth Annual Convention*, 1890, pp. 87–88.

43. For a contemporary survey of plumbing laws, see *Plumbing Problems* (New York: The Sanitary Engineer, 1885), pp. 225–231. On the evolution of plumbing laws, see Ogle, *All the Modern Conveniences;* and May N. Stone, "The

Plumbing Paradox: American Attitudes Toward Late Nineteenth-Century Domestic Sanitary Arrangements," *Winterthur Portfolio* 14 (1979): 283–309. On New York's law, see John Duffy, *A History of Public Health in New York City, 1866–1966* (New York: Russell Sage Foundation, 1974), pp. 231–232.

44. For more detailed overviews of changing bathroom technology and aesthetics, see Ellen Lupton and J. Abbott Miller, *The Bathroom, the Kitchen, and the Aesthetics of Waste* (Cambridge, Mass.: MIT List Visual Arts Center, 1992), esp. pp. 25–33; and Ogle, *All the Modern Conveniences.* On the Dececo toilet, see GW to AFR, Oct. 19, 1881, box 10, folder 381, NA.

45. Sanitary Association, Philadelphia, *Guarding the Home. Essential Conditions of Sanitary Arrangements to Exclude the Germs of Typhoid Fever and Other Zymotic Diseases* (title page lists it as *Skeletons of Our Homes, Tract no. 1*) [1887], insert, p. 1. HC. Entitled *Guarding the Home,* the booklet, whose author was identified only as a "Layman," was originally published in their monthly *Sanitary Magazine.*

46. Ibid., quotations from pp. 2, 3, and title page.

47. Ibid., p. 4.

48. On the growth of the sanitary pottery industry, see Marc J. Stern, *The Pottery Industry of Trenton: A Skilled Trade in Transition, 1850–1929* (New Brunswick, N.J.: Rutgers University Press, 1994). See also Ogle, *All the Modern Conveniences,* esp. Chap. 5.

49. My figures for the living wage are taken from David Montgomery, *The Fall of the House of Labor: The Workplace, the State, and American Labor Activism, 1865–1925* (New York: Cambridge University Press, 1987), p. 136. The prices for sanitary devices are averages based on the price lists usually included in trade catalogues.

50. *Germicide,* p. [17]. Although instructions for homemade ventilators and water filters were commonplace, I never found mention of a do-it-yourself version of the sewer trap.

4. Disciples of the Laboratory

1. Albert Abrams, *Transactions of the Antiseptic Club* (New York: E. B. Treat, 1895), pp. 29–30, 37–38. By the 1920s, Abrams had become a controversial figure due to his advocacy of a healing contraption known as a "dynamizer," which used radio waves to cure and detect various diseases. See James Harvey Young, *The Medical Messiahs: A Social History of Health Quackery in Twentieth-Century America* (Princeton, N.J.: Princeton University Press, 1967), pp. 137–142.

2. "Science," *Independent,* Apr. 14, 1892, p. 512. All the commentary of this sort that I found was written by men.

3. Alan Trachtenberg, *The Incorporation of America: Culture and Society in the Gilded Age* (New York: Hill and Wang, 1982).

4. My conception of the "incorporation of the germ" has been influenced by Stephen S. Morse, "Emerging Viruses: Defining the Rules for Viral Traffic," *Perspectives in Biology and Medicine* 34 (1991): 387–409; and Benedict Anderson, *Imagined Communities: Reflections on the Origin and Spread of Nationalism* (London: Verso, 1983).

5. For an excellent overview of this revolution in technique, see Thomas D. Brock, *Robert Koch: A Life in Medicine and Bacteriology* (Madison, Wisc.: Science Tech, 1988).

6. Whether and when Koch should be given credit for the postulates that bear his name is a matter of long debate among medical historians. See Brock, *Robert Koch,* pp. 179–182.

7. For an overview of the "golden age" of bacteriology, see William Bulloch, *The History of Bacteriology* (London: Oxford University Press, 1960). A still useful summary is Frederick P. Gorham, "The History of Bacteriology and Its Contribution to Public Health Work," in *A Half Century of Public Health,* ed. Mazyck P. Ravenal (1921; reprint, New York: Arno, 1970) pp. 66–93, esp. the chronological listing of the discovery of specific microorganisms on pp. 71–72. On American bacteriologists, see Paul F. Clark, *Pioneer Microbiologists of America* (Madison: University of Wisconsin Press, 1961); and Patricia Peck Gossel, "The Emergence of American Bacteriology, 1875–1900," (Ph.D. diss., Johns Hopkins University, 1988).

8. Edward Trudeau, *An Autobiography* (New York: Doubleday, Page, 1916), pp. 175–176. Despite the many deficiencies in the science writing of the era, which Terra Ziporyn discusses at length in her *Disease in the Popular American Press: The Case of Diphtheria, Typhoid Fever, and Syphilis, 1870–1920* (New York: Greenwood, 1988), I think the authoritativeness of these new methods came across in popular accounts of bacteriology.

9. On Koch and the healthy carrier problem, see Brock, *Robert Koch,* pp. 255–256. On the subsequent evolution of the concept, see Judith Walzer Leavitt, *Typhoid Mary: Captive of the Public's Health* (Boston: Beacon, 1996).

10. There was indeed concern about spitting prior to the 1880s, but for aesthetic and moral reasons. I have surveyed numerous etiquette manuals and health guides for the pre-1880 period and have found virtually no concern about spitting as a mode of spreading infectious diseases. On spitting, see Norbert Elias, *The Civilizing Process: The History of Manners* (1939; reprint, Oxford, Eng: Basil Blackwell, 1978), pp. 153–160; and John F. Kasson, *Rudeness and Civility: Manners in Nineteenth-Century Urban America* (New York: Hill and Wang, 1990), esp. 124–126.

11. Carl Fraenkel, *Textbook of Bacteriology,* 3d ed., trans. and ed. J. H. Linsley (New York: William Wood, 1891), p. 347.

12. Ibid.

13. This new perspective is neatly summed up in Charles Chapin, *Sources and Modes of Infection,* which first appeared in 1910. In the references that follow, I use the revised and expanded second edition (New York: John Wiley and Sons, 1912). For an excellent biography of Chapin and the public health practice of this period, see James H. Cassedy, *Charles V. Chapin and the Public Health Movement* (Cambridge, Mass.: Harvard University Press, 1962).

14. Fraenkel, *Textbook of Bacteriology,* p. 242.

15. For a review of the two theories, see Chapin, *Sources and Modes,* esp. pp. 294–295, 302–305. I discuss the fate of the dust theory in more detail in Chapter 10.

16. T. Mitchell Prudden, *Dust and Its Dangers* (New York: G. P. Putnam's Sons, 1890).

17. Ibid., p. 21.

18. Ibid., pp. 53, 60–61, 70.

19. Ibid., p. 97.

20. Ibid., p. 99. On pp. 74–86, Prudden suggests safeguards to minimize the dust danger.

21. "Do House Flies Convey Infection?" *PSM* 22 (1883): 571. In the late 1860s, Louis-Adolphe Raimbert demonstrated that flies carried the anthrax germ on their bodies. Koch's work on anthrax later proved that other vectors of infection were far more important than flies in spreading that disease, yet the premise that houseflies spread germs remained very popular.

22. The experimental evidence for fly transmission of disease is carefully surveyed in George H. F. Nuttall, *On the Role of Insects, Arachnids, and Myriapods as Carriers in the Spread of Bacterial and Parasitic Diseases of Man and Animals: A Critical and Historical Study* (Baltimore, Md.: The Friedenwald Co., 1899). Nuttall himself thought the emphasis on the fly's role as a disease vector far outstripped the experimental evidence confirming it. For an excellent account of the "pathologization" of the fly, see Naomi Rogers, "Germs with Legs: Flies, Disease, and the New Public Health," *BHM* 63, no. 4 (winter 1989): 599–617. Terra Ziporyn stresses the importance of the Spanish-American War and the determination of Leland O. Howard in popularizing the germ menace. See Ziporyn, *Disease in the Popular American Press,* pp. 76–78, 85–86.

23. See Nuttall, *On the Role of Insects,* pp. 71–75, for a discussion of the cattle fever experiments. Chapin, *Sources and Modes,* pp. 380–447, sums up the remarkable explosion of work on insect vectors completed in the decade after Nuttall's survey appeared. For the larger context of medical thinking about insect vectors, see Victoria A. Harden, *Rocky Mountain Spotted Fever: History of a Twentieth-Century Disease* (Baltimore, Md.: Johns Hopkins University Press, 1990), Chap. 3.

24. Chapin, *Sources and Modes,* pp. 317–339. On the bacteriological study of water and the Lawrence Experiment Station, see Barbara Rosenkrantz, *Public Health and the State: Changing Views in Massachusetts, 1842–1936* (Cambridge, Mass.: Harvard University Press, 1972), pp. 98–107.

25. See Chapin, *Sources and Modes,* pp. 339–342.

26. Ibid., pp. 342–352.

27. Ibid., pp. 352–365. On the problem of bovine tuberculosis, see Barbara Rosenkrantz, "The Trouble with Bovine Tuberculosis," *BHM* 59 (1985): 155–175. On heating milk to remove bacterial contaminants, see Rima D. Apple, *Mothers and Medicine: A Social History of Infant Feeding, 1890–1950* (Madison: University of Wisconsin Press, 1987), esp. pp. 41–44.

28. Chapin, *Sources and Modes,* pp. 371–379. On the history of food poisoning, see Stewart M. Brooks, *Ptomaine: The Story of Food Poisoning* (New York: A. S. Barnes, 1974); and James Harvey Young, *Pure Food: Securing the Federal Food and Drugs Act of 1906* (Princeton, N.J.: Princeton University Press, 1989), esp. pp. 110–113. As Young notes, health concerns about the food supply initially focused on chemical preservatives and adulteration and only gradually broadened to include the dangers of microbial contamination. See also Mitchell Okun, *Fair Play in the Marketplace: The First Battle for Pure Food and Drugs* (Dekalb: Northern Illinois University Press, 1986).

29. Brooks, *Ptomaine,* pp. 19–30; Young, *Pure Food,* pp. 110–113. See also James Harvey Young, "Botulism and the Ripe Olive Scare of 1919–1920," *BHM* 50 (1976): 372–391. The evolution of the ptomaine concept is apparent in the influential text, Victor Vaughan and Frederick G. Novy, *Ptomaïns, Leucomains, Toxins and Antitoxins; or, The Chemical Factors in the Causation of Disease,* 3d ed. (New York: Lea Brothers, 1896), which first appeared in 1888. By the third edition, the authors had replaced the term "ptomaine" with "bacterial poisons."

30. For an overview of the evolution of antiseptic and aseptic surgery, see Owen Wangensteen and Sarah Wangensteen, *The Rise of Surgery from Empiric Craft to Scientific Discipline* (Minneapolis: University of Minnesota Press, 1978); and Ira Rutkow, *The History of Surgery in the United States, 1775–1900,* 2 vols. (San Francisco: Norman, 1988).

31. Carl Beck, *A Manual of the Modern Theory and Technique of Surgical Asepsis* (Philadelphia: W. B. Saunders, 1895), pp. 14–15. As early as the 1840s, the Austrian physician Ignaz Semmelweiss and the American physician Oliver Wendell Holmes had warned that doctors' unwashed hands were spreading childbed fever, but the idea that medical attendants could convey contagion had met with fierce resistance for both scientific and professional reasons. See Judith Walzer Leavitt, *Brought to Bed: Child-Bearing in America, 1750–1950* (New York: Oxford University Press, 1986), esp. pp. 154–157.

32. Hunter Robb, *Aseptic Surgical Technique* (Philadelphia: J. B. Lippincott,

1894), p. 13. On the introduction of the rubber glove, see Curt Proskauer, "Development and Use of the Rubber Glove in Surgery and Gynecology," *JHM* 13 (1958): 373–381.

33. Beck, *A Manual,* p. 242; Edwin Valentine Mitchell, *Concerning Beards* (New York: Dodd, Mead, 1930), p. 42.

34. For a discussion of operating room garb and protocols, see Robb, *Aseptic Surgical Technique,* esp. Chap. 3. See also Wangensteen and Wangensteen, *The Rise of Surgery,* pp. 487–488; and Dan W. Blumhagen, "The Doctor's White Coat: The Image of the Physician in Modern America," *Annals of Internal Medicine* 92 (1979): 111–116. On p. 115, he states that the colors were changed after 1930 to minimize glare.

35. Robb, *Aseptic Surgical Technique,* pp. 10–11, 46. These lapses during operations are cited on pp. 11–12.

36. Nathan Breiter, "The Hand as a Propagator of Microbic Disease—A Medico-Social Question," *Medical Record* (N.Y.) 111 (1897): 813–816, quotation from p. 813.

37. Chapin, *Sources and Modes,* p. 189. Chapin distinguishes between contact infection, which could result from drinking from a cup just used by a sick person, and fomite infection, which implied a much longer interval of time between the shared touch. See pp. 164–258. A fomite was defined by Ezra Hunt as "any porous substance capable of absorbing, retaining, or transmitting the contagious particles" in his "Sanitary Nomenclature," *Public Health Reports and Papers* 11 (1885): 31–37, quotation on p. 35.

38. Chapin, *Sources and Modes,* pp. 189–190. See also pp. 192–193.

39. Prudden, *Dust and Its Dangers,* esp. pp. 76–78; "The Licking of Postage-Stamps," *Literary Digest* 53 (1916): 454–455. On the library book controversy, see Andrew McClary, "Beware the Deadly Books: A Forgotten Episode in Library History," *Journal of Library History* 20 (1985): 427–433.

40. This refinement of fomite theory is already evident in Chapin, *Sources and Modes,* pp. 212–258.

41. On the social history of venereal disease, see Allan M. Brandt, *No Magic Bullet: A Social History of Venereal Disease in the United States since 1880* (New York: Oxford University Press, 1985).

42. Timothy J. Gilfoyle, in his *City of Eros: New York City, Prostitution, and the Commercialization of Sex, 1790–1920* (New York: W. W. Norton, 1992), pp. 62–63, cites Sanger's study on prostitutes and venereal disease. Brandt, *No Magic Bullet,* esp. pp. 12–13, discusses the perception that venereal disease was on the rise in the late 1800s. He cites Morrow's statistic on p. 16.

43. L. Duncan Bulkley, *Syphilis in the Innocent (Syphilis Insontium) Clinically and Historically Considered with a Plan for the Legal Control of the Disease* (New York: Bailey and Fairchild, 1894), p. 54.

44. Ibid., pp. 88–89.

45. Ibid., p. 18.

46. On the difficulties of public education about venereal disease, see Brandt, *No Magic Bullet,* and Ziporyn, *Disease in the Popular American Press,* Chap. 4. I am convinced that fears of venereal disease strongly influenced popular beliefs about casual contact and fomite transmission.

47. Charles Chapin, "The End of the Filth Theory of Disease," *PSM* 60 (1902): 234–239.

48. On these economic changes, especially as they affected consumers, see William Leach, *Land of Desire: Merchants, Power, and the Rise of a New American Culture* (New York: Pantheon, 1993); and Trachtenberg, *Incorporation of America.* On hygiene and the Progressive-era marketplace, see Okun, *Fair Play in the Marketplace.* On the degradation of workers' health, see David Rosner and Gerald Markowitz, eds., *Dying for Work: Workers' Safety and Health in Twentieth-Century America* (Bloomington: Indiana University Press, 1987). On the rise of the "networked" city, see Joel A. Tarr and Gabriel Dupuy, eds., *Technology and the Rise of the Networked City in Europe and America* (Philadelphia: Temple University Press, 1988.)

49. Although distinctions between first- and second-class cars ensured some measure of class and race segregation, railroad travel was often described as a "democratizing" experience. See, for example, H. Roger Grant, *We Took the Train* (Dekalb: Northern Illinois University Press, 1990). On the expansion of public transit, see Charles W. Cheape, *Moving the Masses: Urban Public Transit in New York, Boston, and Philadelphia, 1880–1912* (Cambridge, Mass.: Harvard University Press, 1980).

50. On the new culture of the department store, see Leach, *Land of Desire.* On dining out, see Lewis Erenberg, *Steppin' Out: New York Nightlife and the Transformation of American Culture, 1890–1930* (Chicago: University of Chicago Press, 1984); and Harvey Levenstein, *Revolution at the Table: The Transformation of the American Diet* (New York: Oxford University Press, 1988), pp. 185–187.

51. On changing patterns of entertainment, see Erenberg, *Steppin' Out;* David Nasaw, *Going Out: The Rise and Fall of Public Amusements* (New York: Basic Books, 1993); Kathy Peiss, *Cheap Amusements: Working Women and Leisure in Turn-of-the-Century New York* (Philadelphia: Temple University Press, 1986); and Roy Rosenzweig, *Eight Hours for What We Will: Workers and Leisure in an Industrial City, 1870–1920* (New York: Cambridge University Press, 1983). Peiss discusses reformers' concerns about the sexual dangers of the dance hall in Chap. 4.

52. Lawrence Flick, *The Hygiene of Phthisis* (Philadelphia: William J. Dornan, 1888), pp. 5, 9, CPP.

53. On European immigrants and the health "menace," see Alan M. Kraut, *Silent Travelers: Germs, Genes, and the "Immigrant Menace"* (New York:

Basic Books, 1994); and Howard Markel, "'Knocking Out the Cholera': Cholera, Class, and Quarantines in New York City, 1892," *BHM* 69 (1995): 420–457. On African Americans and urban epidemics, see David McBride, *From TB to AIDS: Epidemics among Urban Blacks since 1900* (Albany: State University of New York Press, 1991). These points are discussed further in Chapter 9.

5. Tuberculosis Religion

1. WMP to Lawrence Flick, Sept. 22, 1893, "Tuberculosis Letters," Flick Papers, CPP.

2. On the contagion versus heredity issue, see Katherine Ott, *Fevered Lives: Tuberculosis in American Culture since 1870* (Cambridge, Mass.: Harvard University Press, 1996); Sheila M. Rothman, *Living in the Shadow of Death: Tuberculosis and the Social Experience of Illness in American History* (New York: Basic Books, 1994), esp. Chaps. 1 and 12; and René Dubos and Jean Dubos, *The White Plague: Tuberculosis, Man, and Society* (1952; reprint, New Brunswick, N.J.: Rutgers University Press, 1987), esp. Chap. 8. John Bunyan is quoted on p. 8. See also Barbara Rosenkrantz, ed., *From Consumption to Tuberculosis: A Documentary History* (New York: Garland, 1994).

3. M. V. Ball, "Preventive Measures against Tuberculosis," p. 3, in Tuberculosis Pamphlets, vol. 106, no. 28, Flick Papers, CPP. On the "religion of science" in the Progressive period, see John Burnham, *How Superstition Won and Science Lost: Popularizing Science and Health in the United States* (New Brunswick, N.J.: Rutgers University Press, 1987). The anti-TB movement illustrates well the general dynamics of science popularization he describes here, esp. in Chap. 2.

4. For older but still useful works on the history of the anti-TB movement, see S. Adolphus Knopf, *A History of the National Tuberculosis Association: The Anti-Tuberculosis Movement in the United States* (New York: NTA, 1922); Richard Harrison Shryock, *National Tuberculosis Association, 1904–1954: A Study of the Voluntary Health Movement in the United States,* Historical Series, no. 8 (New York: NTA, 1957); and Richard Harrison Shryock, "The Historical Significance of the Tuberculosis Movement," in *Medicine in America: Historical Essays* (Baltimore, Md.: Johns Hopkins University Press, 1966). For newer works, see Mark Caldwell, *The Last Crusade: The War on Consumption, 1862–1954* (New York: Atheneum, 1988); Ott, *Fevered Lives;* and Michael E. Teller, *The Tuberculosis Movement* (Westwood, Conn.: Greenwood, 1988). Teller gives statistics on the number of societies on p. 33. On the career of Lawrence Flick, see Barbara Bates, *Bargaining for Life: A Social History of Tuberculosis, 1876–1938* (Philadelphia: University of Pennsylvania Press, 1992). Of the forty-nine officers profiled in Knopf's history, all were white

and only one was a woman (Mabel Boardman of the American Red Cross); thirty-three were physicians.

5. On the scope and significance of the TB movement's publicity methods, see Martin Pernick, "The Ethics of Preventive Medicine: Thomas Edison's Tuberculosis Films; Mass Media and Health Propaganda," *Hastings Center Report* 8 (June 1978): 21–27. There were, of course, many other important health-related crusades in this period, and at least one, the temperance movement, long predated the TB movement's efforts to promote systematic hygiene education. But as Terra Ziporyn has written in her meticulous study of disease in the popular American press, tuberculosis was "perhaps the era's paradigmatic disease"; she excluded it from her analysis, because, in her words, its discussion "could more than fill a book in itself." See Terra Ziporyn, *Disease in the Popular American Press: The Case of Diphtheria, Typhoid Fever, and Syphilis, 1870–1920* (New York: Greenwood, 1988), p. 3. On the temperance movement's effects on health education, see Richard K. Means, *A History of Health Education in the United States* (Philadelphia: Lea and Febiger, 1962).

6. On the decline of tuberculosis, see F. B. Smith, *The Retreat of Tuberculosis, 1850–1950* (New York: Croom Helm, 1988); and Leonard Wilson, "The Historical Decline of Tuberculosis in Europe and America: Its Causes and Significance," *JHM* 45 (1990): 366–396. Reliable nationwide statistics for disease mortality are not available for the United States before 1933. Smith stated on p. 237 that the death rates from tuberculosis began to decline in the United States around 1870 and dropped precipitously after 1916. Looking at the New York City data, which is complete from 1806 onward, Wilson found a steady decline beginning around 1880. On the efficacy attributed to preventive efforts, see Teller, *Tuberculosis Movement*, pp. 134–137. Teller himself doubts that the movement contributed much to the pre-1917 decline, but he notes that many Progressive-era authorities argued to the contrary.

7. On TB mortality rates, see Teller, *Tuberculosis Movement*, p. 3; and Rosenkrantz, *From Consumption to Tuberculosis*, p. 3. Note that by 1900, the death rates from pneumonia and influenza were equal or slightly higher than those from TB. Although tuberculin proved to have no therapeutic efficacy, its diagnostic value was considerable. See Thomas D. Brock, *Robert Koch: A Life in Medicine and Bacteriology* (Madison, Wisc.: Science Tech, 1988), esp. pp. 194–215, 294–295. On the sanitarium movement, see Bates, *Bargaining for Life*, and Rothman, *Living in the Shadow*, esp. Chap. 13.

8. The first nine tracts of the PSPT are reprinted in PSPT, *Report for the Year Ending May 1, 1905*, pp. 13–35. The letters requesting tracts that are preserved among Flick's papers suggest that writers were either consumptives themselves or had relatives suffering from the disease.

9. "The Anti-Tuberculosis Bill Poster Campaign," *JOL* 7, no. 1 (Jan. 1910): 390–393, quotation on p. 390. The NTA's *Monthly Confidential Bulletin*

(ALA), a newsletter circulated among the affiliated societies, documents its active role in sharing good publicity ideas.

10. "Livingston Farrand," *Dictionary of American Biography*, ed. Robert Livingston Schuyler and Edward T. James (New York: Charles Scribner's Sons, 1958), supp. 2, pp. 176–178; "Livingston Farrand, an Appreciation," *BNTA* 25, no. 2 (Dec. 1939): 185–186; and Biographical Files for Charles M. De Forest (T391), Livingston Farrand (P1006), Philip Jacobs (T300), and E. G. Routzahn (1978), Archives, ALA.

11. On the history of American advertising in this period, see Michele H. Bogart, *Artists, Advertising, and the Borders of Art* (Chicago: University of Chicago Press, 1995); Stephen Fox, *The Mirror Makers: A History of American Advertising and Its Creators* (New York: William Morrow,), 1984; Leach, *Land of Desire;* Jackson Lears, *Fables of Abundance: A Cultural History of Advertising in America* (New York: Basic Books, 1994); Roland Marchand, *Advertising the American Dream: Making Way for Modernity, 1920–1940* (Berkeley: University of California Press, 1985); James Norris, *Advertising and the Transformation of American Society, 1865–1920* (New York: Greenwood, 1990); Daniel Pope, *The Making of Modern Advertising* (New York: Basic Books, 1983); Frank Presbrey, *The History and Development of Advertising* (1929; reprint, New York: Greenwood, 1968); and Susan Strasser, *Satisfaction Guaranteed: The Making of the American Mass Market* (New York: Pantheon, 1989).

12. Philip P. Jacobs, *The Tuberculosis Worker* (Baltimore, Md.: Williams and Wilkins, 1923), p. 17. In comparison to post–World War I techniques, which became markedly more psychological in their approach, Progressive-era advertising has an air of innocence and straightforwardness, which helps to explain its appeal to public health workers. Journalism was another important inspiration for these changes, which I neglect here due to limitations of time; for the contemporaneous development of popular science writing, see Burnham, *How Superstition Won;* and Ziporyn, *Disease in the Popular American Press*. Lears, *Fables of Abundance,* pp. 282–283, notes the journalistic influence on ad copywriting.

13. On this process, see especially Leach, *Land of Desire,* and Lears, *Fables of Abundance*.

14. The *JOL* ran a regular feature entitled "Advertising Tuberculosis," which provided many examples of this new kind of display advertising. On the Kansas brick, see Samuel J. Crumbine, *Frontier Doctor* (Philadelphia: Dorrance, 1948), p. 147. On changes in layout of educational material, see Mary Swain Routzahn and Evart G. Routzahn, *Publicity for Social Work* (New York: Russell Sage Foundation, 1928), p. 154.

15. On the double-barred cross, see Knopf, *History of the National Tuberculosis Association,* pp. 152–154. On the development of the trademark, see Presbrey, *History and Development,* esp. pp. 382–386.

16. On the rise of commercial art, see Bogart, *Artists,* and Presbrey, *History and Development,* pp. 356–359, 388–413. On the new visual culture of merchandizing, see Leach, *Land of Desire,* and Lears, *Fables of Abundance.*

17. The story of the Venice poster is told in Lawrence Veiller, "A New Method of Tuberculosis Prevention," *JOL* 5, no. 7 (Aug. 1908): 239–241. Methods for reaching varied ethnic and racial groups were often discussed in the *Monthly Confidential Bulletin.* See, for example, Jan. 1909 and May 1909, ALA.

18. These posters are reproduced in "The Anti-Tuberculosis Bill Poster Campaign," pp. 391–393.

19. "An Effective Fly Poster," *BNTA* 3, no. 9 (June 1917), p. 4; *Public Health Exhibitions: A Catalogue of Necessaries for Their Outfitting and Maintenance,* catalogue B (1910), p. 14, in Tuberculosis Pamphlets, vol. 106, no. 8, Flick Papers, CPP. The illusion had to be based on the world death rate because the U.S. rate would produce a blink only every two minutes, an interval deemed too long to impress the audience.

20. The "doughboy" poster is reprinted in Evart G. Routzahn and Mary Swain Routzahn, *The ABC of Exhibit Planning* (New York: Russell Sage Foundation, 1918), p. 68b; the PSPT poster was reprinted in *The Year Book of the PSPT. . . 1919* (Philadelphia: PSPT, 1919).

21. Reproduced in *The Year Book of the PSPT. . .1919.* On the shift to more "positive" lines of appeal, see Routzahn and Routzahn, *Publicity for Social Work,* esp. pp. 44–56. The cartoons were developed by the Commission for the Prevention of Tuberculosis in France, a Rockefeller Foundation–financed effort to bring American public health methods to war-ravaged France. Some samples are reproduced in Routzahn and Routzahn, *The ABC of Exhibit Planning,* p. 104b. See also Nancy Tomes, "American and European Public Health Reform and the Bacteriological Model, 1880–1930," (paper delivered at the Annual Meeting of the Organization of American Historians, Reno, Nevada, Mar. 1988). For one noteworthy cartoon bacillus, see Harry A. Wilmer, *The Lives and Loves of Huber the Tuber* (New York: NTA, 1942).

22. On the early anti-TB films, see Pernick, "The Ethics of Preventive Medicine." On public health films in general, see Martin Pernick, *The Black Stork: Eugenics and the Death of "Defective" Babies in American Medicine and Motion Pictures since 1915* (New York: Oxford University Press, 1996).

23. On the exhibit, see Knopf, *History of the National Tuberculosis Association,* p. 33, and Jacobs, *Tuberculosis Worker,* pp. 18–19.

24. "Monthly Confidential Bulletin," Feb. 1909, ALA. The exhibit format was also shaped by the social survey methods developed in the Progressive period. See Martin Bulmer, Kevin Bales, and Kathryn Kish Sklar, *The Social Survey in Historical Perspective, 1880–1940* (New York: Cambridge University Press, 1991).

25. On the development of the seal campaign, see Knopf, *History of the National Tuberculosis Association,* pp. 55–66; and Shryock, *National Tuberculosis Association,* pp. 127–134. The dollar amounts raised are given in Leigh Mitchell Hodges, *The People against Tuberculosis: The Story of the Seal Sale* (New York: NTA, 1942), pp. 53–54. The organizational diagram of the 1919 seal campaign suggests how similar to a large-scale advertising campaign it had become. See *BNTA* 5, no. 10 (July 1919): 12.

26. Samuel Hopkins Adams, "Health for Sale: The 'People's Penny Proposition,'" *La Follete's Magazine,* Dec. 7, 1914, pp. 8–9, 29–30, quotation is on p. 8.

27. For the number of children enrolled and a brief history of the Modern Health Crusade, see Knopf, *History of the National Tuberculosis Association,* p. 43. For a fuller account, see Louise Strachan and Elizabeth F. Jordan, *From Pioneer to Partner: Child Health Education in the National Tuberculosis Association, 1917–1945,* Historical Series, no. 4 (New York: NTA, 1947).

28. NTA, "Monthly Confidential Bulletin," Mar. 1909, ALA.

29. F. N. Yeager, "Post Cards as Educational Agents," *JOL* 6, no. 6 (June 1909): 172.

30. Illinois State Board of Health, *Consumption: [Circular] Recommended by the Central Board of Health for General Distribution,* 1908–1910, in vol. 106, no. 24, Flick Papers, CPP; and *BNTA* 4, no. 9 (June 1918), p. 3, emphasis in original. This religious rhetoric was by no means restricted to the anti-TB movement. See, for example, John Ettling, *The Germ of Laziness: Rockefeller Philanthropy and Public Health in the New South* (Cambridge, Mass.: Harvard University Press, 1981), esp. pp. viii–vix. For an insightful reading of the crusade imagery in the anti-TB movement, see JoAnne Brown, "Tuberculosis: A Romance," (paper presented at the Berkshire Conference of Women's History, Poughkeepsie, N.Y., May 1993.)

31. Knopf, *A History of the National Tuberculosis Association,* pp. 154, quotes Philip Jacobs as stating that the NTA chose the particular design of the double-barred cross that was "farthest removed from any design having a religious significance." Different versions of the double-barred cross were known as the "Jerusalem Cross" in the ancient Christian church and the "Lorraine Cross" in the first Crusade; it was also associated with the fraternal order of Masons. This sometimes uneasy marriage of secularism and Protestantism was typical of Progressive reform in general. See Robert M. Crunden, *Ministers of Reform* (New York: Basic Books, 1982).

32. National Tuberculosis Association, *What You Should Know about Tuberculosis* (New York: NTA, 1916), p. 8. For other spit slogans, see, for example, New York City Department of Health, *Do Not Spit: A Tuberculosis (Consumption) Catechism and Primer for School Children* (New York: Board of Health, 1908), p. 10. On earlier critiques of spitting, see John F. Kasson, *Rudeness and Civility:*

Manners in Nineteenth-Century Urban America (New York: Hill and Wang, 1990).

33. New York City Department of Health and the Committee on Prevention of Tuberculosis of the Charity Organization Society, *What You Should Know about Tuberculosis* (New York: Board of Health, 1910), p. 15; and D. E. Salmon, "The Origin and Prevention of Tuberculosis," *The Sanitary Volunteer* 1, no. 4 (Apr. 1889), pp. 81, 82.

34. E. C. Schroeder, *The Relation of the Tuberculous Cow to the Public Health,* reprint from the Bureau of Animal Industry, *Twenty-Fifth Annual Report* (1908) in Tuberculosis Pamphlets, vol. 106, no. 15, Flick Papers, CPP. On the shift from sensible to insensible forms of dirt, see Georges Vigarello, *Concepts of Cleanliness: Changing Attitudes in France since the Middle Ages* (New York: Cambridge University Press, 1988), esp. Chap. 15.

35. Louis Hamman, "The Prevention of Tuberculosis," *JOL* 7, no. 2 (Feb. 1910): 29.

36. Agnes Vietor, "Tuberculosis in Everyday Life," *JOL* 2, no. 7 (Aug. 1905): 163–167, quotation on p. 164; *Little Dangers to Be Avoided in the Daily Fight against Tuberculosis,* reprint from *Pennsylvania Health Bulletin* 7 (1910): 1–5, quotations on pp. 3, 2.

37. See J. C. Wilson to Lawrence Flick, Jan. 17, 1893, Flick TB Papers, 1893, p. 216, CPP. The PSPT evidently prepared a memorial on the subject and forwarded it to the White House. The letter also mentioned that Harrison's granddaughter had recently been nursed through a bout with scarlet fever in the White House, making fumigation all the more urgent. On the Harrisons' health problems, see Harry J. Sievers, *Benjamin Harrison,* 3 vols. (Chicago: H. Regnery Co., 1968), 3: 218–219, 224, 241–42, 250–53. Like Mrs. P, Caroline Harrison was not told she was dying of tuberculosis.

38. *Little Dangers,* p. 4.

39. Edwin F. Bowers, "The Menace of Whiskers," *McClure's* 46 (1916): 90; *Little Dangers,* p. 4.

40. Lawrence Flick, *The Hygiene of Phthisis* (Philadelphia: William J. Dornan, 1888), p. 7. William Osler claimed that he was the first to use the "seed and soil" metaphor for TB in an 1892 lecture, but as Chapter 1 makes evident, it was used for other diseases long before then. See William Osler, "The Home in Its Relation to the Tuberculosis Problem," *First Annual Report of the Henry Phipps Institute* (Philadelphia: Henry Phipps Institute, 1905), p. 146.

41. New York City Department of Health, *What You Should Know,* p. 20; Vietor, "Tuberculosis in Everyday Life," p. 163. More discretely, the anti-TB crusade served as a respectable front for social hygienists' efforts to prevent venereal disease, as discussed in Chapter 5.

42. Cyrus Edson, "The Microbe as a Social Leveller," *North American Review*

161 (1895): 421–426. On Edson's career, see John Duffy, *A History of Public Health in New York City, 1866–1966* (New York: Russell Sage Foundation, 1974), esp. pp. 92–93.

43. New York City Department of Health, *What You Should Know*, p. 16; "Texts for Tuberculosis Sunday," undated mimeograph, Virginia Lung Association Papers, Claude Moore Health Sciences Library, University of Virginia. On the identification of TB with the poor, particularly immigrants, and the varied explanations for their susceptibility, see Rothman, *Living in the Shadow*, pp. 183–185; and Alan M. Kraut, *Silent Travelers: Germs, Genes, and the "Immigrant Menace"* (New York: Basic Books, 1994), esp. pp. 120–123, 155–156.

44. Frank Buffington Vrooman, "Public Health and National Defence," *The Arena* 69 (1895): 425–438, quotation is on p. 434. On the importance of cholera in prompting earlier public health programs, see Charles E. Rosenberg, *The Cholera Years: The United States in 1832, 1849, and 1866* (Chicago: University of Chicago Press, 1962). For the importance of TB to Progressive-era reforms, see Mark Caldwell, *The Last Crusade: The War on Consumption, 1862–1954* (New York: Atheneum, 1988); and Rothman, *Living in the Shadow*, esp. pp. 179–210.

45. Osler, "The Home," pp. 141–154, quotation is on p. 148. See Teller, *The Tuberculosis Movement*, esp. pp. 95–108, for an overview of the Progressive-era debate over TB etiology.

46. Lucius Morse, "Tuberculophobia—Can It Longer Be Excused as an Adult Entity?" *JOL* 16, no. 2 (Feb. 1919): 45–48, quotation on p. 47.

47. "Crusade a Democratic Movement," *BNTA* 5, no. 13 (Oct. 1919): 6.

48. Accounts of antituberculosis parades and exhibits abound in the *JOL* and other TB literature. See, for example, "Memphis Pageant," *BNTA* 5, no. 14 (Nov. 1919): 7. On parades as integrative ceremonies, see Susan G. Davis, *Parades and Power* (Philadelphia: Temple University Press, 1986); and Mary Ryan, "The American Parade: Representations of the Nineteenth-Century Social Order," in *The New Cultural History*, ed. Lynn Hunt (Los Angeles: University of California Press, 1989), pp. 131–153. On the special role accorded children as hygiene reformers, see Naomi Rogers, "Vegetables on Parade: Child Health Education in the 1920s," (paper delivered at the Annual Meeting of the American Association for the History of Medicine, Baltimore, Md., May 1990).

49. "Cincinnati's Death Calendar," *JOL* 6, no. 9 (Sept. 1909): 276. See also Morse, "Tuberculophobia," p. 45. For perceptive discussions of the increasing stigma of TB, see Ott, *Fevered Lives;* and Rothman, *Living in the Shadow*, esp. pp. 211–217.

50. "A Stock Lecture," *JOL* 12, no. 8 (Aug. 1915): 262–263.

51. New York City Department of Health, *What You Should Know*, p. 24; *Little Dangers*, p. 4.

52. Ellen N. La Motte, "The Unteachable Consumptive," *JOL* 6, no. 4 (Apr. 1909): 105–107, quotation on p. 105. Emphasis in original. This article is one of the harshest statements I found in my reading of antituberculosis literature.

53. On the communion cup movement, see Howard S. Anders, "Present Status of the Sanitary Movement for the Adoption of the Individual Drinking Cup," *Transactions of the Medical Society of the State of Pennsylvania* 26 (1895): 187–192; and Howard S. Anders, "The Progress of the Individual Cup Movement, Especially among Churches," *Journal of the American Medical Association* 29 (1897): 789–794.

54. W. M. Parker, "The Hygiene of the Holy Communion," *Medical Record* 41 (1892): 264–265, quotation on p. 265; Minutes of the General Assembly of the Presbyterian Church in the United States of America, n.s., vol. 18 (1895), p. 75, PC.

55. Quoted in Parker, "Hygiene of Holy Communion," p. 265.

56. Ellen A. Wallace, *Dietetic Hygienic Gazette* 12 (1896): 211–212, quotation on p. 212.

57. "To the Session of the Walnut Street Church," Walnut Street Presbyterian Church, Philadelphia, included in Minutes of Session, Apr. 6, 1898, PC; and Walnut Street Presbyterian Church, Minutes of Session, May 15, 1898, PC.

58. The brochure is in "Silver and Pewter Communionware," ser. 9, record group 22, Records of the Presbyterian Historical Society, 1851–1965, PC. The list has check marks by all the Presbyterian churches. The PC collection increased dramatically when congregations began abandoning the common cup in the 1890s, according to Ledlie I. Laughlin in her "Pewter Communion Services of the Presbyterian Historical Society," *Journal of Presbyterian History* 44 (1966): 83–88. The first sanitary communion set, using aluminum cups, was invented in 1894 by the Reverend J. Gethin Thomas, a Presbyterian minister in Lima, Ohio. See "Reverand [sic] J. G. Thomas," MS C735i, PC.

59. Anders, "Progress," p. 790.

6. The Domestication of the Germ

1. Ellen H. Richards, *Sanitation in Daily Life*, 3d ed. (Boston: Whitcomb and Barrows, 1915), p. 3; Ellen H. Richards and Marion Talbot, *Home Sanitation: A Manual for Housekeepers* (Boston: Ticknor, 1887), p. 73. By "Mosaic Code," Richards and Talbot were referring to the large body of religious law derived from God's lengthy instructions to Moses, as recorded in the Old Testament and elaborated in Talmudic commentaries. To keep faith with God, the Jews observed special rules of cleanliness for everyday life; these

included, for example, keeping *kashruth* (kosher) in the kitchen, and taking ritual baths, or *mikves*. On these customs as American Jews observed them in Richards and Talbot's day, see Andrew R. Heinze, *Adapting to Abundance: Jewish Immigrants, Mass Consumption, and the Search for American Identity* (New York: Columbia University Press, 1990), chap. 3.

2. I use the term "domestic science" to cover the broad range of efforts to uplift the home, which united women's voluntary organizations and professions, and "home economics" to refer to the specific profession that grew out of that larger movement. There is a growing literature on women's political culture and professional identities in the Progressive era, of which I found most useful: Ellen Fitzpatrick, *Endless Crusade: Women Social Scientists and Progressive Reform* (New York: Oxford University Press, 1990); Suellen Hoy, *Chasing Dirt: The American Pursuit of Cleanliness* (New York: Oxford University Press, 1995); Robyn Muncy, *Creating a Female Dominion in American Reform, 1890–1935* (New York: Oxford University Press, 1991); Anne Firor Scott, *Natural Allies: Women's Associations in American History* (Urbana: University of Illinois Press, 1993); and Kathyrn Kish Sklar, *Florence Kelley and the Nation's Work: The Rise of Women's Political Culture, 1830–1900* (New Haven, Conn.: Yale University Press, 1995). On the growth of the women's professions in this period, see Joan Jacobs Brumberg and Nancy Tomes, "Women in the Professions: A Research Agenda for American Historians," *Reviews in American History* 10, no. 2 (June 1982): 275–296.

3. On the early home economics movement, see Hoy, *Chasing Dirt*, esp. pp. 153–157; Laura Shapiro, *Perfection Salad: Women and Cooking at the Turn of the Century* (New York: Farrar, Straus and Giroux, 1986); Sarah Stage, "From Domestic Science to Social Housekeeping: The Career of Ellen Richards," in *Power and Responsibility: Case Studies in American Leadership,* eds. David M. Kennedy and Michael E. Parrish (New York: Harcourt Brace Jovanovich, 1986), pp. 211–228; Sarah Stage and Virginia Vincenti, eds., *Rethinking Home Economics: Women and the History of a Profession* (Ithaca, N.Y.: Cornell University Press, 1997); and Emma S. Weigley, "It Might Have Been Euthenics: The Lake Placid Conferences and the Home Economics Movement," *American Quarterly* 26 (1974): 79–96. On the domestic science movement in rural America, see Marilyn Irwin Holt, *Linoleum, Better Babies, and the Modern Farm Woman, 1890–1930* (Albuquerque: University of New Mexico Press, 1995).

4. On women physicians and their commitment to hygiene, see Regina Morantz-Sanchez, *Sympathy and Science: Women Physicians in American Medicine* (New York: Oxford University Press, 1985), pp. 58–63.

5. Harriette M. Plunkett to Lawrence Flick, Apr. 11, 1893, TB Letters, CPP; Richards quoted in Shapiro, *Perfection Salad,* p. 175.

6. On women's higher education in this period, see Lynn D. Gordon, *Gender and Higher Education in the Progressive Period* (New Haven, Conn.: Yale

University Press, 1990); and Rosalind Rosenberg, *Beyond Separate Spheres: Intellectual Roots of Modern Feminism* (New Haven, Conn.: Yale University Press, 1982). On women in the physical sciences, see Margaret Rossiter, *Women Scientists in America: Struggles and Strategies to 1940* (Baltimore, Md.: Johns Hopkins University Press, 1982). On women in the social sciences, see Fitzpatrick, *Endless Crusade*, and Rosenberg, *Beyond Separate Spheres*. Women professionals' conceptions of the "scientific method" were influenced not only by experimentalism but also by the social survey. See Martin Bulmer, Kevin Bales, and Kathryn Kish Sklar, *The Social Survey in Historical Perspective, 1880–1940* (New York: Cambridge University Press, 1991).

7. On the career of Ellen Richards, see George Rosen, "Ellen H. Richards," *American Journal of Public Health* 64 (1974): 816–819; Rossiter, *Women Scientists,* pp. 68–70; and Stage, "From Domestic Science." On the career of Marion Talbot, see Edward James, Janet James, and Paul Boyer, eds., *Notable American Women: A Biographical Dictionary, 1607–1950,* vol. 3 (Cambridge, Mass.: Belknap Press, Harvard University Press, 1971), pp. 423–424; and Fitzpatrick, *Endless Crusade,* esp. pp. 30–32, 84–86.

8. Florence Kelley, "Aims and Principles of the Consumer's League," *American Journal of Sociology* 5, no. 3 (Nov. 1899): 289–304, quotation on p. 296. On Kelley's career, see Sklar, *Florence Kelley.*

9. Charlotte Angstman, "College Women and the New Science," *PSM* 53 (1898): 674–690, quotation on p. 686.

10. On the class and race assumptions embedded in women reformers' conceptions of uplift, see, for example, Rivka Shpak Lissak, *Pluralism and Progressives: Hull House and the New Immigrants, 1890–1919* (Chicago: University of Chicago Press, 1989); and Elizabeth Lasch-Quinn, *Black Neighbors: Race and the Limits of Reform in the American Settlement House Movement, 1890–1945* (Chapel Hill: University of North Carolina Press, 1993).

11. S. Maria Elliott, *Household Bacteriology* (Chicago: American School of Home Economics, 1907), pp. 124, 157. For a good example of how the laboratory and demonstration methods were meshed in home economics, see Ava L. Johnson, *Bacteriology of the Home: A Textbook of Practical Bacteriology* (Peoria, Ill.: Manual Arts Press, 1929). On the tradition of public science, see Jan Golinski, *Science as Public Culture: Chemistry and Enlightenment in Britain, 1760–1820* (New York: Cambridge University Press); and Larry Stewart, *The Rise of Public Science: Rhetoric, Technology, and Natural Philosophy in Newtonian Britain, 1660–1750* (New York: Cambridge University Press, 1992).

12. I base these generalizations on my reading of the Lake Placid Conference on Home Economics, *Proceedings of Annual Conferences,* (Lake Placid, N.Y. and Boston Mass.: Lake Placid Conference on Home Economics, 1899 to 1920). This group subsequently renamed itself the American Home Economics Association.

13. On the *LHJ*, see Helen Damon-Moore, *Magazines for the Millions: Gender and Commerce in the Ladies' Home Journal and the Saturday Evening Post* (Albany: State University of New York Press, 1994); and Jennifer Scanlon, *Inarticulate Longings: The Ladies' Home Journal, Gender, and the Promises of Consumer Culture* (New York: Routledge, 1995). The *LHJ* alone had more than a million subscribers; by the 1910s, one in five American women was a reader. Other popular magazines with large female readerships, including *Harper's Bazaar, Cosmopolitan,* and *House Beautiful,* carried articles on the new domestic science.

14. "Discoveries," *GH* 48 (May 1909): 656.

15. On the role of advertising in mass-circulation magazines, see Damon-Moore, *Magazines for the Millions,* and Scanlon, *Inarticulate Longings,* esp. Chap. 6. Suellen Hoy dates the commercialization of cleanliness to the 1910s; see Hoy, *Chasing Dirt,* pp. 140–149. Chapter 4 suggests that it starts earlier, in the 1880s, as evidenced by products such as the Germicide, and accelerated in the 1890s. In surveying the ads in the *LHJ* and *GH,* I found a marked increase starting in the mid-1890s in ad copy that mentioned germs.

16. On Richards's conception of the science of the controllable environment, see Stage, "From Domestic Science."

17. This incident is reported in Flora Rose, "A Page of Modern Education, 1900–1940: Forty Years of Home Economics at Cornell University," *A Growing College: Home Economics at Cornell University* (Ithaca: New York State College of Ecology, 1969), pp. 22–23. Popular science education itself was not novel, but rather its extension to housewives with limited education. See John Burnham, *How Superstition Won and Science Lost: Popularizing Science and Health in the United States* (New Brunswick, N.J.: Rutgers University Press, 1987).

18. Herbert W. Conn, *Bacteria, Yeasts, and Molds in the Home* (Boston: Ginn, 1903), p. 1.

19. Elliott, *Household Bacteriology,* p. 5; Conn, *Bacteria, Yeasts, and Molds;* "Bacteriology of the Household," *CRCFW,* n.s. 1, no. 4 (Feb. 1909). Elliott discussed the history of the germ theory and Koch's postulates on pp. 109–113. She helped Van Rensselaer prepare this bulletin for the Cornell Series. As I note in Chapter 8, hygiene reformers watered down the teaching of scientific principles when addressing their least educated audiences.

20. Marion Harland, "Little Things That Are No Trifles," *GH* 54 (May 1912): 705–707, quotation on p. 706; Elliott, *Household Bacteriology,* p. 20.

21. Ibid., p. 32.

22. S. Maria Elliott, *Household Hygiene* (Chicago: American School of Home Economics, 1907), p. 3.

23. Ibid., pp. 192–193. Elliott advocates the peppermint test on p. 137.

24. See, for example, Mary Taylor Bissell, *A Manual of Hygiene* (New York: Baker and Taylor, 1894), pp. 96–97.

25. Martha Van Rennselaer (hereafter MVR) to Dora Bird, May 11, 1912, box 24, folder 36, CU. For typical dusting directives, see Elliott, *Household Bacteriology,* pp. 104–105. Prudden's work is listed in the model syllabus on pp. 157–161.

26. MVR, "Decoration in the Farm Home," *CRCFW,* ser. 1, no. 2 (Dec. 1902). See also Claudia Q. Murphy, "Wall Sanitation," Lake Placid Conference, *Proceedings of the Eighth Annual Conference, 15–22 September, 1906* (Lake Placid, N.Y.: Lake Placid Conference, 1906), pp. 47–50.

27. Flora Rose, "The Laundry," *CRCFW,* n.s. 1, no. 3 (Jan. 1909), pp. 42, 67; L. R. Balderston and E. H. Gunther, "Sanitary Precautions in Laundry Work," *GH* 54 (1912): 712–715, quotations on pp. 712, 714.

28. MVR, "Household Insects," 1913, pp. 1–2, box 46, CU. As evidenced by the many receipts for insect repellents to be found in nineteenth-century domestic handbooks, housewives had long shunned moths and stinging insects because of the damage they did to clothing and skin. See, for example, Catharine Beecher and Harriet Beecher Stowe, *American Woman's Home* (1869; reprint, Hartford, Conn.: Stowe-Day Foundation, 1991), p. 377.

29. See, for example, Amy Elizabeth Pope, *Home Care of the Sick* (Chicago: American School of Home Economics, 1907). The discussion of tuberculosis is on pp. 125–144. On the rising use of hospitals, see Charles E. Rosenberg, *The Care of Strangers: The Rise of America's Hospital System* (New York: Basic Books, 1987), esp. Chap. 10.

30. MVR, "Household Bacteriology," pp. 93–94. See also Hibbert Winslow Hill, "Teaching Bacteriology to Mothers," *JHE* 2, no. 6 (Dec. 1910): 635–640, quotation on p. 640.

31. Conn, *Bacteria, Yeasts, and Molds,* pp. 2, 139. See the discussion of cold on pp. 148–156.

32. MVR, "Household Bacteriology," p. 94; Maria Parloa, "To Keep Refrigerators Sweet," *LHJ* 9, no. 8 (July 1892): 21.

33. Conn, *Bacteria, Yeasts, and Molds,* pp. 111–112. These aspects of the American diet are noted in Shapiro, *Perfection Salad,* but with no connection to the issues of disease prevention.

34. MVR, "Household Bacteriology," p. 66. For a typical invocation of the fly leaving the manure pile to walk on the baby's bottle, see "The Home and Its Mistress," *The Metropolitan* 28, no. 2 (winter 1917): 3.

35. "Thursday Program 1913," box 16, file 7, Farmers' Week, CU. Rose also warned against anyone in the household drinking directly from the milk bottle, which would leave bacteria on its rim to be passed along to the next drinker. On concerns about milk, see Charles Chapin, *Sources and Modes of Infection,* 2d ed. (New York: John Wiley and Sons, 1912), pp. 342–365.

36. For a good example of the new precision in kitchen work, see MVR, "Household Bacteriology," and Mary Urie Watson, "Rules for Cleaning,"

CRCFW, n.s., vol. 1, no. 23 (Sept. 1, 1912), pp. 325, 336–337. A contemporary plumbing text also advised women to separate kitchen and laundry processes. "Never make a laundry of the kitchen," the author wrote, because water from boiling clothes contained whatever dirt was on the clothes, and condensed on surfaces of the kitchen, food as well as furniture. "There may be no live germs in this filthy condensation, but we should guard against dead matter in our food." See James J. Lawler, *Lawler's American Sanitary Plumbing* (New York: Excelsior, 1896), pp. 219–220.

37. Ola Powell, *Successful Canning and Preserving,* 3d rev. ed. (Philadelphia: J. B. Lippincott, 1917), pp. vii, 15. On the early bacteriology of canning, see James Harvey Young, *Pure Food: Securing the Federal Food and Drugs Act of 1906* (Princeton, N.J.: Princeton University Press, 1989), esp. pp. 110–113.

38. See, for example, ibid., pp. 15–28; see also Maria Parloa, "Canning and Preserving," *CRCFW* 4, no. 20 (Feb. 1906).

39. Powell, *Successful Canning,* p. 6; Mary B. Hughes, *Everywoman's Canning Book: The ABC's of Safe Home Canning and Preserving by the Cold Pack Method* (Boston: Whitcomb and Barrows, 1918), p. 6.

40. Powell, *Successful Canning,* p. 12. Although cases of botulism were recorded as early as the eighteenth century, the Belgian researcher von Ermengen isolated the microorganism responsible for this form of food poisoning only in 1895. See James Harvey Young, "Botulism and the Ripe Olive Scare of 1919–1920," *BHM* 50 (1976): 372–391.

41. The home is referred to as a "kingdom peculiar to women" in Isabel Bevier, "Household Science in a State University," paper delivered to 1900 Farmers' Institute, Jacksonville, Illinois, Isabel Bevier Papers, RS 8/1/20, box 5, University of Illinois Archives, Urbana, Illinois. On the larger conception of social housekeeping, see Stage, "From Domestic Science."

42. [S. Maria Elliott], *Public Hygiene in Relation to the Housekeeper,* brochure for the Women's Educational and Industrial Union, School of Housekeeping, 1900–1901, box 1, folder 1, SCH. Elliott taught the course with the assistance of the famed sanitary chemist and bacteriologist William Sedgwick. On the central role women played in Progressive-era sanitary reform, see Suellen M. Hoy, "'Municipal Housekeeping': The Role of Women in Improving Urban Sanitation Practices, 1880–1917," in *Pollution and Reform in American Cities, 1870–1930,* ed. Martin V. Melosi (Austin: University of Texas Press, 1980), pp. 173–198.

43. Kelley, "Aims and Principles," pp. 298–299.

44. MVR, "Household Bacteriology," p. 97. In a sense, domestic scientists were trying to "reconstruct the linkages" between commodities and the places and processes that produced them, as described in William Cronon, *Nature's Metropolis: Chicago and the Great West* (New York: W. W. Norton, 1991). He uses that phrase on p. xv.

45. MVR, "Household Bacteriology," p. 95. For the advice about wearing gloves, see Elliott, *Household Bacteriology,* p. 122.

46. "Cheap Food, or Clean Food?" *Literary Digest* 50 (May 29, 1915): 1272–1273, quotation on p. 1272.

47. MVR, "Household Bacteriology," p. 97.

7. Antisepticonscious America

1. "Eight Rainy Day Suits," *NYT,* Mar. 4, 1898.

2. "A Plea for Long Skirts," *Harper's Bazaar* 33 (1900): 2066.

3. On the new conceptions of modernity, see especially William Leach, *Land of Desire: Merchants, Power, and the Rise of a New American Culture* (New York: Pantheon, 1993); Jackson Lears, *Fables of Abundance: A Cultural History of Advertising in America* (New York: Basic Books, 1994); Roland Marchand, *Advertising the American Dream: Making Way for Modernity, 1920–1940* (Berkeley: University of California Press, 1985); and Susan Strasser, *Satisfaction Guaranteed: The Making of the American Mass Market* (New York: Pantheon, 1989).

4. William W. Bauer, "Antisepticonscious America," *American Mercury* 29 (July 1933): 323–326.

5. "The Passing of the Beard," *Harper's Weekly* 47 (1903): 102.

6. William Inglis, "The Revolt against the Whisker," *Harper's Weekly* 51 (1907): 612–613, quotation on p. 612. On Gillette, see Russell B. Adams, Jr., *King C. Gillette: The Man and His Wonderful Shaving Device* (Boston: Little, Brown, 1978), quotation from p. 56. For photographs of NTA officers, see Knopf, *History of the National Tuberculosis Association* (New York: NTA, 1922), "Biographies of the Officers of the Association," pp. 273–464. Of the other officers, two had small goatees; twenty-nine had mustaches, almost all of them small and closely trimmed; and twelve were completely clean shaven. On the larger redefinition of "manliness" in this period, see Gail Bederman, *Manliness and Civilization: A Cultural History of Gender and Race in the United States, 1880–1917* (Chicago: University of Chicago Press, 1995).

7. Emily A. Bruce, "The Rational Dress Movement: A Physician's View," *The Arena* 51 (Feb. 1894): 317–319, quotation on p. 318; "A Dirty Fashion," *Puck,* Aug. 8, 1900, p. 7. See also "Septic Skirts," *Scientific American,* Aug. 18, 1900, p. 108.

8. Elizabeth Ewing, *History of Twentieth-Century Fashion,* rev. ed. (Totowa, N.J.: Barnes and Noble, 1986), pp. 10, 23–24, 42–43, 62–63, 81, 90–91. Hemlines went above the ankle around 1915 and never descended again. American women apparently wore their skirts shorter than did their European counterparts. See Jane Mulvagh, *Vogue History of Twentieth-Century Fashion* (New York: Viking, 1988), p. 17.

9. Ellen Richards removed dust-catching furnishings from her own home in the 1880s, according to Caroline L. Hunt in her *Life of Ellen H. Richards, 1842–1911* (Washington, D.C.: American Home Economics Association, 1958), pp. 59–60. On the "artistic" home, see Clifford Edward Clark, Jr., *The American Family Home, 1800–1960* (Chapel Hill: University of North Carolina Press, 1986), pp. 103–130; Bradley C. Brooks, "Clarity, Contrast, and Simplicity: Changes in American Interiors, 1880–1930," in *The Arts and the American Home, 1890–1930*, ed. Jessica H. Foy and Karal Ann Marling (Knoxville: University of Tennessee Press, 1994), pp. 14–43; and Katherine C. Grier, *Culture and Comfort: People, Parlors, and Upholstery, 1850–1930* (Amherst: University of Massachusetts Press, 1988).

10. See especially Clark, *American Family Home*, pp. 131–192; and Brooks, "Clarity, Contrast, and Simplicity." See also Ellen Lupton and J. Abbott Miller, *The Bathroom, the Kitchen, and the Aesthetics of Waste* (Cambridge, Mass.: MIT List Visual Arts Center, 1992).

11. Brooks, in his "Clarity, Contrast, and Simplicity," pp. 34–37, suggests the link between the new styles and the "new immigration." Clark, in *American Family Home*, pp. 156–157, emphasizes the importance of hygienic issues.

12. The fullest accounts of this growing trend toward using germs to sell products are Suellen Hoy, *Chasing Dirt: The American Pursuit of Cleanliness* (New York: Oxford University Press, 1995), esp. pp. 123–149; and Vincent Vinikas, *Soft Soap, Hard Sell: American Hygiene in an Age of Advertisement* (Ames: Iowa State University Press, 1992). See also the interesting discussion in Lears, *Fables of Abundance*, Chap. 6, esp. pp. 171–174.

13. On immigrants and consumer goods, see Lisabeth Cohen, *Making a New Deal: Industrial Workers in Chicago, 1919–1939* (New York: Cambridge University Press, 1990), esp. Chap. 3; and Andrew Heinze, *Adapting to Abundance: Jewish Immigrants, Mass Consumption, and the Search for American Identity* (New York: Columbia University Press, 1990). On sanitary Americanization, see Hoy, *Chasing Dirt*, esp. Chap. 4.

14. Earnest Elmo Calkins, "The Influence of Advertising," *GH* 48 (May 1909): 643–645, quotation on p. 644. On Calkins's career, see Lears, *Fables of Abundance*, esp. pp. 308–314.

15. Potteries Selling Co., *Household Health* (ca. 1915), p. 7. HC. Charles-Edward Amory Winslow's definitive report discrediting the sewer gas fear appeared in the National Association of Master Plumbers, *Report of the Sanitary Committee, 1907–1908–1909* (Boston, Mass.: National Association of Master Plumbers, 1909).

16. "Advertising Section," *GH* 43 (July–Dec. 1906), and *GH* 41 (July–Dec. 1905). The advertising section was a regular feature at the back of *GH;* the pages were not numbered, so in these and the citations to follow, I simply indicate the volume number.

17. *McConnell Germ-Proof Water Filters, Illustrated Catalogue, 1894,* CPP; The Pasteur-Chamberland Filter Co., *Illustrated Types of Pasteur Water Filters* (1900), HC. Note that the use of Pasteur's name was not a mere honorific; he actually held the patent on this filter. See Gerald L. Geison, *The Private Science of Louis Pasteur* (Princeton, N.J.: Princeton University Press, 1995), p. 41.

18. Trenner-Lee Formaldehyde Disinfector, J. Ellwood Lee Co., n.d., p. 6, in Related Pamphlets; American and Continental "Sanitas" Co., "How to Disinfect: A Guide to Practical Disinfection in Everyday Life"; and Germ-a-Thol, Pratt Food Co.; all in Disinfectants, box 1, WC. Household pets, or more specifically their fur, were often cited as a potential source of infection; early tuberculosis tracts made occasional references to tuberculosis-carrying cats and dogs. The new recognition of the role of insect vectors probably increased these anxieties. Even though the common pet flea was never implicated as a disease carrier, it most likely suffered guilt by association.

19. C. N. West Disinfecting Co., "Message to Good Housewives," Disinfectants, box 1, WC. I suspect that the often strident tone of disinfectant advertisements attests to the struggle that companies faced in getting housewives to pay for expensive soaps and disinfectants when less expensive alternatives were available.

20. Oakland Chemical Company, *Monograph on Hydrogen Dioxide,* 1897, pp. 3, 23, NLM; Lambert Pharmacal Co., *Summer Complaints of Infants and Children,* back cover. In medical journals and brochures, Listerine's manufacturer, the Lambert Pharmacal Company of St. Louis, continued the appeal to male medical authority that was typical of its advertising in the 1880s. Listerine advertisements in the mass circulation magazines stressed its generic uses as an antiseptic mouthwash and skin cleanser.

21. "Hoymei," *LHJ* 58 (Nov. 1899): 33.

22. Mildred Maddocks, "The Cleaner versus the Broom," *GH* 64 (May 1917): 68, 158, 161, 162, quotation on p. 68; Charles J. Clarke, "Do You Use a Vacuum Cleaner?" *GH* 68 (Mar. 1919): 28. By 1926, 80 percent of affluent households owned a vacuum cleaner. For the effect on housework of this new household technology, see Ruth Schwartz Cowan, *More Work for Mother: The Ironies of Household Technology from the Open Hearth to the Microwave* (New York: Basic Books, 1983), statistic on p. 173; and Susan Strasser, *Never Done: A History of American Housework* (New York: Pantheon, 1982).

23. *Cosmopolitan Magazine* 54 (Dec. 1912): 168a; *Harper's Bazaar* 43 (July–Dec. 1909): unpaginated back matter.

24. Sanitary Manufacturing Company, *The Passing of the Carpet* (1902), p. 12, HC; *Harper's Bazaar* 43 (July–Dec. 1909): 4.

25. Standard Textile Products Company, "Sanitas Modern Wall Covering and Its Uses," n.d., p. 9, HC; Carbola Chemical Company, "Carbola, the

Disinfectant That Paints," n.d., unpaginated, Disinfectants, box 1, WC; E. I. du Pont de Nemours & Co., *Saniflat,* HC; *SEP* June 5, 1926, p. 160.

26. *GH* 43 (July–Dec. 1906).

27. *GH* 48 (1909). On the development of the refrigerator, see Cowan, *More Work,* pp. 128–145.

28. *Harper's Bazaar* 43 (July–Dec. 1909); *GH* 48 (1909); *GH* 41 (July–Dec. 1905).

29. *GH* 32 (1902). Emphasis in original.

30. See Thomas Hine, *The Total Package: The Evolution and Secret Meanings of Boxes, Bottles, Cans, and Tubes* (New York: Little, Brown, 1995); and Strasser, *Satisfaction Guaranteed,* pp. 252–285. I base my generalizations about use of words such as "air-tight" and "sanitary" on my own reading of turn-of-the-century advertisements.

31. "The Development of Cellophane," Reference Files, Public Relations Department, E. I. du Pont de Nemours & Co., box 3, folder 328, Accession no. 1410, HD; *GH* 43 (July–Dec. 1906).

32. "Supplement to Baked Goods Survey Report," Apr. 1, 1928, box 502, HD; "Food-Candy, 1917–1929 misc.," Competitive Advertisements Files, JWT.

33. "Canned Foods," Competitive Advertisements Files, JWT. On the adoption of the glass bottle, see Eric Lampard, *The Rise of the Dairy Industry in Wisconsin: A Study in Agricultural Change, 1820–1920,* (Madison: State Historical Society of Wisconsin, 1963), pp. 228–229, 405.

34. Marion Harland, *The Story of Canning* (n.p.: National Canners Association, 1910), pp. 5–6. Harland's real name was Mary Virginia Hawes Terhune.

35. *Harper's Bazaar* 43 (July–Dec. 1909). For the canning industry's efforts to stress its sanitary credentials, see, for example, Max Ams Machine Company, *The Seal of Safety* (New York: Max Ams Machine Publicity Department, 1915). Although the first textbook on the bacteriology of canning appeared in 1899, the canning industry did not begin systematically to implement bacteriologically based principles until the 1910s, largely due to its "dispersed and fragmented" state, according to James Harvey Young, *Pure Food: Securing the Federal Food and Drugs Act of 1906* (Princeton, N.J.: Princeton University Press, 1989), p. 113.

36. *Butcher's Advocate* 77 (Sept. 3, 1924 and Apr. 16, 1924).

37. On the new corporate culture, see Olivier Zunz, *Making America Corporate, 1870–1920* (Chicago: University of Chicago Press, 1990), esp. Chap. 7, "Drummers and Salesmen."

38. On the discipline exercised over workers in these service industries, see Matthew Josephson, *Union House, Union Bar: The History of the Hotel and Restaurant Employees and Bartenders International Union, AFL-CIO* (New York: Random House, 1956), esp. p. 86.

39. On the development of hotels as purveyors of fine service, see the interesting discussion in Leach, *Land of Desire*, Chap. 5. Good background information on hotels and restaurants is also available in Josephson, *Union House, Union Bar.*

40. "How Hotel Keepers Can Aid in Preventing the Spread of Tuberculosis," PSPT, *Report for the Year Ending May 1, 1906* (Philadelphia: PSPT, 1906). pp. 20–21. These generalizations are based on my reading of the *NYT* and *HM*. Frank H. Hamilton, in his "Sewer Gas," *PSM* 22 (1882): 1–20, esp. pp. 17–18, refers to the fact that many hotels removed water basins to prevent sewer gas from entering sleeping chambers.

41. "The Practical Hotel Housekeeper," *HM* 5, no. 49 (Apr. 1897): 28–29, quotation on p. 28. The same advice was being given decades later; see Jane C. Van Ness, *The Housekeeper's Primer* (Chicago: The Hotel Monthly Press, 1940), CHS.

42. *HM* 8, no. 89 (Aug. 1900): 11. On hotels as purchasers of sanitary equipment, see, for example, American Aromatic Disinfector Company, "Disinfectants and Disinfecting Appliances," [ca. 1910], HC; and Albert Pick and Co., Chicago, *General Catalog. . .,* 1911–1912, CHS.

43. The ad for "Acme Hygienic Couches, Mattresses, and Pillows" appears in the *HM* 7, no. 74 (1899). Advertisements in the *HM* for refrigerators, vacuum cleaners, and the like invoke the same themes as do the ads in *GH*.

44. *NYT,* Jan. 23, 1910. For a typical warning about hotel blankets, see "Little Dangers to Be Avoided in the Daily Fight against Tuberculosis," *Pennsylvania Health Bulletin* 7 (Jan. 1910): 1–5.

45. *NYT,* Jan. 16, 1912; *NYT,* Feb. 13, 1912. The law stated that "each sheet shall be ninety-one inches long" and wide enough to cover the whole mattress. Virginia also passed a law requiring hotels to use eight-foot sheets, but the courts found it unconstitutional (this ruling was appealed). See *NYT,* Dec. 14, 1911.

46. Edward Hungerford, "Sleeping-Cars and Microbes," *Harper's Weekly,* Feb. 1, 1914, pp. 20–22. This article gives the statistic on the number of passengers using sleeping cars and mentions the lines to "tubercular resorts" being a particular point of concern. On the history of the Pullman car, see Joseph Husband, *The Story of the Pullman Car* (Chicago: A. C. McClurg, 1917). As of 1914, the Pullman Company supplied 95 percent of all sleeping cars used in the United States. In the early 1900s, public health officials took seriously the issue of railway sanitation. See, for example, the discussions of railroad car sanitation in U.S. Public Health Service, *Transactions of the Third Annual Conference of State and Territorial Health Officers with the United States Public Health and Marine Hospital Service, May 15, 1905* (Washington, D.C.: Government Printing Office, 1906), pp. 24–29.

47. Husband, *Story of the Pullman Car,* p. 153. Crowder's appointment is

mentioned in U.S. Public Health Service, *Transactions of the Third Annual Conference,* p. 27.

48. Hungerford, "Sleeping-Cars and Microbes," p. 22. Whether the company actually complied with these exacting regulations deserves further investigation.

49. On dining out, see Lewis Erenberg, *Steppin' Out: New York Nightlife and the Transformation of American Culture, 1890–1930* (Chicago: University of Chicago Press, 1984). On the restaurant industry's growing emphasis on cleanliness, see Harvey Levenstein, *Revolution at the Table: The Transformation of the American Diet* (New York: Oxford University Press, 1988), p. 186.

50. "Editor's Diary," *North American Review* 183 (1906): 699; *HM* 9, no. 103 (Oct. 1901): 17.

51. *HM* 5, no. 49 (Apr. 1897): 28–29, quotation on p. 29; John Goins, *The American Colored Waiter,* n.d., Hotel School, CU; I. S. Anoff, "The Albert Pick and Company Story," unpublished typescript, p. 12, CHS.

52. On Typhoid Mary and the healthy carrier concept, see Judith Walzer Leavitt, *Typhoid Mary: Captive of the Public's Health* (Boston: Beacon, 1996).

53. See *NYT,* Dec. 24, 1910, and Dec. 3, 1915. The city's sanitary code was amended to require bacteriological exams of city food handlers; restaurant workers needed a certificate to work. The 1915 article reported that after the first round of required examinations, approximately 3.5 percent of the 40,000 people tested were found to be "in such a state of health as to make their employment unlawful." For further discussion of sanitary regulation of food handlers, see Leavitt, *Typhoid Mary,* pp. 52–54.

54. In their organizing campaigns, union activists tried to capitalize on the fact that although service workers were forced to observe sanitary niceties while serving affluent customers, their working conditions were often extremely unsanitary. See, for example, Pasquale Russo, *Twelve O'Clock Lunch* (Chicago: the author, 1923), CHS.

55. See, for example, the discussion of food inspection in Robert S. Lynd and Helen Merrell Lynd, *Middletown: A Study in American Culture* (New York: Harvest/Harcourt Brace, 1957), pp. 449–450. On the politics of packaged goods, see Strasser, *Satisfaction Guaranteed,* Chap. 8. On the entertainment and service industries serving the urban working classes, see Erenberg, *Steppin' Out;* and Levenstein, *Revolution at the Table.*

56. The articles and commentaries provided in the *American Journal of Public Health* provide a sense of this widening process of state hygienic regulation in the 1910s and 1920s. For an in-depth survey of one concrete area of change, namely the provision of individual eating utensils, see David Fromson, comp., *Regulatory Measures Concerning the Prohibition of the Common Drinking Cup and the Sterilization of Eating and Drinking Utensils in Public Places* (New York: Cup and Container Institute, 1936).

57. "Individual Drinking Cups," *JOL* 6, no. 8 (Aug. 1909): 237 mentions that railroads installed cup vendors. Public health officers also discussed experiments with paper cups on various railway lines in U.S. Public Health Service, *Transactions of the Fourth Annual Conference. . . 1906*, pp. 31–45. Ads for collapsible drinking cups can be found in many periodicals in this period, including *Scientific American* and the *JOL.*

58. Quoted in John H. White, Jr., *The American Railroad Passenger Car* (Baltimore, Md.: Johns Hopkins University Press, 1978), p. 432.

59. "A Drink of Cold Water," *Survey* 29 (Oct. 12, 1912): 54–55, quotation from p. 55. See also the temperance argument stated in *The Independent* 71 (Oct. 12, 1911): 830–831.

60. For an overview of the water fountain's history, see Louis V. Dieter, "The Relative Sanitary Values of Different Types of Drinking Fountains," *The American City* 21, no. 5 (Nov. 1919): 452–457, and 21, no. 6 (Dec. 1919): 549–554. A state-by-state compendium of drinking-cup laws is given in Fromson, *Regulatory Measures.*

61. "A Drink of Cold Water," pp. 54–55.

62. *The Independent* 74 (May 22, 1913): 1118–1119.

63. Marie Correll, *Sanitary Drinking Facilities, with Special Reference to Drinking Fountains,* Bulletin of the Women's Bureau, no. 87, 1931, (Washington, D.C.: Government Printing Office, 1931), p. 6.

64. JBSC, *Bulletin* 1, no. 3 (Oct. 1919): back cover, box 10, folder 12, ILR.

8. The Wages of Dirt Were Death

1. Marion Harland, "Little Things That Are No Trifles," *GH* 54 (May 1912): 705–707, quotation on p. 705.

2. Harriette M. Plunkett, *Women, Plumbers, and Doctors* (New York: D. Appleton, 1885), p. 203. Note that besides having a mother who died of typhoid, Theodore Roosevelt had a sister who contracted TB as a child. See David McCullough, *Mornings on Horseback* (New York: Simon and Schuster, 1981), pp. 32–34.

3. On class and race differentials in mortality rates, see S. K. Kleinberg, *The Shadow of the Mills: Working-Class Families in Pittsburgh, 1870–1907* (Pittsburgh: University of Pittsburgh Press, 1989); Barbara Bates, *Bargaining for Life: A Social History of Tuberculosis, 1876–1938* (Philadelphia: University of Pennsylvania Press, 1992), pp. 313–327; Georgina D. Feldberg, *Disease and Class: Tuberculosis and the Shaping of Modern North American Society* (New Brunswick, N.J.: Rutgers University Press, 1995); David McBride, *From Tuberculosis to AIDS: Epidemics among Urban Blacks since 1900* (Albany, N.Y.: State University of New York Press, 1991); and Katherine Ott, *Fevered Lives: Tuberculosis in*

American Culture since 1870 (Cambridge, Mass.: Harvard University Press, 1996).

4. For broad overviews of these crusades, see Feldberg, *Disease and Class,* esp. Chap. 3; Suellen Hoy, *Chasing Dirt: The American Pursuit of Cleanliness* (New York: Oxford University Press, 1995), esp. Chap. 4; Ott, *Fevered Lives;* and Michael E. Teller, *The Tuberculosis Movement* (Westwood, Conn.: Greenwood, 1988).

5. On the larger expansion of the female-dominated professions in this period, see Joan Jacobs Brumberg and Nancy Tomes, "Women in the Professions: A Research Agenda for American Historians," in *Reviews in American History* 10 (1982): 275–296.

6. See Kleinberg, *Shadow of the Mills,* esp. pp. 94–99, and pp. 347–348, nn. 125, 126. Based on new evidence from the 1900 census, Samuel H. Preston and Michael R. Haines, in their *Fatal Years: Child Mortality in Late Nineteenth-Century America* (Princeton, N.J.: Princeton University Press, 1991), argue that as of 1900, infant mortality in urban areas remained very high even among privileged social groups.

7. John Duffy, *The Sanitarians: A History of American Public Health* (Chicago: University of Illinois Press, 1990), pp. 178–179; Kleinberg, *The Shadow of the Mills,* pp. 97–98; and Hoy, *Chasing Dirt,* Chap. 3. On the crusade for tenement reform, see Roy Lubove, *The Progressives and the Slums: Tenement Housing Reform in New York City* (Pittsburgh: University of Pittsburgh Press, 1962).

8. Edith Abbott et al., *The Tenements of Chicago, 1908–1935* (Chicago: University of Chicago Press, 1936), p. 206; Kleinberg, *Shadow of the Mills,* pp. 97, 92.

9. On the strengthening of public health powers in this period, see Duffy, *Sanitarians,* esp. Chap. 13; and Judith Walzer Leavitt, *The Healthiest City: Milwaukee and the Politics of Health Reform* (Princeton, N.J.: Princeton University Press, 1982). On the campaign to reduce infant mortality, see Richard A. Meckel, *Save the Babies: American Public Health Reform and the Prevention of Infant Mortality, 1850–1929* (Baltimore, Md.: The Johns Hopkins University Press, 1990). On the Metropolitan program, see Diane Hamilton, "The Cost of Caring: The Metropolitan Life Insurance Company's Visiting Nurse Service, 1909–1953," *BHM* 63 (1989): 419–426; and Elizabeth Toon, "Corporations, Consumers, and Classrooms: Women and Health Education at the Metropolitan Life Insurance Company, 1920–1940," (paper presented at the Berkshire Conference on Women's History, Chapel Hill, N.C., 1996).

10. Bureau of Child Hygiene, New York City Department of Health, *Ten Commandments for Keeping Baby Well* (New York: privately printed, 1916). On the general tendency to "dumb down" health education, see John Burnham, *How Superstition Won and Science Lost: Popularizing Science and Health in the*

United States (New Brunswick, N.J.: Rutgers University Press, 1987), esp. pp. 45–84. He argues that this process accelerated sharply in the 1930s.

11. *Report of the Henry Street Settlement* (New York: privately printed, 1926), p. 3. On the history of visiting nursing, see Karen Buhler Wilkerson, "False Dawn: The Rise and Decline of Public Health Nursing in America, 1900–1930," in *Nursing History: New Perspectives, New Possibilities,* ed. Ellen C. Lagemann (New York: Teachers College Press, 1983), pp. 89–106; "Left Carrying the Bag: Experiments in Visiting Nursing, 1877–1909," *Nursing Research* 36 (1987): 42–47; and Hamilton, "Cost of Caring."

12. "Nursing Technique," box 1, folder 2, SCP.

13. My conception of resistance and adaptation is much indebted to Lizabeth Cohen, "Embellishing a Life of Labor: An Interpretation of the Material Culture of American Working-Class Homes, 1885–1915," in *American Material Culture: The Shape of Things around Us,* ed. Edith Mayo (Bowling Green, Ohio: Bowling Green State University Popular Press, 1984), pp. 158–181, quotation from p. 178; and Lisabeth Cohen, *Making a New Deal: Industrial Workers in Chicago, 1919–1939* (New York: Cambridge University Press, 1990).

14. On poor housing conditions, see Kleinberg, *Shadow of the Mills,* pp. 65–99; Elizabeth Ewen, *Immigrant Women in the Land of Dollars: Life and Culture on the Lower East Side, 1890–1925* (New York: Monthly Review Press, 1985), esp. 148–163; and Laura Anker Schwartz, "Immigrant Voices from Home, Work, and Community: Women and Family in the Migration Process, 1890–1938" (Ph.D. diss., State University of New York at Stony Brook, 1983), esp. Chaps. 9 and 10.

15. Henry Street Settlement, Visiting Nurse Society Brochure, Oct. 1936, p. 13, in box 86, Henry Street Settlement Papers, Social Welfare History Archives, University of Minnesota.

16. Florence Larrabee Lattimore, "Three Studies in Housing and Responsibility," in *The Pittsburgh Survey,* ed. Paul U. Kellogg, vol. 5: *The Pittsburgh District: Civic Frontage* (New York: Survey Associates, 1914), pp. 124–138, quotation on p. 128; and *Kingsley Record* 22, no. 2 (Mar. 1920): 1, 3, quotation on p. 3.

17. "Miss O'Donnell, Personal Experiences during the Epidemic," box 8, folder 1, SCP; and Mary Simkhovitch, quoted in Ewen, *Immigrant Women,* p. 97. In one of the oral histories done by Corinne Krause, a Slavic woman said it was a mark of "mother love" to sleep with a sick child and criticized a relative for leaving her small children to sleep in a separate room because "that boy and girl could be dead in the morning, you don't know." Interview S-2-A, p. 32, KC. The Krause Collection, upon which I relied heavily in writing this chapter, consists of approximately 225 oral histories. Interviews were done with three generations of ethnic women in the same family

(grandmother, mother, and daughter) in 1975 and 1976 as part of a study of ethnic women's mental health. As part of a general effort to assess sources of self-esteem, the interview schedule included a number of questions relating to housework and child care. The participants were also asked about the family's medical history, including experiences with disease and attitudes toward doctors. Interviewees are identified only by initials to protect their privacy. The prefix *I* designates an Italian family; *J* is Jewish; *S* is for Slavic. The letter *A* was used for the grandmother, *B* for the mother, and *C* for the daughter.

18. Peter Roberts, "The New Pittsburghers: Slavs and Kindred Immigrants in Pittsburgh," *Charities and Commons* 21 (1909): 533–552, quotation on p. 543; Lattimore, "Three Studies in Housing and Responsibility," p. 132; National Council of Jewish Women, *By Myself I'm a Book! An Oral History of the Immigrant Jewish Experience in Pittsburgh* (Waltham, Mass.: American Jewish Historical Society, 1972), pp. 102–103.

19. Lattimore, "Three Studies in Housing and Responsibility," p. 125; and Interview S-1-A, p. 16, KC.

20. Mabel Hyde Kittredge, "The Need of the Immigrant," *JHE* 5, no. 4 (Oct. 1913): 307–316, quotations on pp. 307, 308.

21. Cohen, "Embellishing a Life of Labor," esp. pp. 164–167.

22. Social investigators often commented on the contrast between the meticulous housekeeping standards women imposed on their homes and the filth and disorder of the streets as a whole. See Ewen, *Immigrant Women,* pp. 136, 137. An Italian woman who came to the United States in 1920, when she was about thirty, explained it in terms of a gendered division of labor: "Him responsible for outside, I'm inside." Interview I-1-A, p. 22, KC.

23. Interviews S-1-B, p. 8; S-13-B, p. 4; I-14-B, p. 5; and J-9-B, p. 24, all KC. On the rigors of cleaning, see also Interviews I-1-B, pp. 4–5; I-3-A, p. 63; I-4-B, pp. 33–34; and S-3-A, p. 19, all KC. See also Ewen, *Immigrant Women,* pp. 95, 100–101, 154–157; and Schwartz, "Immigrant Voices," pp. 636–638. One woman recalled that a child was scalded to death by falling in his mother's wash bucket. See Interview S-7-A, pp. 10–11, KC.

24. On cleaning customs, see Ewen, *Immigrant Women,* pp. 32, 148–149.

25. For a discussion of advertising in immigrant newspapers, see Andrew Heinze, *Adapting to Abundance: Jewish Immigrants, Mass Consumption, and the Search for American Identity* (New York: Columbia University Press, 1990), esp. Chaps. 6, 9, and 10. He notes that Yiddish papers were quicker than Italian papers to feature ads for American products such as soap. For mentions of linoleum, see, for example, Interviews 1-5-A, p. 63; and S-4-A, p. 8, KC. Linoleum was also popular among rural women. See Marilyn Irwin Holt, *Linoleum, Better Babies, and the Modern Farm Woman, 1890–1930* (Albuquerque: University of New Mexico Press, 1995), pp. 88–89.

26. Interview S-1-B, p. 13; and I-22-A, p. 6, KC.

27. Interviews S-8-A, p. 13; S-14-A, p. 21; S-14-B, p. 13; and I-24-B, p. 75, KC.

28. Interviews S-1-B, p. 39; and S-3-B, p. 13, KC.

29. Interviews J-6-A, p. 14; and J-4-B, n.p., KC.

30. Interview I-7-B, p. 10, KC.

31. Frank Persons to F. E. Crowell, Nov. 27, 1916, "Tuberculosis, 1916–1917," box 180, CSS.

32. Interview I-23-B, p. 26, KC; Robert S. Lynd and Helen Merrell Lynd, *Middletown: A Study in American Culture* (New York: Harcourt, Brace and World, 1929), p. 452.

33. Persons to Crowell, Nov. 27, 1916, CSS; "Twentieth Century Follies," p. 8, box 1, folder 2, SCP.

34. Interview I-11-B, p. 25, KC.

35. On the association of cleanliness and assimilation, see Hoy, *Chasing Dirt,* esp. Chap. 4. On the notion of "whiteness," see David R. Roediger, *The Wages of Whiteness: Race and the Making of the American Working Class* (New York: Verso, 1991).

36. In the Krause oral histories, deaths of infants and children are frequently mentioned side by side with comments on the obsessive cleanliness of immigrant mothers and daughters.

37. On the country life movement, see William L. Bowers, *The Country Life Movement in America, 1900–1920* (Port Washington, N.Y.: Kennikat, 1974). For a perceptive account of the extension movement among rural women, see Holt, *Linoleum.* On the history of rural home economics at Cornell, see Flora Rose, *A Growing College: Home Economics at Cornell University* (Ithaca: New York State College of Human Ecology, 1969).

38. The extension movement mirrored the gendered division of labor on the farm. For an overview of the family economy in rural New York, see Nancy Grey Osterud, *Bonds of Community: The Lives of Farm Women in Nineteenth-Century New York* (Ithaca, N.Y.: Cornell University Press, 1991), esp. pt. 3.

39. Rose, *Growing College,* p. 16. For background on Bailey and early home economics, see pp. 10–23.

40. "Report, Cornell Study Clubs, 1914," box 24, folder 44, CU. I have found no direct evidence that Bailey and Van Rensselaer modeled the Cornell Reading Course on the Chautauqua circles, but the similarity of their methods is striking. My thanks to Paul Pedisich for acquainting me with the Chautauqua publications.

41. Details on the Farmers' Institute programs are in box 16, folder 7, CU; for reports on the kitchen conferences, see CUE. On the opening of the cafeteria laboratory, see Rose, *Growing College,* pp. 48–49.

42. Martha Van Rensselaer (hereafter MVR) to Mrs. R. W. Potter, Nov. 23, 1903, Letterpress, p. 605, CU; Louise Montgomery to MVR, Mar. 27, 1913, box 24, folder 36, CU; and Isabel Parsons to MVR, Apr. 16, 1910, box 24, folder 42, CU. This is not to say that no one ever challenged the home economists' authority. See, for example, MVR to Mr. Alva Agee, Dec. 17, 1903, Letterpress, pp. 642–643, box 49, CU.

43. "Survey of Results of Extension," Chenango County, Oct. 24–25, 1923, CUE; Mrs. E. P. Ellinwood to MVR, Mar. 1, 1909, box 24, folder 33, CU.

44. Louise Montgomery to MVR, Mar. 27, 1913, box 24, folder 36, CU; and Estelle Cole to MVR, Jan. 31, 1910, box 24, folder 31, CU.

45. Eppie E. Yantis to MVR, Feb. 13, 1915, box 23, folder 40, CU; and Geneva Watson to MVR, n.d., box 24, folder 41, CU.

46. Mrs. E. P. Ellinwood to MVR, Mar. 1, 1909, box 24, folder 33, CU; and Mrs. E. H. Warren to Farmer's Wives Reading Club, Oct. 1, 1907, box 24, folder 41, CU.

47. Martha Van Rensselaer, "Suggestions on Home Sanitation," *CRCFW*, ser. 3, no. 11 (Nov. 1904): 205. Emphasis in original.

48. Ibid., p. 207. On her resolve to emphasize male responsibility in the sanitation leaflet, see MVR to Mrs. John H. McClure, Mar. 6, 1901, Letterpress, p. 56, CU.

49. "Reading Course Testimonials," box 25, folder 16, CU. The author is not identified. She also wrote that compared to her, "The colts are taken care of, and are not worked to death," and concluded, "Your paper I look for with pleasure and enjoy and keep them. They are good."

50. MVR to Mr. Alva Agee, Dec. 17, 1903, Letterpress, pp. 642–43, box 49, CU; and Mrs. Rufus Stanley to MVR, Dec. 26 1909, box 16, folder 7, CU.

51. Elizabeth Ecker to MVR, n.d., box 24, folder 34, CU; and Mrs. John L. Fuller to MVR, n.d., box 24, folder 39, CU.

52. Discussion paper on household bacteriology submitted by Kathrea Edes, box 24, folder 34; MVR to Mrs. S. W. Terry, Oct. 25, 1909, box 24, folder 40; MVR to Mrs. D. F. Boutwell, Mar. 21, 1901, Letterpress p. 87, box 49, all CU.

53. Mrs. W. S. Peck to MVR, Jan. 25, 1909, box 24, folder 42, CU; and Ella Cushman, "Notice of Trip Arrangements," Oct. 3, 1934, Chenango County, box 1, record group 919, CUE.

54. "Reading Course Testimonials," box 24, folder 9, CU; Ella Cushman, "Report of Meeting on Household Management," Cattaraugus County, box 1, CUE.

55. Ella Cushman, "Report on Kitchen Tour" [1934?], Chenango County, box 1, CUE. On the use of boiling water, see "Summary of Eighty-one Records," Dec. 1932, Jefferson County, box 2, CUE. Only twenty-four of the homes surveyed in Jefferson County had hot running water.

56. Kathrea Edes, "Household Bacteriology: Discussion Paper," Apr. 1, 1913, box 24, folder 34, CU; and MVR to Mrs. C. V. Zelley, Apr. 29, 1912, box 24, folder 42, CU.

57. Mrs. George Bancroft to MVR, n.d. [answered Mar. 1909], box 24, folder 32, CU; and Helen Boden to Miss White, [1910?], box 24, folder 35, CU.

58. May E. Abbuhl to Flora Rose, Jan. 16, 1915, box 16, folder 8, CU; Ella Cushman, "Trip Report," Apr. 6, 1934, Cattaraugus County, box 1, CUE.

59. Ella Cushman, "Trip Report," Oct. 20–21, 1931, Cattaraugus County, box 1, CUE. See also MVR to Mrs. James Swan, May 4, 1904, Letterpress, p. 727, box 49, CU; MVR to Mrs. George B. Ward, Mar. 18, 1901, Letterpress, pp. 82–83, box 49, CU.

60. Mrs. A. C. Abbuhl to MVR, July 9, 1920, box 24, folder 36, CU.

61. On the demographic debates, see for example, Douglas C. Ewbank and Samuel H. Preston, "Personal Health Behaviour and the Decline in Infant and Child Mortality: The United States, 1900–1930," in John Caldwell et al., eds., *What We Know about Health Transition: The Cultural, Social and Behavioural Determinants of Health* (Canberra: Australian National University Press, 1990), pp. 116–149. The "makework" interpretation is well illustrated by Barbara Ehrenreich and Deirdre English, *For Her Own Good: 150 Years of the Experts' Advice to Women* (New York: Doubleday, 1978), esp. p. 159.

9. The Two-Edged Sword

1. Maud Nathan, *The Story of an Epoch-Making Movement* (New York: Doubleday, Page, 1926), pp. 61, 62, 63.

2. Cyrus Edson, "The Microbe as a Social Leveller," *North American Review* 161 (1895): 421–426.

3. Ernest Poole, *The Plague in Its Stronghold* (New York: Charities Organization Society, 1903). My discussion of the garment industry relies on Steven Frazer, "Combined and Uneven Development in the Men's Clothing Industry," *Business History Review* 57 (winter 1983): 522–547; Steven Fraser, *Labor Will Rule: Sidney Hillman and the Rise of American Labor* (New York: Free Press, 1991); and David Montgomery, *The Fall of the House of Labor* (New York: Cambridge University Press, 1987). On working conditions among immigrant garment workers, see Alan M. Kraut, *Silent Travelers: Genes, Germs, and the "Immigrant Menace"* (New York: Basic Books, 1994), esp. pp. 180–182, 187.

4. On the garment union movement, Fraser, *Labor Will Rule;* Gus Tyler, *Look for the Union Label: A History of the International Ladies' Garment Workers' Union* (Armonk, N.Y.: M. E. Sharpe, 1995); and Joan M. Jensen, "The Great Uprisings: 1900–1920," in *A Needle, a Bobbin, a Strike: Women Needleworkers in America,* ed. Joan M. Jensen and Sue Davidson (Philadelphia: Temple Uni-

versity Press, 1984), pp. 88–89. By 1900, Illinois, Massachusetts, and New York all had factory inspection laws.

5. Nathan, *Story,* p. 61. On the emergence of consumers' leagues, see Kathryn Kish Sklar, *Florence Kelley and the Nation's Work: The Rise of Women's Political Culture, 1830–1900* (New Haven, Conn.: Yale University Press, 1995), esp. Chap. 12.

6. New York City Consumer's League, *The Menace to the Home from Sweatshop and Tenement-Made Clothing* (New York: The League, 1901), p. 3. The pamphlet reprinted testimony given in 1900 before the Tenement-House Commission of New York State.

7. Nathan, *Story,* pp. 60–67, quotations from p. 66. On the national movement, see Sklar, *Florence Kelley,* pp. 308–311. The New York City league had previously used a "White List" to designate shops that observed its "Standard of a Fair House," which included decent wages, reasonable hours of work, and a willingness to "conform in all respects to the present sanitary laws." But the garment industry was too decentralized for this strategy to work; department stores simply had no idea if the garments they sold had been made in tenement houses. On the history of the union label, see Ernest R. Spedden, *The Trade Union Label,* Johns Hopkins University Studies in Historical and Political Science, ser. 28, no. 2 (Baltimore, Md.: Johns Hopkins University Press, 1910). The term "white label" was first used by union cigar makers in San Francisco to distinguish their goods from those made by Chinese workers. But as employed by the Consumers' League, the term "white label" referred to a product's moral and hygienic status rather than the race of the worker who made it. Most garment workers were Russian Jews or Italians, groups that many Anglo-American Protestants did not consider members of the "white" race.

8. Nathan, *Story,* pp. 67–69.

9. Quoted in Spedden, *Trade Union Label,* p. 67.

10. In the 1880s, Gompers already had a firm command of sanitarian language and quoted extensively from physicians and public health officials. See Samuel Gompers, "Reports on Tenement-House Cigar Manufacture: Some Implications," in *The Samuel Gompers Papers,* ed. Stuart B. Kaufman, vol 1: *The Making of a Union Leader, 1850–86* (Chicago: University of Illinois Press, 1986), p. 200. Gompers was president of the American Federation of Labor from 1886 to 1924, with the exception of one year, 1895. On the "white label" used by San Francisco cigar makers, see Spedden, *Trade Union Label,* p. 63. On workers' long-standing concerns about health conditions, see, for example, Anthony F. C. Wallace, *St. Clair: A Nineteenth-Century Coal Town's Experience with a Disaster-Prone Industry* (Ithaca, N.Y.: Cornell University Press, 1988), esp. Chap. 5. On *The Jungle,* see James Harvey Young, *Pure Food: Securing the Federal Food and Drugs Act of 1906* (Princeton, N.J.: Princeton

University Press, 1989), Chap. 10; on the efforts by International Workers of the World to organize hotel and restaurant workers, see Pasquale Russo, *Twelve O'Clock Lunch* (Chicago: Pasquale Russo, 1923).

11. Those records of the Chicago Lung Association, the Atlanta Lung Association, the PSPT, and the Charity Organization Society of the City of New York that I have used all document the formation of labor committees during the early 1900s. Boston evidently had a similar committee, according to Richard Shryock in his *National Tuberculosis Association, 1904–1954: A Study of the Voluntary Health Movement in the United States,* NTA Historical Series, no. 8 (New York: NTA, 1957), p. 91. For a contemporary account of union work, see "Labor's Fight against Tuberculosis," *JOL* 6, no. 6 (June 1909): 172–173. For a union paper's coverage of the TB crusade, see the printers' *Typographical Journal* in the early 1900s.

12. American Federation of Labor, *Tuberculosis* (Washington, D.C.: AFL, 1906), George Meany Memorial Archives, Silver Spring, Md., quotations from cover, pp. 2, 3.

13. Reprinted in Central Labor Union of Philadelphia, Minutes, 1908–1914, Oct. 9, 1910, p. 228, UA.

14. Central Labor Union of Philadelphia, Minutes, 1908–1914, Sept. 25, 1910, p. 222, UA; Kennaday's speech is reprinted in American Federation of Labor, *Tuberculosis,* pp. 5–7, quotation on p. 7.

15. American Federation of Labor, *Tuberculosis,* p. 7. I suspect that the male-dominated labor unions found the male-dominated anti-TB societies to be more palatable allies than were the female-dominated consumer leagues.

16. On the 1909–1910 strikes, see Jensen, "The Great Uprisings," and Tyler, *Look for the Union Label,* Chaps. 4 and 5.

17. On its founding, see JBSC, *Fifteen Years of Industrial Sanitary Self Control,* Fifteenth Anniversary Report of the JBSC. . .1926, pp. 26–27, box 5, folder 13, ILR.

18. Ibid., pp. 28–29. The sanitary provisions of the protocol were extended to the dress and waist manufacturers and unions in 1913. On the appeal of the JBSC to both sides, see Tyler, *Look for the Union Label,* p. 127.

19. JBSC, *Fifteen Years,* p. 7.

20. The standards are reprinted in *BJBSC,* no. 1 (June 1911), n.p. box 10, folder 4, ILR.

21. JBSC, *Six Years' Work and Progress of the JBSC. . .1916* (New York: JBSC, 1916), p. 7, box 5, folder 6, ILR; JBSC, *Ten Years of Industrial Sanitary Self Control: Tenth Annual Report of the JBSC. . .* (New York: JBSC, 1921), p. 7; and E. Packard, "Impressions of a Temporary Inspector," *BJBSC* 2, no. 1 (May 1914): 12.

22. For biographical information on William Jay Schieffelin, see his obituary in *NYT,* May 1, 1955, p. 88; on Lillian Wald, see Edward James, Janet

James, and Paul Boyer, eds., *Notable American Women: A Biographical Dictionary, 1607–1950* (Cambridge, Mass.: Belknap Press, Harvard University Press, 1971), vol. 3, pp. 526–529; on Henry Moskowitz, see Elizabeth Israels Perry, *Belle Moskowitz: Feminine Politics and the Exercise of Power in the Age of Alfred E. Smith* (New York: Oxford University Press, 1987), pp. 99–107; on George Price, see "Industrial Medicine's Hall of Fame: George M. Price," *Industrial Medicine and Surgery* 22, no. 1 (Jan. 1953): 37–38.

23. See biographical sources in previous note, as well as Tyler, *Look for the Union Label*, pp. 127–128. Tyler emphasizes the critical role played by Price; although not an active Socialist, he definitely believed in the principle of socialized medicine. On the careers of Newman and Schneiderman, see Annelise Orleck, *Common Sense and a Little Fire: Women and Working-Class Politics in the United States, 1900–1965* (Chapel Hill: University of North Carolina Press, 1995). In an oral history deposited at the ILR, Newman noted that her brother, who was an Orthodox Jew, offered to pay for her schooling if she would stop reading influential Socialist daily, *The Forward*, but she refused. See Henoch Mendelsund, "Interview with Pauline Newman," 1973, p. 42.

24. "The Sanitary Label," *BJBSC*, no. 5 (Jan. 1912): n.p., box 10, folder 5, ILR. The JBSC's philosophy was very similar in this regard to that of Sidney Hillman, the fiery leader of the Amalgamated Clothing Workers' of America, as analyzed in Fraser, *Labor Will Rule.*

25. George M. Price, "Factory Introspection," *The Survey* 26 (1911): 219–228, quotations on pp. 219, 220.

26. JBSC, *Workers' Health Bulletin* (New York: JBSC, 1915), quotation on p. 24. The sanitary division of responsibility is evident in *Sanitary Control Monthly Bulletin* 1, no. 3 (Oct. 1919): 8–9, which gives two categories of sanitary defects, those "for which shop owner is responsible" and those "for which shop owners and the workers are jointly responsible."

27. *BJBSC*, no. 3 (Aug. 1911): 1; "The Impressions of Miss Rose Schneiderman," *BJBSC*, no. 6 (May 1912): n.p., box 10, folder 5, ILR.

28. "Impressions of Miss Rose Schneiderman," unpaginated; Oral History, Pauline Newman, 1973, pp. 4–5, Oral History 4, ILR.

29. "The Sanitary Label," *BJBSC*, no. 5 (Jan. 1912), unpaginated, box 10, folder 5, ILR.

30. JBSC, *Fifteen Years*, p. 30, simply noted that the convention approved the label idea, but that its implementation failed "for many reasons"; it didn't give any specifics. The coverage of the label issue in *The Ladies' Garment Worker* is more useful in reconstructing opposition to the idea. The journal itself was in favor of the Protocol label. An editorial, "The White Protocol Label," *The Ladies' Garment Worker* (Aug. 1914): 9–10, argued that previous attempts to develop a union label had failed due to "the erroneous belief in the inferiority

of label goods." The Protocol label not only would help erase that idea, but would also allow the union to improve conditions "in a much quicker way and in a more businesslike manner than we could hope to attain by mere agitation." Another article, "Convention Echoes," *The Ladies' Garment Worker* (June 1913): 27, identifies the "militant" element in the ILGWU as the source of opposition to the label idea. Discussing a similar disagreement among cigar workers, Patricia Cooper writes, "Some agreed with an Industrial Workers of the World charge that supporting the label basically meant drumming up business for capitalists." See Patricia Cooper, *Once a Cigar Maker: Men, Women, and Work Culture in American Cigar Factories, 1900–1919* (Chicago: University of Illinois Press, 1987), pp. 138–139, quotation on p. 139.

31. Henry Moskowitz, "The Prosanis Label of the Joint Board of Sanitary Control," in JBSC, *Fifteen Years*, pp. 2–16, quotations on pp. 3, 2, 7. The resolution was reprinted as part of Moskowitz's report.

32. Ibid., p. 4.

33. Newman is quoted in Tyler, *Look for the Union Label*, p. 292. For an insightful discussion of the label movement and its failure to win support among working-class women, see Dana Frank, *Purchasing Power: Consumer Organizing, Gender, and the Seattle Labor Movement, 1919–1929* (New York: Cambridge University Press, 1994). Her work suggests why the sanitary label attracted more support among middle-class women than the union label did among working-class women. See esp. pp. 216–224. At first glance, it might be tempting to dismiss the JBSC as a conservative precursor of 1920s welfare capitalism. But I believe it represents a considerably more complex and interesting example of what David Montgomery terms "constructive Socialism," a labor philosophy that "linked union struggles over job conditions to community reforms of desperate importance to workers." As Montgomery writes, "Nothing could be more misleading than to identify 'sewer socialism' with bourgeois influence on the party." See Montgomery, *Fall of the House of Labor*, p. 286. For a similar argument about the career of Sidney Hillman, see Fraser, *Labor Will Rule*.

34. Moskowitz, "The Prosanis Label," p. 6.

35. George M. Price, "Labor's Fight against Tuberculosis," *Transactions of the Twenty-First Annual Meeting of the National Tuberculosis Association*, pp. 457–460. For a sense of its varied activities, see Union Health Center, *Report for 1926*, box 5, folder 14, ILR. See also Tyler, *Look for the Union Label*, Chap. 9.

36. [Pauline Newman], untitled MS, box 11, folder 21, ILR.

37. Charles Wertenbaker, "My Experience in Organizing Negro Anti-Tuberculosis Leagues," in *The Call of the New South: Addresses Delivered at the Southern Sociological Congress . . . 1912*, ed. James E. McCulloch (Nashville, Tenn.: Southern Sociological Congress, 1912), pp. 216–220, quotation on p.

216. For an introduction to race and public health issues in this period, see Vanessa Gamble, *Germs Have No Color Line: Blacks and American Medicine, 1900–1940* (New York: Garland, 1989); David McBride, *From Tuberculosis to AIDS: Epidemics among Urban Blacks since 1900* (Albany: State University of New York Press, 1991), esp. Chaps. 1 and 2; Susan L. Smith, *Sick and Tired of Being Sick and Tired: Black Women's Health Activism in America, 1890–1950* (Philadelphia: University of Pennsylvania Press, 1995), esp. the Introduction and Chaps. 1 and 2; and Marion M. Torchia, "Tuberculosis among American Negroes: Medical Research on a Racial Disease, 1830–1950," *JHM* 32 (1977): 252–279. For two perceptive accounts of the challenges faced by black leaders in this period, see Kevin K. Gaines, *Uplifting the Race: Black Leadership, Politics, and Culture in the Twentieth Century* (Chapel Hill: University of North Carolina Press, 1996); and Evelyn Brooks Higginbotham, *Righteous Discontent: The Women's Movement in the Black Baptist Church, 1880–1920* (Cambridge, Mass.: Harvard University Press, 1993).

38. For a good portrait of African-American life in Atlanta, as well as of the events of the 1906 race riot, see Jacqueline Anne Rouse, *Lugenia Burns Hope: Black Southern Reformer* (Athens: University of Georgia Press, 1989), pp. 41–45, 57–67. For a discussion of segregation and public health in Atlanta, see the fine article by Stuart Galishoff, "Germs Know No Color Line: Black Health and Public Policy in Atlanta, 1900–1918," *JHM* 40 (1985): 22–41. The U.S. Department of the Census gave Atlanta's population in 1900 as 89,872, of which 39.8 percent were categorized as "negro."

39. See Minutes of the Executive Committee [later Board of Directors], June 1, Oct. 12, and Nov. 9, 1909, box 2, folder 3, AHS. The head of the "Negro Race Committee," a Dr. Campbell, took the initiative in meeting with various black groups; once he stopped coming to meetings, the effort at biracial cooperation seems to have ceased. For a general history of the Atlanta Anti-Tuberculosis and Visiting Nurse Association, see Margaret Kidd Parsons, "White Plague and Double-Barred Cross in Atlanta," (Ph.D. diss., Emory University, 1985).

40. See Scrapbook, 1909, box 7, folder 40, AHS, quotations from pp. 17, 26. Emphasis in the original. On African-American domestic workers, see Tera Hunter, "Household Workers in the Making: Afro-American Women in Atlanta and the New South, 1861 to 1920," (Ph.D. diss., Yale University, 1990). She analyzes the laundress controversy in Chap. 5.

41. Minutes, Executive Committee, Apr. 12, 1910, box 2, folder 32, AHS. On the politics surrounding the laundress controversy, see Hunter, "Household Workers," pp. 212–228.

42. See Rouse, *Lugenia Burns Hope*, esp. pp. 11–19, 26–27, 41–46, 65–68. Quotation is from "The Constitution of the Neighborhood Union," typescript, box 1, NU.

43. See Marion Torchia, "The Tuberculosis Movement and the Race Question, 1890–1950," *BHM* 49 (1975): 152–168. For a contemporary account, see Kelly Miller, "The Negro Anti-Tuberculosis Society of Washington," *JOL* 6, no. 5 (May 1909): 129–130. The papers of Charles Wertenbaker at the University of Virginia Library contain extensive information about the formation of black leagues.

44. Dr. Henry Rutherford Butler, "Negligence a Cause of Mortality," *Mortality among Negroes in Cities: Proceedings of the Conference for Investigation of City Problems, Held at Atlanta University, May 26–27, 1898,* Atlanta University Publications, no. 1, 1896, pp. 15–18, 20–25, quotation is on pp. 24–25.

45. Lugenia Burns Hope (hereafter LBH), Draft of speech, ca. 1908–1909, NU.

46. Parsons, "White Plague," pp. 83–86, 115–116. Parsons writes that prior to Hope's visit, Lowe had followed a suggestion made by Alice Carey, an African-American teacher, and had begun meeting informally with six black women to teach them the basics of tuberculosis prevention so that they might spread that information in their neighborhoods. If Hope knew of Lowe's group, it may well have encouraged her to make this overture to the Atlanta Association.

47. Rosa Lowe (hereafter RL) to LBH, May 23, 1914, Correspondence, box 1, folder 4, AHS.

48. RL, "City Tuberculosis Program for Negroes," 1914, Reports, Negro Program, box 3, folder 32, AHS.

49. Ibid.

50. The name was changed because it was felt that the former designation did "not identify the negroes with the tuberculosis work." See Minutes, Negro Anti-Tuberculosis Association (hereafter NATA), Jan. 8, 1915, box 3, folder 6, AHS. I suspect that the African-American participants preferred the name "Negro Association" to the "Colored Branch." On the zone system used by the Neighborhood Union, see Rouse, *Lugenia Burns Hope,* pp. 67–69.

51. The range of support for NATA is evident from its minutes, collected in box 3, folders 6 and 32, AHS.

52. See, for example, threats to report dance-hall owners, Minutes of NATA, Jan. 8, 1915, box 3, folder 6, AHS. The NATA's approach closely resembles the "politics of respectability" described by Higginbotham in her *Righteous Discontent.*

53. A sample "Sanitary Survey" from 1925 is preserved in box 2, folder 3, NU. For an interesting account of the sanitary work done by the NU, see "Work of the Neighborhood Union," *Spellman Messenger,* Nov. 1916, pp. 5–6, in box 1, NU.

54. Minutes, NATA, June 8, 1915, box 3, folder 6, AHS. On the referral of nuisances to the white association, see *Minutes of the Negro Meeting Held on April*

18, 1917, box 3, folder 6, AHS; and *Minutes of the Educational-Medical Campaign Meeting. . . April 20, 1917,* box 3, folder 6, AHS. A list sent to the Chamber of Commerce of 103 homes with sanitary defects suggests how home conditions were linked directly with disease: for example, one note identified a house "next to dumping ground, 3 people died here of TB in December." See *Chamber of Commerce List,* Apr. 1921, box 2, folder 1, NU. On the extension of the surveys to stores, see Rouse, *Lugenia Burns Hope,* pp. 81–82.

55. "Partial Report of the Work of the NU," 1919, box 2, NU. The award is reported in Minutes, NATA, July 24, 1917, box 3, folder 6, AHS. According to the U.S. census, Atlanta's African-American population went from 154,839 in 1910 to 270,366 in 1920.

56. Rosa Lowe reported the city's promise that the trash would be collected in *Minutes of the Negro Meeting Held on April 18, 1917;* Lugenia Burns Hope complained about their lack of compliance in *Report of the Educational Dept., AATA* [Atlanta Anti-Tuberculosis Association], June 12 to July 17, 1919, box 2, NU.

57. Minutes, NATA, Oct. 17, 1916, box 3, folder 6, AHS. In 1917, Lowe and Hope visited Negro schools to make a survey of overcrowding, lack of playgrounds, and the like. See Minutes, NATA, Nov. 14, 1917, box 3, folder 5, AHS. For the introduction of the Modern Health Crusade, see "Report of Colored Dept. for Year 1919," box 2, folder 18, AHS.

58. Minutes, NATA, Jan 1, 1917, and Mar. 13, 1917, box 3, folder 6, AHS, quotation is from Jan. 1, 1917. Henry H. Pace was often identified by his nickname, Harry. For further information on the library controversy, see Donald L. Grant, *The Way It Was in the South: The Black Experience in Georgia,* ed. Jonathan Grant (New York: Carol Publishing, 1993), pp. 217–218.

59. Minutes, NATA, Apr. 20, 1916, box 3, folder 6, AHS. I do not know if this park was ever built.

60. On the migration northward, see James R. Grossman, *Land of Hope: Chicago, Black Southerners, and the Great Migration* (Chicago: University of Chicago Press, 1989); and Daniel M. Johnson and Rex R. Campbell, *Black Migration in America: A Social Demographic History* (Durham, N.C.: Duke University Press, 1981).

61. *Minutes of the Joint Meeting Held on July 27th, 1917,* box 3, folder 6, AHS.

62. Ibid.

63. On the insurance companies' support, see Minutes, NATA, Mar. 19 and Apr. 11, 1919 (hiring black agent); and Minutes, NATA, Jan. 16, 1917 (printing material), both in box 3, folder 6, AHS.

64. *Anti-Tuberculosis Association, Colored Department, Annual Report,* [1921], box 3, folder 32, AHS; Minutes, NATA, May 19, 1921, box 3, folder 6, AHS. Further details of the church campaign are in Minutes, NATA, June to Sept. 1921, in box 3, folder 6, AHS.

65. Minutes of the Annual Meeting, Feb. 18, 1926, box 2, folder 19, AHS. Weisiger described the NATA as a group that "meets regularly and has within its organization outstanding leaders who are working largely from an educational and informational standpoint, with Atlanta's Colored Population." On the institutes, see Rouse, *Lugenia Burns Hope,* pp. 82–85.

66. On National Negro Health Week, see Smith, *Sick and Tired,* pp. 33–57.

67. Minutes, NATA, Sept. 21, 1915, box 3, folder 6, AHS; Minutes, Feb. 21, 1924, box 3, folder 7, AHS. In 1920, the Atlanta Association took the bold step of sending John Hope as one of its delegates to the Southern Tuberculosis Association. See Parsons, "White Plague," p. 125.

68. Leet Myers to LBH, Aug. 24, 1920, box 2, NU. The children's drawings can be seen in the files of the National Negro Health Week, TU.

69. Mary Antin to LBH, Apr. 19, 1922, box 2, NU.

70. Monroe Work Files, "Women's Work," clipping from *Savannah Journal,* July 23, 1921, TU. Stuart Galishoff makes this same argument in his article about Atlanta, especially regarding the success of the 1910 bond issue, which financed improvement of the city's water supply. See Galishoff, "Germs Know No Color Line."

10. The Waning of Enthusiasm

1. Charles-Edward Amory Winslow, "Man and the Microbe," *PSM* 85 (1914): 5–20, quotations on p. 9.

2. On the importance of the healthy carrier concept, see Judith Walzer Leavitt, *Typhoid Mary: Captive of the Public's Health* (Boston: Beacon, 1996).

3. See Charles Chapin, "The End of the Filth Theory of Disease," *PSM* 60 (Jan. 1902): 234–239; and "The Fetich of Disinfection," *Journal of the American Medical Association* 47 (Aug. 1906): 574–577. Winslow's experimental studies of sewer gas appeared in National Association of Master Plumbers, *Report of the Sanitary Committee 1907–1908–1909* (Boston, Mass.: National Association of Master Plumbers, 1909), pp. 4–22, 39–85. On Walter Reed's experiments and the evolving understanding of yellow fever, see Margaret Humphreys, *Yellow Fever and the South* (New Brunswick, N.J.: Rutgers University Press, 1992), pp. 38–39, 153–157.

4. Quoted in Livingston Farrand to General Director, AICP [Association for Improving the Condition of the Poor], Dec. 9, 1914, box 52, CSS. For a general overview of the "winnowing" process, see Charles Chapin, *Sources and Modes of Infection,* 2d ed. (New York: John Wiley and Sons, 1912).

5. Bailey Burritt to Lawrence Veiller, Feb. 11, 1916, Veiller Correspondence, CSS. Burritt worked for the Association for Improving the Condition of the Poor.

6. The increasing skepticism toward the dust theory can be traced by

comparing succeeding editions of Milton J. Rosenau, *Preventive Medicine and Hygiene* (New York: D. Appleton, 1913). One factor that kept the dust theory alive was the discovery that workers in the so-called dusty trades, such as quarrying and metal polishing, had particularly high rates of TB. See Gerald Rosner and Gerald Markowitz, *Deadly Dust: Silicosis and the Politics of Occupational Disease in Twentieth-Century America* (Princeton, N.J.: Princeton University Press, 1991).

7. Hibbert Winslow Hill, *The New Public Health* (Minneapolis: Press of the Journal Lancet, 1913), p. 5. On the new public health, see Elizabeth Fee, *Disease and Discovery: A History of the Johns Hopkins School of Hygiene and Public Health, 1916–1939* (Baltimore, Md.: Johns Hopkins University Press, 1987); Judith Walzer Leavitt, *The Healthiest City: Milwaukee and the Politics of Health Reform* (Princeton, N.J.: Princeton University Press, 1982); Leavitt, *Typhoid Mary;* and Barbara Gutmann Rosenkrantz, *Public Health and the State: Changing Views in Massachusetts, 1842–1936* (Cambridge, Mass.: Harvard University Press, 1972).

8. On the new diagnostic resources, see George Rosen, *A History of Public Health,* expanded ed. (Baltimore, Md.: Johns Hopkins University Press, 1993), esp. pp. 307–314; Terra Ziporyn, *Disease in the Popular American Press: The Case of Diphtheria, Typhoid Fever, and Syphilis, 1870–1920* (New York: Greenwood, 1988); René Dubos and Jean Dubos, *The White Plague: Tuberculosis, Man, and Society* (1952; reprint, New Brunswick, N.J.: Rutgers University Press, 1987), pp. 120–122. The Widal test became available in the late 1890s.

9. Hill, *New Public Health,* p. 5; Winslow, "Man and the Microbe," p. 20.

10. Iago Gladston, "Debunking Health Education," *Journal of the American Medical Association* 91, no. 14 (Oct. 6, 1928), pp. 1056; Sinclair Lewis, *Arrowsmith* (1924; reprint, New York: Harcourt Brace Jovanovich, 1952); Paul de Kruif, *Microbe Hunters* (1926; reprint, New York: Harcourt, Brace and World, 1953), pp. 282, 300. On the writing of *Arrowsmith,* see Charles E. Rosenberg, "Martin Arrowsmith: The Scientist as Hero," in *No Other Gods: On Science and American Social Thought,* ed. Charles Rosenberg (Baltimore, Md.: Johns Hopkins University Press, 1976), pp. 123–131.

11. Lewis, *Arrowsmith,* pp. 189, 242, 244, 216, 219.

12. De Kruif, *Microbe Hunters,* pp. 282, 300.

13. For overviews of the twenties, see Stanley Coben, *Rebellion against Victorianism: The Impetus for Cultural Change in 1920s America* (New York: Oxford University Press, 1991); and Ellis Hawley, *The Great War and the Search for a Modern Order: A History of the American People and Their Institutions, 1917–1933* (New York: St. Martin's, 1992). On changes in the public health movement, see Paul Starr, *The Social Transformation of American Medicine* (New York: Basic Books, 1982), esp. Bk. 1, Chap. 5, and Bk. 2, Chap. 1. On the

infant welfare movement, see Richard A. Meckel, *Save the Babies: American Public Health Reform and the Prevention of Infant Mortality, 1850–1929* (Baltimore, Md.: Johns Hopkins University Press, 1990). On the turn to science and the retreat from social and political advocacy, see Christopher C. Sellers, *Hazards of the Job: From Industrial Disease to Environmental Health Hazard* (Chapel Hill: University of North Carolina Press, 1997); John Burnham, "American Physicians and Tobacco Use," *BHM* 63 (1989): 1–31; and Philip J. Pauly, "The Struggle for Ignorance about Alcohol," *BHM* 64 (1990): 366–392. The retreat from social issues argument should not be taken too far, however; Georgina Feldberg argues persuasively that despite the turn to the laboratory, a powerful "therapeutic of social reform" continued to shape public health policy toward tuberculosis control. See Georgina D. Feldberg, *Disease and Class: Tuberculosis and the Shaping of Modern North American Society* (New Brunswick, N.J.: Rutgers University Press, 1995).

14. On mass health education in the interwar period, see Martin Pernick, *The Black Stork: Eugenics and the Death of "Defective" Babies in American Medicine and Motion Pictures since 1915* (New York: Oxford University Press, 1996); Martin Pernick, "The Ethics of Preventive Medicine: Thomas Edison's Tuberculosis Films; Mass Media and Health Propaganda," *Hastings Center Report* 8 (June 1978): 21–27; and Elizabeth Toon, "Managing the Conduct of the Individual Life: Public Health Education and American Public Health, 1910–1940" (Ph.D. diss., University of Pennsylvania, forthcoming Dec. 1997). The work of Pernick and Toon will be of great use in exploring how the explosion of public health films and radio programs helped reshape the interwar gospel of germs. On women in bacteriology, see Margaret Rossiter, *Women Scientists in America: Struggles and Strategies to 1940* (Baltimore, Md.: Johns Hopkins University Press, 1982), esp. pp. 238–243. On the early development of health education, see Elizabeth Toon, "Selling Health: Consumer Education, Public Health, and Public Relations in the Interwar Period" (paper delivered at the Berkshire Conference on Women's History, Chapel Hill, N.C., June 1996).

15. For a brief overview of the changing mortality statistics, see Judith Walzer Leavitt and Ronald L. Numbers, "Sickness and Health in America: An Overview," in *Sickness and Health in America: Readings in the History of Medicine and Public Health* (Madison: University of Wisconsin Press, 1985). For more detailed studies of declining child mortality rates, see Meckel, *Save the Babies,* and Samuel H. Preston and Michael R. Haines, *Fatal Years: Child Mortality in Late Nineteenth-Century America* (Princeton, N.J.: Princeton University Press, 1991). On vitamins, see Rima D. Apple, *Vitamania: Vitamins in American Culture* (New Brunswick, N.J.: Rutgers University Press, 1996). The ideal of positive health is defined in Martha Koehne, "The Health Education Program and the Home Economist," *JHE* 16, no. 7 (July 1924): 373–380.

16. Stanhope Bayne-Jones, *Man and Microbes* (Baltimore, Md.: Williams and Wilkins, 1932), p. 128. On the increasing effectiveness of public health regulation in areas such as water and sanitation, see John Duffy, *A History of Public Health in New York City, 1866–1966* (New York: Russell Sage Foundation, 1974); and Leavitt, *Healthiest City*. On the food industry, see Harvey Levenstein, *Revolution at the Table: The Transformation of the American Diet* (New York: Oxford University Press, 1988). On packaging, see Thomas Hine, *The Total Package: The Evolution and Secret Meanings of Boxes, Bottles, Cans, and Tubes* (New York: Little, Brown, 1995).

17. On the evolution of the anti-TB societies, see Michael E. Teller, *The Tuberculosis Movement* (Westwood, Conn.: Greenwood, 1988), and Richard H. Shryock, *National Tuberculosis Association, 1904–1954: A Study of the Voluntary Health Movement in the United States,* Historical Series, no. 8 (New York: NTA, 1957). On the crusade against lung cancer, see James T. Patterson, *The Dread Disease: Cancer and Modern American Culture* (Cambridge, Mass.: Harvard University Press, 1987).

18. See, for example, Estelle D. Buchanan and Robert Earle Buchanan, *Bacteriology for Students in General and Household Science,* rev. ed. (New York: Macmillan, 1922); Ava L. Johnson, *Bacteriology of the Home: A Textbook of Practical Bacteriology* (Peoria, Ill.: Manual Arts Press, 1929); and Paul W. Allen, D. Frank Holtman, and Louise Allen McBee, *Microbes Which Help or Destroy Us* (St. Louis, Mo: C. V. Mosby, 1941). The title pages indicate that Estelle Buchanan had an M.S. and had formerly been an assistant professor of botany at Iowa State College; Robert Buchanan was a Ph.D. and professor of bacteriology. Johnson had both a B.Home Ec. and an M.S. and had been director of practical bacteriology at the Pratt Institute. Allen and Holtman were both Ph.D.'s and professors of bacteriology at the University of Tennessee. McBee held an M.S. and was formerly an assistant at the same university. Note the continued strong association between household bacteriology and state universities.

19. These generalizations are based on my reading of the *Journal of Home Economics* in the 1910s and 1920s. Surveys of interwar job opportunities for home economists emphasized nutrition and textiles; the bacteriology of food is conspicuously absent. Articles on health education in the 1920s suggest that home economists were feeling the competition from other professions, particularly nursing, and that they saw nutrition as their best hope of aligning with the "new public health." See also the historical essays in Sarah Stage and Virginia Vincenti, eds., *Rethinking Home Economics: Women and the History of a Profession* (Ithaca, N.Y.: Cornell University Press, 1997).

20. None of the texts cited in n. 17 had a discussion of household plumbing. Allen, Holtman, and McBee, in their *Microbes Which Help,* pp. 147–149,

154, 164, include a discussion of the dust theory of infection that T. Mitchell Prudden could hardly have faulted.

21. *Scouting for Girls: Official Handbook of the Girl Scouts,* 3d ed. (New York: Girl Scouts, 1922), p. 121. The American Red Cross nursing courses also became an important conduit for teaching household hygiene in the interwar period. See Jane A. Delano and Isabel McIsaac, *American Red Cross Textbook on Elementary Hygiene and Home Care of the Sick* (Philadelphia: P. Blakiston's Son, 1913).

22. On changes in housework, see Ruth Schwartz Cowan, *More Work for Mother: The Ironies of Household Technology from the Open Hearth to the Microwave* (New York: Basic, 1983); and Suellen Hoy, *Chasing Dirt: The American Pursuit of Cleanliness* (New York: Oxford University Press, 1995). On hospital expansion, see Rosemary Stevens, *In Sickness and in Wealth: American Hospitals in the Twentieth Century* (New York: Basic Books, 1989).

23. See, for example, the discussion of terminology in Charles F. Bolduan and Nils W. Bolduan, *Applied Microbiology and Immunology for Nurses* (Philadelphia: W. B. Saunders, 1940), p. 13.

24. On the history of virology, see Sally Smith Hughes, *The Virus: A History of the Concept* (New York: Science History Publications, 1977).

25. On the influenza epidemic, see Alfred W. Crosby, *America's Forgotten Pandemic: The Influenza of 1918* (New York: Cambridge University Press, 1989). Statistics are on p. 206. Vivid accounts of the influenza epidemic can be found in the papers of Simmons College's School of Public Health Nursing, box 8, folder 1, SCP.

26. My discussion of polio is based on the excellent study of Naomi Rogers, *Dirt and Disease: Polio before FDR* (New Brunswick, N.J.: Rutgers University Press, 1992).

27. See Ibid., pp. 1–2.

28. Ibid., esp. pp. 138–190. On the continued association of cleanliness and polio avoidance, see Emily Martin, *Flexible Bodies: Tracking Immunity in American Culture from the Days of Polio to the Age of AIDS* (Boston: Beacon, 1994), pp. 24–25; and Jane S. Smith, *Patenting the Sun: Polio and the Salk Vaccine* (New York: William Morrow, 1990), pp. 155–157.

29. On the history of child health education, see Richard K. Means, *A History of Health Education in the United States* (Philadelphia: Lea and Febiger, 1962).

30. For an overview of the black health movement, see Susan L. Smith, *Sick and Tired of Being Sick and Tired: Black Women's Health Activism in America, 1890–1950* (Philadelphia: University of Pennsylvania Press, 1995). The work at Tuskegee is discussed on pp. 33–57. For a contemporary account of rural health extension work, see Thomas Monroe Campbell, *The Moveable School Goes to the Negro Farmer* (1936; reprint, Tuskegee, Ala.: Tuskegee Institute,

Tuskegee University Archives, 1992). On black health care in general, see David McBride, *Integrating the City of Medicine: Blacks in Philadelphia Health Care, 1910–1965* (Philadelphia: Temple University Press, 1989); and McBride, *From TB to AIDS*.

31. For general accounts of advertising and hygiene in the interwar period, see Vincent Vinikas, *Soft Soap, Hard Sell: American Hygiene in an Age of Advertisement* (Ames: Iowa State University Press, 1992); and Hoy, *Chasing Dirt*, pp. 123–149. See also Jackson Lears, *Fables of Abundance: A Cultural History of Advertising in America* (New York: Basic Books, 1994), Chap. 6.

32. For an insightful account of the Listerine campaign, see Roland Marchand, *Advertising the American Dream: Making Way for Modernity, 1920–1940* (Berkeley: University of California Press, 1985), pp. 18–21.

33. My analysis here and in the next few paragraphs is based on the Listerine advertisements for the 1920s and 1930s found in Domestic Advertisements, Warner-Lambert Company, JWT.

34. "Just about All about Cellophane," pp. 74–102 [clipping, source unidentified], in vol. 81, Scrapbook 1932–1933, HD, quotation on p. 74.

35. "Just about All," p. 75; "Lost and Found: A Profit Maker for the Baking Industry," item 509, HD. Expenditures on cellophane marketing are given in Minutes of Directors' Meetings, Du Pont Cellophane Company, vols. 18, 20, HD. My generalizations about Du Pont marketing strategies are based on my reading of the company archives. As one of Du Pont's own chemists noted in 1924, cellophane was not a marked hygienic improvement over wrappings already in wide use, such as glassine or tinfoil. See H. R. Whitaker to O. F. Benz, Aug. 19, 1924, box 502, "Defendant's Exhibits," no. 299, HD.

36. On the radio show, see *Emily Post*, brochure in Scrapbook, 1932–1933, vol. 81, HD. For a survey of the major advertising campaigns, see "Making American Cellophane Conscious," Scrapbook, 1932–1933, vol. 81, HD. The scrapbook also contains samples of advertisements run in the mass-circulation magazines.

37. Oliver Benz, form letter, May 26, 1933, vol. 82, Scrapbook, 1933, HD.

38. "The Public Cup," *Independent* 71 (1911): 830–831, quotation on p. 831. Du Pont tried to soften the waste by promoting the idea of cellophane crafts among young girls, who were encouraged to make hats, belts, and purses from folded squares of cellophane. On cellophane "homecraft," see *It Started with Belts*, vol. 81, Scrapbook, 1932–1933, HD.

39. On sanitary pads, see Jane Farrell-Beck and Laura Klosterman Kidd, "The Roles of Health Professionals in the Development and Dissemination of Women's Sanitary Products, 1880–1940," *JHM* 51 (1996): 325–352. On the history of toilet paper, see Walter T. Hughes, "A Tribute to Toilet Paper," *Reviews of Infectious Diseases* 10, no. 1 (Jan.–Feb. 1988): 218–222.

40. Homer N. Calver, "Foreword," *Regulatory Measures Concerning the Prohi-*

bition of the Common Drinking Cup and the Sterilization of Eating and Drinking Utensils in Public Places (New York: Public Health Committee, Cup and Container Institute, 1936), p. 3.

41. "Kimberly-Clark Corporation," in Adele Hast, ed., *International Directory of Company Histories,* vol. 3 (Chicago: St. James, 1991), pp. 40–41. On the history of sanitary products and menstrual hygiene, see Farrell-Beck and Kidd, "Roles of Health Professionals"; and Joan Jacobs Brumberg, *Body Projects* (New York: Random House, 1997), Chap. 2.

42. "Kimberly-Clark Corporation," pp. 40–41; "The Introduction of Kleenex Tissues," Archives, Kimberly-Clark Corporation, Neenah, Wisconsin.

43. For accounts of the history of antibiotics, see Harry F. Dowling, *Fighting Infection: Conquests of the Twentieth Century* (Cambridge, Mass.: Harvard University Press, 1977); Stuart B. Levy, *The Antibiotic Paradox: How Miracle Drugs Are Destroying the Miracle* (New York: Plenum, 1992); and Frank Ryan, *Forgotten Plague: How the Battle against Tuberculosis Was Won and Lost* (Boston: Little, Brown, 1992).

44. On Ehrlich, see Ryan, *Forgotten Plague,* pp. 88–89.

45. Ibid., pp. 96–109.

46. Dowling, *Fighting Infection,* pp. 125–157.

47. See Ryan, *Forgotten Plague,* esp. pp. 209–223; and Levy, *Antibiotic Paradox,* esp. pp. 42–46. On p. 47 of his book, Levy provides a useful chronology of the discovery and clinical introduction of antimicrobial drugs from 1919 to 1972.

48. Early problems associated with antibiotic use are discussed in Ryan, *Forgotten Plague,* esp. Chap. 19; and Dowling, *Fighting Infection,* esp. Chap. 11.

49. J. D. Ratcliff, "Yellow Magic of Penicillin," *Reader's Digest* 43 (1943): 47–51, quotation on p. 48; and J. D. Ratcliff, "Bugs Are Their Employees," *Nation's Business* 37 (1949): 32–34, 58–59, quotations on pp. 33, 34.

50. Quoted in Ryan, *Forgotten Plague,* p. 296. Unfortunately, the euphoria faded quickly because it soon became evident that TB was far harder to cure than first anticipated. Effective drug treatment eventually required a combined dose of several different antibiotics. See Ryan, *Forgotten Plague,* esp. pp. 377–384.

51. On the development and introduction of the polio vaccine, see Smith, *Patenting the Sun.*

52. "The Killers All Around," *Time,* Sept. 12, 1994, p. 65; the surgeon general's quote is cited in Barry Bloom and Christopher J. L. Murray, "Tuberculosis: Commentary on a Reemergent Killer," *Science* 257 (Aug. 21, 1992): 1055. *Pax antibiotica* is a term used by John D. Arras in his "Fragile Web of Responsibility," *Hastings Center Report* 18, no. 2 (Apr.–May 1988): supp., p. 10

53. Interview I-4-A, KC.

Epilogue

1. Ryan White and Ann Marie Cunningham, *Ryan White: My Own Story* (New York: Dial, 1991).

2. For a historical overview of the AIDS epidemic, see Mirko Grmek, *History of Aids: Emergence and Origin of a Modern Pandemic* (Princeton, N.J.: Princeton University Press, 1990). See also Randy Shilts, *And the Band Played On: Politics, People and the AIDS Epidemic* (New York: St. Martin's, 1987); and Elinor Burkett, *The Gravest Show on Earth: America in the Age of AIDS* (New York: Houghton Mifflin, 1995). Like Shilts, many AIDS activists felt that the medical establishment moved too slowly in tracking the AIDS virus. But most medical historians would agree with Grmek's argument that both laboratory researchers and the Centers for Disease Control responded to the challenge of the epidemic with remarkable efficiency given the elusive nature of the retrovirus and its modes of transmission. On the larger context of the public health problems raised by the AIDS epidemic, as well as of other issues raised in this chapter, see the thoughtful discussion in Joshua Lederberg, Robert E. Shope, and Stanley C. Oaks, Jr., eds., *Emerging Infections: Microbial Threats to Health in the United States* (Washington, D.C.: National Academy Press, 1992).

3. See Shilts, *And the Band Played On,* and Brandt, *No Magic Bullet: A Social History of Venereal Disease in the United States since 1880,* exp. ed. (New York: Oxford University Press, 1987), esp. Chap. 6.

4. On the importance of negative education, see, for example, Steven Kappel et al., "AIDS Knowledge and Attitudes among Adults in Vermont," *Public Health Reports* 104 (1989): 388–391.

5. On popular perceptions of the immune system, see Emily Martin, *Flexible Bodies: Tracking Immunity in American Culture from the Days of Polio to the Age of AIDS* (Boston: Beacon, 1994).

6. On the preventive strategies recommended for AIDS patients and their caretakers, see John G. Bartlett and Anne K. Finkbeiner, *The Guide to Living with HIV Infection,* rev. ed. (Baltimore, Md.: Johns Hopkins University Press, 1993), pp. 42–55.

7. Marion Harland, "Little Things That Are No Trifles," *GH* 54 (May 1912): 705–707. For popular accounts of the emerging viruses, see Laurie Garrett, *The Coming Plague: Newly Emerging Diseases in a World Out of Balance* (New York: Farrar, Straus, and Giroux, 1994). and Richard Preston, *The Hot Zone* (New York: Random House, 1994). For a scientific treatment of their emergence, see Lederberg, Shope, and Oaks, *Emerging Infections,* esp. Chap. 2.

8. See Garrett, *Coming Plague,* esp. pp. 80–84; and Preston, *Hot Zone,* esp. pp. 80–83.

9. Garrett, *Coming Plague,* p. 27.

10. On Hantaviruses, see Garrett, *Coming Plague,* Chap. 15. On Lyme

disease, see Alan Barbour, *Lyme Disease* (Baltimore, Md.: Johns Hopkins University Press, 1996). On dengue, see Robin Marantz Henig, "The New Mosquito Menace," *NYT,* Sept. 13, 1995, p. A23; and Larry Rohter, "U.S. Is Now Threatened by Epidemic of Dengue," *NYT,* Sept. 23, 1995, p. A5. On equine encephalitis, see John Rather, "Mosquito Threat Draws Swift Action," *NYT,* Sept. 1, 1996, sec. 13, pp. 1, 14.

11. "Neglected for Years, TB Is Back with Strains That Are Deadlier," *NYT,* Oct. 11, 1992, pp. A1, A44. On the return of TB, see Katherine Ott, *Fevered Lives: Tuberculosis in American Culture since 1870* (Cambridge, Mass.: Harvard University Press, 1996), Chap. 9; and Frank Ryan, *The Forgotten Plague: How the Battle against Tuberculosis Was Won and Lost* (Boston: Little, Brown, 1992). As Ott points out, the idea of TB's "return" is misleading, in that the disease has remained a major killer in developing nations throughout this century.

12. "Common Bacteria Said to Be Turning Untreatable," *NYT,* Feb. 20, 1994, p. 24. On the new *Streptococcus A* outbreaks, see "Severe Infection Cited in Queens Boy's Death," *NYT,* Apr. 5, 1995, pp. B1, B6. See also "Fears Growing over Bacteria Resistant to Antibiotics," *NYT,* Sept. 12, 1995, pp. C1, C3. The strep infections' virulence appears to be the result of genetic variance, not drug-resistance, and they are treatable with penicillin.

13. See Lederberg, Shope, and Oaks, *Emerging Infections,* esp. Chap. 2, for a good overview of this process.

14. "Outbreak of Disease in Milwaukee Undercuts Confidence in Water," *NYT,* Apr. 20, 1993, p. C3; and "Bacterial Taint in Water Supply Baffles Experts," *NYT,* July 29, 1993, p. A1. Concern about the water supply is a primary reason why sales of bottled water have soared in the last ten years. See Suzanne Hamlin, "Behind American's Love of Bottled Water," *NYT,* July 24, 1996, pp. C1, C6.

15. "Lessons Are Sought in Outbreak of Illness from Tainted Meat," *NYT,* Feb. 9, 1993, p. C3. For the text of the safe handling instructions adopted by the Agriculture Department, see "Coming Soon to a Brisket near You," *NYT* Aug. 15, 1993, sec. 4, p. 2. In July 1996, this same strain of *E. coli* affected eight thousand Japanese, killing seven people. See "A Food Infection Alarms Japanese," *NYT,* July 25, 1996, pp. A1, A8.

16. Keith Bradsher, "Gap in Wealth in U.S. Called Widest in West," *NYT,* Apr. 17, 1995, pp. A1, D4.

17. John Cushman, "Report Says Global Warming Poses Threat to Public Health," *NYT,* July 8, 1996. p. A2.

18. These comments are based on my reading of materials available from the federal government, such as *The Inside Story: A Guide to Indoor Air Quality* (Washington, D.C.: U.S. Environmental Protection Agency, 1988). A comprehensive guide to home health hazards can be found in Arthur C. Upton and Eden Graber, *Staying Healthy in a Risky Environment* (New York: Simon

and Schuster, 1993). For the return of the germ sell, see "Why Germs Make Us Squirm," *Atlanta Journal/Constitution,* July 20, 1993, pp. B1, B7.

19. Tim Weiner, "Finding New Reasons to Dread the Unknown," *NYT,* Mar. 26, 1995, sec. 4, p. 1.

20. For a recent scientific assessment of the risks posed by emergent diseases, see Lederberg, Shope, and Oaks, *Emerging Infections.*

21. "Civilization and the Microbe," *Scientific American* 130 (Mar. 1924): 172.

22. Garrett, *Coming Plague,* pp. 618, 620.

23. Alfred Crosby, *America's Forgotten Pandemic: The Influenza of 1918* (New York: Cambridge University Press, 1989), p. xi.

24. On Typhoid Mary, see Judith Walzer Leavitt, *Typhoid Mary: Captive of the Public's Health* (Boston: Beacon, 1996). On Tuskegee, see James Jones, *Bad Blood: The Tuskegee Syphilis Experiment,* rev. ed. (New York: Free Press, 1993).

25. On the celebrity campaign against sweatshop labor, see "Labor Pains," *People,* June 10, 1996, pp. 55–60, 67–68.

Acknowledgments

In the nine years that it has taken me to write this book, I have accumulated more debts than I can ever properly acknowledge. The project has been generously supported by fellowships from the American Council of Learned Societies and the Rockefeller Foundation; by grants from the National Endowment for the Humanities (Grant no. RH-21055-92) and the National Library of Medicine (Grant no. R01 LM0579-01); and by the State University of New York at Stony Brook. The book's contents are solely my responsibility and do not necessarily represent the views of these institutions or agencies.

I want to thank the many archivists and librarians who have helped me over the years. For service beyond the call of duty, particularly heartfelt thanks are due my old friend Ellen Gartrell of the Hartman Center at Duke University and the staffs of the Atlanta History Center, the College of Physicians of Philadelphia, the University Archives and the Kheel Center at Cornell University, the Hagley Museum and Library, and the Tuskegee University Archives.

Portions of Chapters 2, 3, 5, and 6, in slightly different form, appeared in the following publications, and I am grateful to the publishers for permission to include the following material here: "The Private Side of Public Health: Sanitary Science, Domestic Hygiene, and the Germ Theory," *Bulletin of the History of Medicine* 64:4 (Winter 1990), 598–599, by permission of the Johns Hopkins University Press; "Moralizing the Microbe: The Germ Theory and the Moral Construction of Behavior in the Late Nineteenth Century Tuberculosis Movement," in *Morality and Health,* edited by Allan Brandt and Paul Rozin (New York: Routledge, 1997), by permission of the publisher; and "Spreading the Germ Theory: Sanitary Science and the Home Economics Movement," in *Rethinking Home Economics: Women and the History of a Profession,* edited by Sarah Stage and Virginia B. Vincenti, copyright © 1997 by Cornell University and used by permission of the publisher, Cornell University Press.

In conducting the research for this book, I was ably assisted by Hilary Aquino, Dianne Creagh, Amanda Frisken, Dianne Glave, Dawn Greeley, and Paula Viterbo, all graduate students at Stony Brook. In addition, Mandy read a draft of the manuscript and gave me some excellent suggestions for revision, and Paula played a particularly important role in shaping my understanding of the germ theory. Many other friends and colleagues have shared stories and given me research leads, among them Stuart Galishoff, Robert Joy, Susan Lederer, Sandra Moss, Katherine Ott, Charles Rosenberg, Alice Ross, Marc Stern, James Strick, John Harley Warner, and James Harvey Young. For help in finding illustrations, I want to thank Bert Hansen, who sent me a wonderful *Puck* cartoon, and William Helfand, who gave me examples from his collection of AIDS posters. Others have generously shared with me their work in progress, including Emily Abel, Patricia Cooper, Jon Harkness, Tera Hunter, Judith Walzer Leavitt, Jacqueline Litt, Joel Mokyr, Maureen Ogle, Samuel Preston, Sherrill Redmon, and Edward Russell III.

Joan Jacobs Brumberg not only shared archival sources and ideas, but also read the manuscript at a critical point in its development, setting me on the path to writing a much better book. Robert Zussman performed the same service at a later stage, forcing me to sharpen the book's basic arguments. Jack Coulehan calmed my anxieties about making scientific faux pas with a last-minute reading of the manuscript. My colleagues in the History Department at Stony Brook have given me continual support and encouragement as I have toiled through the seemingly endless effort of completion. In particular, the "sisters of the supper club"—Young-Sun Hong, Temma Kaplan, Brooke Larson, Iona Man-Cheong, Barbara Weinstein, and Kathleen Wilson—have cheered me on at every step. The editorial staff at Harvard University Press, especially Aïda Donald and Elizabeth Suttell, have been wonderful to work with. Julie Carlson did a superb job of copyediting with tact, humor, and efficiency.

Finally, I want to thank my family for their love and encouragement, including my mother's stories about her early experiences with Listerine, my sister Linda's clippings about my beloved University of Kentucky basketball team, and my brother-in-law Charlie's enthusiasm for the subject of microbe hunting. On the other side of the family, my in-laws, Julia and Philip Sellers, have treated me like one of their own children, truly a privileged status, and have provided free meals and quality babysitting when I needed them most. My brothers- and sisters-in-law have provided an enthusiastic audience for my ramblings about germs, and my aunt and uncle-in-law Marilyn and Quentin Shokes lent me the use of a cabin in the Smoky Mountains, where much of the first draft was written.

The last debt, to my husband, Christopher Sellers, is the greatest. He has been in the trenches with me every step of the way, reading drafts, finding references, and bolstering my flagging confidence. Our daughter, Anne Camlin, was born during the last stages of the book's writing, and although her fevers and rashes have been mercifully brief, they have helped me understand the anxieties parents feel for the safety of their children. Most of all, Annie and her father have continually reminded me that there is more to life than writing books. Their love and support gave me the greatest incentive to finish.

Index

347